THE ORIGINAL
HORSE BIBLE
THE DEFINITIVE SOURCE FOR ALL THINGS HORSE

By Moira C. Reeve and Sharon Biggs

Photographs by Bob Langrish

BOWTIE
PRESS®
Irvine, California
A Division of BowTie, Inc.

Lead Editor: Jarelle S. Stein
Associate Editor: Lindsay Hanks
Assistant Editor: Elizabeth L. McCaughey
Consulting Editor: Elizabeth Moyer
Art Director: Cindy Kassebaum
Production Supervisor: Jessica Jaensch
Production Coordinator: Tracy Vogtman
Book Project Specialist: Karen Julian
Indexer: Melody Englund

Vice President, Chief Content Officer: June Kikuchi
Vice President, Kennel Club Books: Andrew DePrisco
BowTie Press: Jennifer Calvert, Amy Deputato, Lindsay Hanks,
Karen Julian, Elizabeth L. McCaughey, Roger Sipe, Jarelle S. Stein

Library of Congress Cataloging-in-Publication Data

Reeve, Moira C.
  The original horse bible : the definitive source for all things horse / by Moira C. Reeve and Sharon Biggs.
      p. cm.
  Includes bibliographical references and index.
  ISBN 978-1-933958-75-0
  1. Horses. 2. Horsemanship. I. Biggs, Sharon, 1966- II. Title.
  SF285.R44 2011
  636.1--dc22
                              2011007902

BowTie Press®
A Division of BowTie, Inc.
3 Burroughts
Irvine, California 92618

Printed and bound in China
15 14 13 12 11   1 2 3 4 5 6 7 8 9 10

For Our Parents

Moira: Elizabeth and Arnold, who made my dreams of horses a reality

Sharon: Richard and Carol, who scraped the money together to send me to horse camp all those years ago

Contents

Preface

Peruse the shelves of your local bookstores and libraries and check out online book sites, and you'll discover thousands of books on the subject of horses. From caring for a newborn foal to making the most of the lives of senior horses, from learning to ride in a weekend to competitive insights from Olympians, there is no shortage of information on the numerous facets of the equine world.

The Original Horse Bible is different from all of those other books, however, in that it is a comprehensive resource about all things "horsey." This volume corrals all facets of horsekeeping to produce one handy resource for readers.

Whatever your skill level with horses may be, you will find something in this book for you: If you are new to horses, the book will serve as a practical guide for your first forays into the horse world. If you are a seasoned equestrian, the book will serve as a great reference tool and be an invaluable addition to your library or tack room.

Although no piece of writing can take the place of hands-on experience, knowledge about horses comes in many different forms, including books. Let *The Original Horse Bible* provide you with information about, and a better understanding of, horses so that your time spent with them is as enjoyable and as rewarding as possible.

Introduction

Look back on our struggle for freedom;
trace our present day's strength to its source.
And you'll find that man's pathway to glory
is strewn with the bones of a horse.
—Anonymous

Horses and humans—our history together is a long and storied one. No other animal has helped shape humanity quite as the horse has. Throughout the centuries, we have domesticated and developed horses for many purposes, including raising sustenance, transportation, recreation, and sport. They have been beasts of burden, partners in work and in war, loyal companions, and cherished family members. The roles horses play in our lives continue to evolve based on our needs, as well as on our desires. For this, we owe horses a debt of gratitude.

Over the millennia, these magnificent creatures have continued to captivate humankind, as evidenced in their portrayal in our art, our literature, and our music. Horses have always inspired; their very name is synonymous with power.

For all their strength and all their beauty, however, horses do have basic needs. They require a tremendous commitment of our time and financial resources to meet those needs. Proper attention to nutrition, health management, training, and overall well-being are but a few of the responsibilities of horse care; none of these responsibilities can be taken lightly. These duties are part of good horsemanship, which is a learning process that never ends. Each horse is special in his own way, with unique characteristics, personality, needs, likes, dislikes, fears, abilities, and talents.

It has been said that patience is a virtue. When it comes to horses, that adage is spot on. From the very basics of primary care to the intricacies of training and riding, we must take a slow, educated approach to instilling the necessary knowledge. For this reason, we hope that all equestrians out there will practice patience in their pursuits with horses. After all, there's no need to rush—the journey is far too enjoyable and rewarding not to be savored.

History, Physiology, and Behavior

Section 1

O ver time, the horse has been shaped by two very different forces: nature and man. Nature, of course, got there first and did the heavy work— taking a small, stubby, short-snouted forest creature, and crafting it into a tall, fast, plains-dwelling animal, with a regal profile. When people finally turned their attention to the horse, they began to tinker with the basic design, controlling reproduction to bring out the traits that would best serve individual purposes. Those purposes differed depending on location and kind of society. Thus the various types and breeds of horses came about. Understanding the horse should begin with learning how the modern horse evolved and why, as well as learning about the horse's basic physiology and behavior.

Chapter 1

The History of the Horse

The idea that equine evolution occurred in a straight line has been debunked by contemporary science. Most researchers believe the horse's evolutionary model is much more complex, with dozens of different equid types making their appearances on earth at given periods through the ages, some concurrently. Changes in environment led to evolutionary changes, creating new genera better adapted to those changes. Some of the new genera proved more successful than others. Ultimately, all but one of the genera died off, many of them within a few million years of their first appearance, leaving *Equus* (the genus of all modern equines) to carry on.

Just as changing environments have shaped the horse, so has humankind, albeit over a much shorter period of time (a few thousand years versus many millions). People began domesticating horses about 6,000 years ago (around 4000 BC), first as sources of food and then as working animals. Horses helped humanity to spread across the globe, which led to the development of distinctive breeds, as different parts of the world called for different types of horses.

NATURAL HISTORY

The earliest scientific evidence of equids dates back at least 55 million years, to the tiny *Hyracotherium* (also known as *Eohippus*, or "dawn horse"). About the size of a large dog, this multi-toed animal scampered through the forests of North America, browsing (rather than grazing) on the leaves of the tropical plants that were so abundant there during the Eocene epoch.

As millions of years passed, other genera evolved from the *Eohippus*. Approximately 30 million years ago, during the height of the Oligocene epoch, the genus type *Mesohippus* appeared. The *Mesohippus* was slightly larger and longer legged than its predecessors, with a slightly longer snout, presenting a more horselike appearance.

Other forest-foraging equids appeared during the late Oligocene and early Miocene epochs (approximately 25 million years ago), including the genus types *Miohippus*, *Parahippus*, and *Megahippus*. Although each genus carried its unique traits, these early equids remained relatively similar to their ancestors. Changing climate conditions, however, would radically transform their descendants.

Over the next few millennia, global temperatures dropped, and the North American continent became cooler and less tropical. In many areas, lush forests were transformed into grasslands. To survive, the Equidae family (including horses and horselike animals) became grass grazers, with several new genera, including *Merychippus*, *Pliohippus*, *Neohipparion*, *Dinohippus*, and *Parahipparion*. Yet life wasn't easy for the grass grazers either. These animals were very visible on the open plains, and ultimately only one genus within the entire Equidae family survived to modern day: *Equus*.

At the Kentucky Horse Park, the sketetal frame of a *Mesohippus* shows the three-toed feet many early equids had.

J. W. Gidley, discoverer of the *Neohipparion* genus, sets out a horse evolution exhibit at the National Museum in 1925.

Believed to have evolved from the *Pliohippus* during the Pleistocene epoch (which began about 2 million years ago), the genus *Equus* comprises all the modern-day horses, asses, and zebras. The horse, which is known by its scientific name *Equus caballus* (*Equus* is the genus, *caballus* the species), includes all of today's domesticated horse breeds. Many scientists think that all contemporary horses can be grouped into *Equus caballus*, but agreement is not universal. Some scientists, for example, consider the primitive Asiatic wild horse, or Przewalski horse, to be a variety of *Equus caballus*, while others call it a separate species, known by the scientific name *Equus przewalskii*.

Adaptation

Scientists often turn to the rich fossil record of the horse to show how evolution works. As discussed above, as early equids adapted to a changing world, new genus types, with slightly different characteristics, came into existence, often during the same period. Some died out, while others evolved further. Certain genetic traits, such as body size, even reversed themselves in a number of the genus types.

Still, the overall evolution of equids from small forest dwellers to large grass grazers can be clearly seen by looking at certain anatomical characteristics of the modern horse and comparing them with those of early equid genera. Today's *Equus*, for example,

The genus *Hipparion*, pictured in this German illustration by Heinrich Harder, was a three-toed ancestor of the modern horse.

is hooved, but early ancestors were multi-toed. Over the course of millions of years, equids lost their side toes: the middle toe evolved into a single large hoof, while the other toes shrank and became functionless. The remains of toes from *Mesohippus* and *Miohippus* are evidenced in the modern-day horse's metacarpal and metatarsal bones, or splint bones, which bracket the cannon bone on all four legs. The ergot, which is a hard calloused growth on the back of the fetlocks, and the chestnut, which is a round calloused disk above the knees of the front legs, may also be evidence of vestigial toes from older four-toed ancestors.

This anatomical change was most likely helped along by the climate change that transformed forests into drier grasslands. As some members of Equidae family evolved to move out of the forests and onto the prairies, they became visible targets for predators. To survive, these early equids had to flee, and flee quickly. The evolutionary development of hooves, longer legs, and larger body types gave equids the ability to use speed—as well as a keen sense of balance and strong flight instinct—to escape hungry carnivores.

Environmental forces required other adaptations as well. The sparse nutrition of the dry grasslands was a stark contrast to the protein-rich plants found in the lush forests. Grass is not only tough; it is indigestible for most mammals. To survive, the mouths and teeth of early horses evolved to become more efficient at grazing, and their digestive systems adapted to utilize a high-fiber/low-nutrient grass diet.

Migration

Scientists believe that equids began life in North America and migrated to other continents throughout the ages. Scientific evidence shows, for example, that about 11 million years ago, horses of the *Hipparion* genus began migrating from North America to Asia and South America by way of ancient land bridges, and by about 3 million years ago, *Equus* could be found on several continents.

Then about 10,000 years ago, horses became extinct in North and South America within a short period of time. Researchers speculate that the end of the horse's almost 55-million-year existence in North America was caused by climatic change, disease, and overhunting by humans, who killed the animals for food.

When horses finally made their return to North America, they came not by land bridge but by ship.

The Wild One: Przewalski Horse

There are many untrained horses in the world; some people may argue that these horses are wild, but that's simply not the case. A truly wild horse is a species or subspecies that has no domesticated ancestors. Only one true wild horse species is alive today: the Przewalski horse, named after the Russian explorer who discovered them, Nikolai Przhevalsky (original Polish spelling is Przewalski). Scientists argue about whether this horse is a member of *Equus caballus* or a member of a separate species, *Equus przewalskii*.

Called Takai by its native people, the Przewalski still exists in the wild in Mongolia and in zoos around the world, but its survival has been on the brink in recent times. The species was considered extinct in the wild between 1969 and 1992, but a small breeding population has been reestablished, thanks to conservation efforts to save the Przewalski.

The Przewalski is a small horse, standing only 12 to 14 hands high (each hand represents four inches) and is mostly dun in color (brownish beige) with black points, a light tan stomach, and a dorsal stripe. The Przewalskis are tough little horses that can subsist on meager rations. The Przewalski horse differs from the modern-day horse in that it possesses sixty-six chromosomes instead of sixty-four. Other truly wild equids include the zebra and the wild ass.

Early European explorers imported horses to breeding ranches on Cuba, Haiti, and other Caribbean islands, as well as on the mainland of the New World, beginning with Columbus's second voyage in 1493. Subsequent explorers brought even more horses—some of the animals were commissioned directly from Europe (mostly Spain), while others were shipped over from the breeding establishments situated in the Caribbean.

DOMESTICATION

Given that horses are prey animals, it is amazing that people have had such success in domesticating them. That success may be due in part to the fact that horses have inherent traits that make them more easily domesticated. They are, for example, highly adaptable and able to survive and propagate in a wide range of environments, making them easier to keep in captivity. In addition, although horses are large and strong, their threat to people is limited because they aren't carnivores and don't possess claws, horns, or antlers.

The most compelling reason for the ease of domestication, however, is that horses are social herd animals. In the natural order of the equine herd setting, there is a hierarchy, or pecking order, from alpha horse down to the most submissive herd member. Although some horses are more domineering than others, most have the potential to be somewhat submissive—a trait that worked quite nicely in our favor for domestication: we became the leaders, and the horses followed.

The modern horse was first domesticated for its meat around 4000 BC. Neolithic humans evolved from hunter-gatherers to farmers, and they tamed horses in herds to make them manageable and compliant. But humans soon discovered that the horse could be trained for other uses, which changed the animal's role in human society dramatically. Early farmers found that sitting atop a horse made it simpler to manage the herd. Early riders found they could control and direct each horse with a sinew wrapped around the nose or around the lower jaw; with that control, they could maneuver the horses to carry them over great distances.

The horse is believed to have been completely domesticated by 3000 BC. A thousand years later, horses were being used extensively across Eurasia. These dates are supported by archaeological finds of horse remains that bear dental pathologies associated with bitting (the use of a bit). Other evidence includes changes in economies and human settlement patterns, the depiction of equids in artifacts, and the appearance of horse bones in human graves.

In the late 1800s, a Central Asian displays his horse's decorative saddle gear. Horses are still vitally important in the region.

Horses began to develop into very distinct types as humans started to delve into the science of horse breeding. The environment and climate, along with work purposes, had an impact on how different horse breeds developed in different parts of the world. Some of the oldest breeds of horses were developed in the area of Persia (Iran). These hot-blooded horses were well suited to traveling long distances in desert conditions, making them prized under saddle and in harness. Many of today's horse breeds can trace their lineage back to horses from the Middle East and North Africa.

Historically, domesticated horses were tools to be utilized for building new societies. Only the very privileged used horses for more recreational endeavors, such as sport. Over the centuries, horses worked in harness, under saddle, and as beasts of burden.

By the turn of the twentieth century, however, the role for horses in the Western world had drastically changed. Industrialization brought the advent of machines for transportation and farmwork. The automobile replaced the buggy, and the tractor supplanted the plow. As a result, equines became less prized for their pulling brawn, and horses fell out of favor as work partners. Yet there were people who saw something different to value in the horses—their athleticism, which could be utilized for sporting pursuits.

In nonindustrialized countries, equines still serve as plow animals on farms and as modes of transportation for goods and people. In the United States and other industrialized countries, some horses do still work for a living as mounts for police officers and park rangers, as entertainers and circus animals, as carriage horses in big cities, as transportation sources in rural Amish communities, and as plow animals for those who prefer horsepower to machine power. These jobs, though, are not the norm for horses in the developed world. The great majority are used for pleasure riding, showing, and racing.

HISTORY IN A HUMAN WORLD

As humankind domesticated the horse, the world opened up for us. Once the horse could be controlled, people on horseback saw the opportunity to expand civilization. They migrated southward to the Fertile Crescent area of the Near East. Horses found roles both in hunting and in warfare. By the middle of the second millennium, chariot horses were used throughout Greece, Egypt, Mesopotamia, and China.

NAMES BY AGE AND SEX

A horse is a horse, of course. But there are many kinds of horses and many terms to describe them. Learning the following terms will help you "talk horse."

■ **Broodmare:** A mare used for breeding, not riding. A mare does not necessarily have to be carrying a foal to be considered a broodmare.

■ **Colt:** A young male equine, up to three years of age.

■ **Dam:** A female parent of a foal.

■ **Filly:** A young female equine, up to three years of age.

■ **Foal:** A baby equine, male or female, still at his or her dam's side.

■ **Gelding:** A castrated male equine.

■ **Mare:** An adult female equine.

■ **Pony:** A type of small equine with particular conformation traits. A pony is not a baby horse.

■ **Sire:** A male parent of a foal.

■ **Stallion:** An uncastrated male equine.

■ **Weanling:** A male or female equine no longer nursing from his or her dam.

■ **Yearling:** A year-old male or female equine.

Over the ages, people have mythologized the horse. Human cultures from around the world have long depicted equines in their art, their writings, and their religions. These works are evidence of our reliance on and our worship of the horse through history.

Feral Horses

A common misconception among many people is that Mustangs or other free-roaming horses found in the wild are truly wild. Actually, these horses are feral. The feral horses are born and live in the wild, but they are descended from domesticated animals that escaped captivity or were turned loose.

Evidence has shown that some US free-roaming horses carry bloodlines that date back centuries, to the days when early European explorers brought horses to the New World. Not all free-roaming horses, however, have bloodlines going back that far. As the early colonists settled America, they developed new horse breeds. Over time and for various reasons, some of these horses became feral.

By 1900, there were an estimated 2 million free-roaming horses in North America. Over the following decades, their numbers fell dramatically as people freely hunted, captured, sold, or slaughtered the horses. Mounting pressures by animal-welfare advocates led the US Congress, in 1971, to pass what is known as the Wild Free-Roaming Horse and Burro Act. It established protections for feral herds of horses and burros, and today only the US Bureau of Land Management (BLM) has the legal right to remove free-roaming horses from public rangelands.

Several populations of feral horses exist throughout the world, some protected, some not. The Camargue horse, for example, roams the marshlands of southern France *(below)* under the protection of the government. Australia's Brumby, by contrast, is constantly under threat of extermination, as it has no legal protection and is considered a pest by many residents.

Ramses II charges into battle in his chariot, in this rendering of an ancient relief. Horses played vital roles in war and work.

It can be argued that the horse has done more to shape human culture than has any other creature on the earth. It's no wonder people have so feverishly romanticized their relationships with horses.

Roles in Human Society

The horse has had many responsibilities and positions within the human world. It has been a partner at work, at war, in sport, and in ceremony. Through the millennia, humans have found many ways to utilize the abilities of the horse, including for work and war, for ceremony and sport, for travel and pleasure.

WORKING HORSES

The term *horsepower* is synonymous with strength, and for hundreds of years, the horse was the primary and most versatile source of power in the world. Specially bred heavy horses, known as drafts, were used for tilling, plowing, logging, transporting goods, and drawing carts and wagons. Lighter horses worked in harnesses and transported passengers, in conveyances from common stagecoaches to luxury carriages to hansom cabs. Ponies carried a successful day's hunt from the mountain and coal from underground mines. The horse helped move civilization into modern times. Although today the machine has mostly replaced the working horse, there are many places around the globe that still use horses in traditional roles.

During the Middle Ages (ca. AD 500–1500), horses played important roles in war, agriculture, and transportation. Nobles, the main owners of horses, used them as warhorses and as riding mounts. Horses were also used in farming, but they became very expensive in this era, costing as much as a house. A peasant family was unlikely to be able to own a valuable horse, but a village's people might pool their resources to buy one or two and then rotate use of the animals. Because of the high price of horses, oxen became more prevalent as plow animals during this period.

When European colonization of the Americas began at the end of the fifteenth century, Spain, France, England, and other nations raced to stake out their territories. They brought hundreds of horses with them to serve mainly as transportation. In the 1600s, the Americas had a proliferation of horses, brought by settlers and missionaries. Native Americans, used as laborers (often slaves) by the Spanish, learned about horses while working on ranches. Although the Spanish made it a crime for them to own horses, Native Americans quickly learned the nature of horses and became expert horsemen. They used horses not only to ride but also to carry packs and drag travois.

In 1680, Pueblo Indians revolted against the rule of the Spanish, driving them back into Old Mexico. As a result, many Spanish horses were abandoned. When the Spanish returned fourteen years later, they found Pueblo Indians raising large herds of horses. The Pueblo Indians began selling and trading their horses to other Native Americans, such as the Kiowa and the Comanche, and teaching them how to ride and how to breed horses.

Horses quickly spread across the southern plains. French traders noted that the Cheyenne Indians in Kansas received their first horses in 1745. That introduction of the horse changed life greatly for the Plains Indians. Before the coming of the horse, they had had to hunt buffalo on foot, a difficult feat. On horseback, hunters could keep up with the stampeding buffalo.

In Europe, urbanization had begun, with more and more country people flocking to London, Frankfurt, Paris, and other cities. The rise in city populations meant a need for more passenger transport. Cab operations became big business. Unfortunately, life was grueling for the nineteenth-century urban cab horse, as owners sought to get the most out of their charges. Horses had to adapt to intense traffic congestion, meager food, and long hours in harness. They were worked hard—sometimes to death.

Horse-drawn vehicles were a common sight along docks, as they served as the primary mode for transporting goods from ports into cities and towns. The horses working the canals were known as boaters; they were powerful, yet small and compact enough to fit under bridges and navigate narrow pathways, known as towpaths. Some had to jump stiles in the paths.

Thousands of ponies worked and spent their lives in coal mines during the nineteenth and twentieth

A Shire is used for logging at the Westonbirt Arboretum in England. Shires are still used for heavy lifting and other chores.

centuries. Known as pit ponies, these small, hardy equines hauled coal through the network of underground shafts and only came up to the surface when the pit was closed for holidays or when their working days were over.

Ford's Model A automobile, built in 1903, marked the beginning of the end of the horse as city transport—the single horsepower carriage being no match for the internal combustion engine. Yet for a time, horses and automobiles shared the roads, two very different types of passenger vehicles.

Although many jobs that horses once held are today performed by machines, horses can still be found hard at work both in Western society and in nonindustrialized countries. They are used in harness to till the land, bring in crops, log, and more. They serve as pack animals and also carry deer from stalking expeditions. Horses remain the preferred mode of transport for South American gauchos, North American cowboys, Australian drovers, and others when they work in areas that are inaccessible even by all-terrain vehicles.

WARHORSES

Civilizations the world over can attribute their successful conquests to the use of horses in battle. Our history books are full of accounts of domesticated horses that helped to further human expansion.

Eurasians were using horses to pull their chariots as early as 3000 BC. The invention of such wheeled vehicles made a significant change in warfare, as men could have greater speed and mobility during battle.

From as early as 3000 BC, armies found that the power of the horse could make their efforts to conquer other nations more effective. From the Near East to the Mediterranean, the horse's stamina and strength were used as new weapons. Horsemanship, spurred on by the needs of warring nations, became a necessary skill for battle. Methods of training and breeding horses were initially created to serve military purposes.

Ancient texts extolled the benefits of domesticating horses in the expanding civilizations. In one of the earliest writings on horses, *The Kikkuli Text* (1345 BC), a Hittite horseman named Kikkuli describes the care and training of the warhorse. Xenophon, an Athenian soldier, wrote *On the Art of Horsemanship* (ca. 400 BC), a foundation for classical riding that bears up today.

The horse-drawn chariot made its way into Egypt relatively late in its history, in the 1700s BC, when the Hyksos introduced it. Chariots were pulled by a pair of horses and were used in war, as well as in hunting and in ceremony. Because horses were uncommon, people considered them signs of prestige and wealth. They lived in fine stables and received excellent care and feed—better than many humans did.

Farther east, China's warhorses of the Qin Dynasty (221–206 BC) pulled individual drivers of two-wheeled chariots into battle. The Chinese also commonly used their horses under saddle. The Chinese have been credited with creating the horseshoe to make the hoof more durable and the stirrup to bring comfort and stability to riding.

In the ancient Near East, warriors of the Parthian Empire (247 BC–AD 224) were famed for a battle tactic called the Parthian shot. In battle, Parthian warriors on horseback would ride away from the enemy, as if in retreat, thus inviting pursuit. As the enemy approached, the Parthians would twist around on horseback and let loose their arrows. The tactic required great skill in horsemanship, as the rider used

Bronze horses and chariots like these formed part of China's First Emperor's terra-cotta army, long hidden underground.

only his legs to steer his steed while attacking with weapon in hand. The Parthians performed this feat of horsemanship without stirrups.

In Japan, during the twelfth century, a group of elite warriors known as samurai arose. They became expert fighters on horseback and on foot. Early samurai fought mainly with bow and arrow from horseback. Only samurai could ride horses into battle.

In Europe, during the Middle Ages, the need arose for immense, powerful horses that could carry a knight in full chain mail along with their own protective equine armor. Big draft horses (placid and obedient), from which knights could do battle, were needed, and European breeders met that need. During medieval competitions, when knights would face each other, horses had to be just as equipped in armor as their riders in armor; these horses were valuable and had to be protected. If a knight lost a tournament, he often lost his horse as well—and therefore part of his honor. The losing knight might have to pay large sums to get his mount back from the winner.

World War I (the Great War) would be the last major conflict in which military forces utilized mounted cavalry. Even so, horses could only be used in a more limited capacity than before. Barbed wire strung across the battlefield made the traditional cavalry charge nearly impossible, and the machine gun could cut down man and horse alike with alarming effectiveness. So horses, as well as mules, were used mostly for transporting soldiers and supplies and for hauling artillery.

Even in this more limited role, draft horses, light horses, and mules died in large numbers. In the four-year duration of the war (1914–1918), the United States exported nearly a million horses to Europe. Approximately 6 million horses served in WWI, and a substantial number of them died in it. When the war was over, most of the surviving horses ended up at slaughterhouses in France.

Although very few horses appeared on the battlefields during World War II, the Polish, Russian, German, and British armies did maintain equestrian cavalry units. The armies utilized horses mainly in logistical support, transporting troops and supplies. Even today, the military keeps horses for transport and ceremonial purposes. India's army maintains the last operationally ready, fully horse-mounted regiment: the Sixty-First Cavalry, the only remaining nonceremonial horse-mounted cavalry in the world.

Two armies of samurai clash in this mid-1800s woodcut of one of the fierce battles of Kawanakajim, fought 1553–1564.

LAW-ENFORCEMENT HORSES

Horses have served as partners in service since 1758, when the English established the first police patrol in London's Bow Street. Since then, mounted branches have become a worldwide necessity for metropolitan police forces. Horses also serve in state and federal parks and border patrols and in military and royal ceremonies. Many horses do not pass the training or lameness evaluations they must go through to become service horses; those that do are highly regarded. An ideal service horse is levelheaded under pressure, sturdy, sound, patient, and brave—an uncommon combination. The riders often see these horses as partners and compatriot officers.

Mounted police units are often pressed into service to patrol parks, wilderness areas, and other places where using a vehicle would be impractical or too noisy. In tourist areas, police horses also serve as ambassadors for their police forces or for their cities. Mounted police units still find horses very effective for riot control due to the animals' imposing size and their ability to place the officers above the crowd. The horse's agility

These US Customs and Border Protection horses carry their agents through environments regular vehicles can't negotiate.

allows the officers to respond quickly in areas where cars or motorbikes would be unable to pass.

Unlike vehicles, however, horses are not impervious to danger when working in riot or crowd control. Like modern-day gladiators, horses sent into these situations are equipped for a fight, wearing equipment such as visors, leather nose guards, rain gear, and special boots for knee and leg protection. Riot-patrol horses must also go through extreme desensitization training in order to ignore stimuli that would normally engage their flight responses.

The US Customs and Border Protection also uses horses because they can quickly—and quietly—get agents through mountainous terrain and trails that are difficult to traverse by motor vehicle. In addition, the size of the horse gives his agent a good vantage point over the difficult passages that illegal border crossers often take. Horses' keen sight and hearing also make them active partners in detecting illegal migrant workers and smugglers. A horse's reaction to a twig snapping will alert agents that someone is nearby. The horse's impressive stature also helps create control.

SPORTING HORSES

After the age of mechanization, horses found new roles in sport, with most competitive disciplines derived from working, agriculture, or military use. Dressage, for instance, has its origins in army training, while western-style events are based on ranch work. Today, riders compete from childhood to adulthood, and in most countries, male and female riders (and horses) compete on an equal playing field.

The top level of equine sport is governed by the Fédération Equestre Internationale, which recognizes the disciplines of eventing, show jumping, dressage, endurance, vaulting, driving, and reining. Equine organizations strive to make sports safe for horse and rider, yet as with any athletic endeavor, injuries occur. Special shoeing, feeding, conditioning, and equipment are used to optimize performance. In some cultures, equine welfare still has far to go for horses in sport, as prevailing attitudes toward animals figure closely into competitive activities.

RACING HORSES

The horse is one of the fastest land mammals on earth, a fact that people have noted and used to their advantage for many centuries. Since the beginning of recorded time, horse racing has been a sport of nearly every major civilization. A number of records suggest that nomadic tribesmen of Central Asia may have raced as early as 2500 BC.

In Greek society, the horse embodied speed and competition, and the people respected horses for their versatility. During the sixth century BC, the ruling aristocratic families in Athens demonstrated their equine appreciation by beginning or ending their names with the word *hippos*, Greek for "horse." Greek horses were bred for racing, as well as for riding; chariot racing was one of the Greeks' most popular contests. Some historians cite these competitions as being the foundation of the Olympic Games.

JOCKEYS

Don't judge jockeys by their small stature. Despite their size, jockeys are tremendous athletes. Because speed is everything in horse racing—and because excess weight slows a horse down—jockeys must maintain a low weight yet still have the strength to balance their bodies over (rather than sit on) their horses. They must have the presence of mind to pilot their 1,200-pound animals at full gallop for at least two minutes through a packed field and know exactly when to make a bid for the lead. if they finish in the money, jockeys typically collect 10 percent of their horses' winnings.

In seventeen-century Europe, horse racing became (and remained) popular among the nobility. Eventually it became known as "the sport of kings."

Today, horse racing takes on many forms. Flat racing (originally match races) is now the most popular style of racing found worldwide. The most prestigious Thoroughbred races in the world are: the United States' Kentucky Derby, the United Kingdom's Epsom Derby, Australia's Melbourne Cup, France's Prix de l'Arc de Triomphe, and the Dubai World Cup. At $10 million, the Dubai's purse is the world's largest. There are also special sprint races for Quarter Horses, which can sprint up to 47.5 miles per hour. Other forms of racing include steeplechases, in which horses negotiate jumps along the course, and harness racing, in which the horses pull lightweight racing carts, called sulkies, at a trot or a pace.

Although humans have been racing each other for long distances for centuries, organized endurance racing did not begin until 1955. It began in America with Nevada's Tevis Cup, a 100-mile ride. Endurance racing is now embraced worldwide and sanctioned by the Fédération Equestre Internationale. Races usually cover 25, 50, or 100 miles, often over difficult terrain. An award is given to the horse that arrives first and passes all veterinary checks, as well as the horse that finishes in the best condition.

The Horse in Words and Art

Figuring prominently in myth and reality, the horse has captured humankind's imagination down the ages. Most people point to the horse's outward beauty as inspiration, but there is more to the majesty than that: an intangible spirit of freedom that exists deep under hide and mane. The horses represent strength and speed, but they are also symbolic of liberty and, in a sense, vulnerability. The partnership forged with this powerful yet willing and obedient creature has inspired us to memorialize the horse in fable, legend, and other literary and artistic endeavors.

Perhaps one of the most perplexing and ancient relics of equine homage is that of the Uffington White Horse. In an area rife with stone circles, barrows, and henges lies the great horse, a giant abstract figure in crushed white chalk measuring 374 feet (110 meters) long that was cut into a trench on a hillside in Oxfordshire, England. The vast form is only visible from overhead. The horse was originally thought to be an emblem of the people who built the Uffington Castle, an early Iron Age hill fort. But contemporary dating performed in the 1990s suggests the figure is from the Bronze Age, sometime between 1400 and 600 BC. The exact purpose of the horse is unknown, but some researchers point to religious symbolism. The Celts in Gaul worshipped Epona, the horse goddess; her

Thoroughbreds round a turn at California's Santa Anita Park, fighting to break out of the pack and take the lead.

Welsh counterpart was Rhiannon, who dressed in glittering gold and rode a white horse.

Over time, the people of the British Isles continued to pay homage to the horse in other ways. Judging from the harnesses and carts used to adorn them, the horses of Britain's Iron Age (ca. 800 BC–AD 100) were most likely regarded well within the society. Other Iron Age artifacts depict the horse's importance, including coins bearing an image similar to that of the Uffington White Horse. In the art of the medieval era, horses were often shown as the transport that enabled knights to do battle, in everything from enacting jousting matches to slaying dragons. Horses were bedecked in armor and the flowing colors of their riders. The period's less-prominent art depicts horses plowing fields for the landowners of the time.

Ancient Greece's refined society has also left us with many equine relics to treasure, among them a horse-care manual. As noted on page 20, an Athenian soldier named Xenophon wrote *On the Art of Horsemanship*, a work regarded as the first text detailing the fundamentals for dressage and what later became known as natural horsemanship. In the book, he discusses the care, training, and selection of the military horse. Xenophon had such insight on the horse's nature that the book is still used today, primarily by hobbyists interested in classical riding styles such as dressage.

An outstanding illustration of ancient horse art was produced around 353 BC for King Mausolus's tomb at Halicarnassus (the coast of Asia Minor) and is considered one of the Seven Wonders of the ancient world. Sadly, an earthquake during the Middle Ages destroyed the four-horse chariot sculpture and the tomb was later plundered. But the remains of the head and neck of a horse in a bridle have survived.

About the same time farther east, Asian artists were depicting horses in pottery and statues, as well as on tapestries. From a tomb of the First Emperor of China, Qin Dynasty (221–207 BC), came an amazing archaeological discovery (in the 1970s) of a vast life-size terra-cotta army, estimated to comprise about 8,000 warriors, 400 horses, and 100 chariots (*see page 20*).

Renaissance art is rife with horse imagery, often used to help illustrate religious and moral teachings.

In 1878, Eadweard Muybridge's series of still photos of a galloping horse led to a new art form for capturing the horse—motion pictures.

A great example of equine art in Renaissance times shows the legend of St. George and the Dragon, which was rendered by several different artists at the time to help teach important virtues to the illiterate people. According to lore, George was a Christian soldier born in Asia Minor (modern-day Turkey) more than 1,600 years before. It was said that the Syrian king's daughter was to be sacrificed to the dragon that was terrorizing the village of Silene. In the legend, St. George (almost always depicted riding a white charger) killed the dragon, protecting the villagers and converting them to Christianity. Allegorically, St. George embodied bravery, chivalry, and purity, while the dragon represented evil. Therefore, the art taught that St. George protected the town from evil, and the legend inspired the people to live good, pure lives.

During the Renaissance, Leonardo da Vinci (1452–1519) studied the movement and locomotion of the horse, as well as that of other animals. He made countless sketches of horses for the Gran Cavallo, an unrealized bronze sculpture.

The Baroque age, which began in the late sixteenth century, following the Renaissance, saw a renewal of the art of equitation; in fact, depicting the equine in art became more popular than ever before. Artistic masters, including Rubens, Van Dyck, and Velázquez, were commissioned to paint their subjects mounted on elegant horses. Lavishness and wealth were common themes in these paintings, and the artists always depicted the horses as heavy and muscular, with long manes and tails.

In the eighteenth century, the painting of animal and sporting images became popular. British painter George Stubbs (1724–1806) is arguably the most famous equine artist in history. Obsessed with the animal's anatomy, he spent eighteen months dissecting horses to study their physiology before producing the book *The Anatomy of the Horse,* in 1766. People saw his work as being far more accurate than that of other famed equine artists of the time—including James Seymour, Peter Tillemans, and John Wootton—and Stubbs began to receive commissions from wealthy and aristocratic horse owners. His most famous painting, *Whistlejacket*, portrays an early Thoroughbred racehorse from around 1762; it currently hangs in London's National Gallery.

For centuries, the horse's beauty has inspired artists the world over, from equine depictions in France's Lascaux caves dating 15,000 to 20,000 years

UNICORNS

The fabled unicorn is not a "one-horned horse," as is typically assumed. It is a bearded horse-goat-like creature, with cloven hooves and a lion's tail. References to unicorns can be found in the Bible, and ancient Greeks believed that the creature actually existed in India. In Medieval religious art, unicorns were often depicted alongside maidens and were symbolic of chaste love and fidelity. It was thought that a virgin could lure and trap a unicorn.

ago, to the nineteenth-century motion-picture pioneer Eadweard J. Muybridge's use of a galloping horse to develop the first moving images, to today's talented equine photographers and artists. Quite simply, people can't take their eyes off these majestic creatures.

The Birth of Equine Welfare

Horses today enjoy more protection than ever in history, and their welfare is enforced by many laws, both nationally and internationally. But animal cruelty remains an important issue throughout the world, and many organizations fight to improve conditions for horses. Humankind's values have shifted over the years to include more regard for the animal's welfare, but there are always issues to improve or resolve.

Some societies and cultures have nurtured the horse through the ages. For the Islamic people, the Arabian horse was considered a gift from Allah, to be revered and treasured. Some historical accounts say that Bedouins treated their horses as part of the family and would allow the horses to stay in their tents.

Most other cultures, however, have viewed horses as strictly utilitarian and often disposable. Working horses in the seventeenth and eighteenth centuries were often worked to death. From pit ponies in Britain to cart horses throughout Europe, there were no laws and no protection for the working animal. If one horse fell ill, another took his place.

WELFARE EFFORTS IN BRITAIN

Anna Sewell's famous book, *Black Beauty*, published in 1877, is considered the first novel to discuss the subject of equine welfare. Sewell tells the story of a horse in Victorian London whose fortunes changed from promising to tragic at the hands of various cruel or

In July 1911, New York street-cab horses drink eagerly from a half-barrel of water provided by concerned ASPCA members.

ignorant owners. (Black Beauty also meets with kindness and ends his days in the country.) Sewell found it particularly difficult to understand how upper-class society could be so cruel. One practice that angered her was the use of a bearing rein. This was a barbaric piece of tack that forced the horse to carry his head and neck in a high-arched position. It may have been fashionable to see carriage horses with their heads held artificially high, but the position made breathing difficult and often led to respiratory conditions resulting in death.

Another prominent figure in British equine welfare, Ada Cole, is considered one of the first activists for the cause. At the beginning of the twentieth century, as the numbers of out-of-work horses increased due to the rising use of machinery, so did the numbers that were exported for meat to foreign slaughterhouses. In 1911, Cole witnessed several British draft horses being unloaded and whipped for four miles to slaughter in Belgium. Appalled by the callous treatment the horses received in their last days on earth, she lobbied politicians, fund raised, and worked tirelessly over decades to heighten awareness. Cole founded the International League for the Protection of Horses (ILPH) in 1927 as a campaigning organization to prevent the export of live British horses for slaughter. A decade later, through the efforts of ILPH and other committed activists, the

British Parliament adopted an act to help protect horses from exportation to slaughter.

Later renamed the World Horse Welfare, Cole's organization has continued to campaign for equine protection, expanding its activities to include welfare and protection around the world.

WELFARE EFFORTS IN AMERICA

Britain had Ada Cole; America had Velma Bronn Johnston, also known as Wild Horse Annie. Born in 1912, Johnston was an animal rights activist who campaigned to stop the removal of Mustangs and burros from federal public lands for slaughter.

Johnston's activist career began in 1950, when she noticed blood dripping from the back of a truck she was driving behind. She discovered that the truck was full of Mustangs on their way to a slaughterhouse. Appalled, she investigated further and began publicly campaigning against the cruel roundup and transport practices. She collected evidence and began speaking to ranchers, businessmen, and politicians; she started a children's letter-writing campaign to the government. In 1959, the campaign resulted in federal legislature, Public Law 86-234, which banned air and land vehicles from hunting and capturing wild horses on state land. It became known as the Wild Horse Annie Act.

Not satisfied with the results, Johnston continued her quest for equine protection. As result of her efforts and those of other activists, the Ninety-Second US Congress passed the Wild Free-Roaming Horse and Burro Act of 1971, which was signed into law by President Richard Nixon. This act prohibited capture, injury, or disturbance of wild (feral) horses and burros, and it allowed for their relocation when their numbers grew too large in any particular area.

Over the past sixty years, dozens of groups have risen to aid in equine welfare. The mission of the American Horse Defense Fund (AHDF), founded in 2000, is "to facilitate the protection, conservation, and humane treatment of members of all Equine species." The organization "works to address inhumane treatment of horses, ponies, donkeys, mules and burros, both wild and domesticated through education, advocacy and litigation when necessary in the state, federal and international arenas." AHDF's work concentrates in the fields of horse slaughter, horse tripping, racing, feral horses, the PMU industry, and nurse mare foals (see Welfare Issues below).

The Hooved Animal Humane Society (HAHS) was founded in 1971 by six concerned and committed citizens. According to the HAHS, its mission is "to promote the humane treatment of hooved animals through education, legislation, investigation and if necessary, legal intervention (impoundment)." The group also provides "physical rehabilitation to animals that have endured severe neglect and abuse and then adopt[s] them out to compassionate forever homes."

The Unwanted Horse Coalition, created in 2005, is a broad alliance of equine organizations that have joined together under the umbrella of the American Horse Council. The coalition's mission is to reduce the number of unwanted horses and to improve their welfare through education and the efforts of organizations committed to the health, safety, and responsible care and disposition of these horses.

WELFARE ISSUES

Equine advocates have had their work cut out for them for quite some time and in a variety of practices. The many welfare concerns plaguing equines ranged from the use of horses for food to the use of mares to nurse orphaned foals, while the mares' own foals were killed.

Slaughter: In most countries where horses are slaughtered for food, they are processed like cattle, in large-scale factory slaughterhouses (called abattoirs). The horse is shot in the forehead with a metal rod using a captive bolt stunner, rendering the animal unconscious. He is then killed by exsanguination (bleeding out)—cutting the jugular vein or carotid artery.

Horse slaughter in America used to flourish at several abattoirs nationwide. However, public protest and legislation eliminated these plants. In 2007, the remaining three slaughterhouses closed: Beltex Corporation and Dallas Crown in Texas, and Cavel International in Illinois.

Efforts persist in the United States to ban horse slaughter for good. On September 8, 2006, the US House of Representatives passed the American Horse Slaughter Prevention Act, a bill designed to stop the slaughter of horses for human consumption. Had the legislation also passed the Senate and been signed by the president, it would have made killing horses for human consumption an illegal practice in America. As of March 2011, this bill has yet to become law.

The latest issue in the practice of horse slaughter lies in the transportation and treatment of the animals going to slaughter outside the United States. Since the closure of the US facilities, more American horses are now sent to Canada and Mexico. Groups advocating the slaughter of American horses have called for the reopening of US horse slaughter plants, saying the horses are better protected by US Department of Agriculture laws than by laws in Canada and Mexico.

A thirty-month investigation conducted by the nonprofit group Animal Angels, which ended in 2009, revealed that abuse and inhumane treatment are still occurring for American horses on the way to slaughter in other countries. The abuse has included beating horses and jabbing them in the eyes and using a cable winch to drag them with a wire wrapped around a back leg.

The US Department of Transportation has officers at the enforcement points to ensure proper transportation of the horses, but it has no jurisdiction beyond transportation matters. Horses are transported in double-decker trailers with low ceilings meant for cows and pigs, making it impossible for the average-size horse to stand properly. Horses not only become injured while they are packed on these trailers but also can freeze to death. Food and water and basic care are denied on what are often long journeys.

American horses continue to be slaughtered—just not in America.

British equine-welfare advocate Ada Cole, founder of what is now called the World Horse League, reviews documents.

Wild Horse Annie, seen relaxing with her dog and horse, worked ceaselessly for the federal protection of Mustangs.

PMU horses: PMU is an acronym for pregnant mare urine, which is collected and used for the production of Premarin, a hormone-replacement drug for women. Although Premarin was first developed in 1942, equine-welfare advocates only gained knowledge of how it was collected decades later, in the 1990s. They soon became concerned about the well-being of the mares, which were kept indoors for up to six months at a time, as well as the welfare of their resulting foals, which often ended up in slaughterhouses. PMU ranchers, who managed the mares that produced the urine, were also criticized for restricting water intake through intermittent watering and for keeping the mares in tie stalls without adequate turnout or the ability to lie down.

In 2005, studies began to show that the use of Premarin had caused cancer in women, and subsequently, the drug's use waned, lessening the need for the PMU ranches. More than 300 PMU ranches closed from 2005–2010; about 70 remained in operation in 2010. This industry has been located primarily in Manitoba, Canada, close to the Wyeth-Ayerst pharmaceutical company, which produces Premarin. Because horse slaughter is still legal in Canada, many mares (who have outlived their usefulness) and their foals (who are considered a by-product of the industry) are slaughtered. Many horse rescues specialize in rescuing and rehoming these horses.

Nurse mare: The nurse mare industry, which probably is centuries old, exists so that an orphan foal of quality has a surrogate dam to nurse from. These mares are usually pressed into service by the Thoroughbred racing industry or performance horse industry. When the nurse mare is sent to raise the Thoroughbred, she must leave her foal behind. The nurse mare's foal is essentially a by-product of the mare's milk industry.

Historically, these abandoned foals have been killed because it proved difficult to raise them, especially in large numbers. Welfare advocates allege that the foals have gained value in the fashion and fur industry, as their hides are sometimes used for pony skin leather goods.

Horse tripping: Horse tripping occurs regularly in charreadas (Mexican-style rodeos), which began in 1921. There are three events in which the front or hind legs of a galloping horse are roped by a charro cowboy on foot or horseback; this action causes the horse to trip and fall to the ground, which gains points for the cowboy.

Feral Mustangs are herded into BLM holding pens to be auctioned off in an effort to control overpopulation on federal land.

In the United States, horse tripping has been banned in California, Florida, Illinois, Maine, New Mexico, Oklahoma, and Texas. The Professional Rodeo Cowboys Association has also banned horse tripping at its sanctioned events. Horse tripping differs from the popular rodeo event of calf roping. Because horses are tall and long legged, they have a higher center of gravity than do calves, which are more compact and travel at slower speeds on much shorter, sturdier limbs.

Film and TV's equine actors: In 1940, the American Humane Association's Film and Television unit began monitoring filmmaking; AHA created the unit after a horse plummeted to his death over a cliff in a stunt for the 1939 movie *Jesse James*. The unit was not welcome on sets until the 1980s, when federal legislation, oversight sanctioning was written into the Screen Actors Guild (SAG) contract. Today American Humane's certified animal-safety representatives work worldwide to protect equine actors. Productions that meet their strict guidelines receive this end credit: No Animals Were Harmed.

Feral horses: For decades, America's feral horses have been under scrutiny from many groups, including the government, ranchers, the livestock industry, state wildlife agencies, and others who do not support the protection of horses on federal lands. In 2004, then-senator Conrad Burns (R–Montana) engineered a backdoor congressional rider that basically gutted the protections afforded by the Wild Free-Roaming Horse and Burro Act of 1971. In this change, the Bureau of Land Management, the agency responsible for protecting feral horses, was charged with dispersing at auction any horses ten years of age or older or not adopted after three tries. As a result, feral horses were sold at auction, and many, in a roundabout way, ended up at slaughterhouses.

The BLM announced in 2008 that it was considering eliminating Mustangs currently in holding facilities in huge numbers. Then in October 2009, Secretary of the Interior Ken Salazar offered a new proposal to create additional holding facilities in the Midwest and the eastern United States. Although adoption through the BLM is still a popular option to control overpopulation on federally managed lands, equine advocates want better range management, such as birth control for mares, as well as more freedom for the horses and burros to roam over public rangeland.

There are countless other issues, some great and some small, still negatively affecting horses throughout the globe. But as shifting attitudes about animal welfare continue to take positive steps, the horse can look toward a better future throughout the world.

Chapter 2

Biology, Intelligence, and Behavior

The horse is an intricate combination of senses, intelligence, and instinct. This blend results in a prey animal with a high degree of self-preservation, yet one armed with enough weaponry—teeth, hooves, and strength—to fend off a would-be attacker if cornered. The horse's main defense, however, is speed. With a typical horse's galloping speeds rated between 35 and 40 miles an hour, a predator would need to be fast and focused to keep up.

Yet a horse is not driven by instinct alone. Although most people try to compare horses to dogs and cats, centuries of being prey have hardwired horses differently. Whereas dogs and cats think like hunters, horses must think like the hunted, learning and adapting quickly to changing circumstances to survive. That also means that horses are highly intelligent.

To better understand what horses are able to do physically, what they are capable of mentally, and how they are likely to respond in various situations, begin by learning about the horse from the inside out.

ANATOMY AND PHYSIOLOGY

Anatomy encompasses the physical structures of the body: the organs, the bones, the tissues, and the other physical elements. Physiology refers to how the horse's organs function, and how the organs, in turn, make the horse function. The following sections take a look at how the horse is put together.

Bone Structure

The horse's skeleton is composed of approximately 205 bones. *(See illustration on page 32.)* The skull alone is made up of 34 bones. The spine consists of seven cervical (neck) vertebrae, usually eighteen thoracic (connected to the rib cage) vertebrae, five or six lumbar vertebrae, and five fused sacral vertebrae called the sacrum. An average of eighteen coccygeal vertebrae make up the tail.

A horse's forelegs are made up of the shoulder blade (scapula); the humerus; the radius; eight carpal bones that compose the knee, cannon, splint bones, and long and short pastern bones; and the coffin bone in the foot. Because of the design of the horse—with the heavy head and neck—the front legs carry at least 60 percent of the animal's total weight.

The hind legs comprise the pelvis, the femur, the tibia, and the fibula. From there are the hock joint, the cannon and splint bones, the long and short pastern bones, and the coffin bone in the foot. The horse's stifle joint has a patella (kneecap). The horse's hind legs support only 40 percent of the horse, but they are

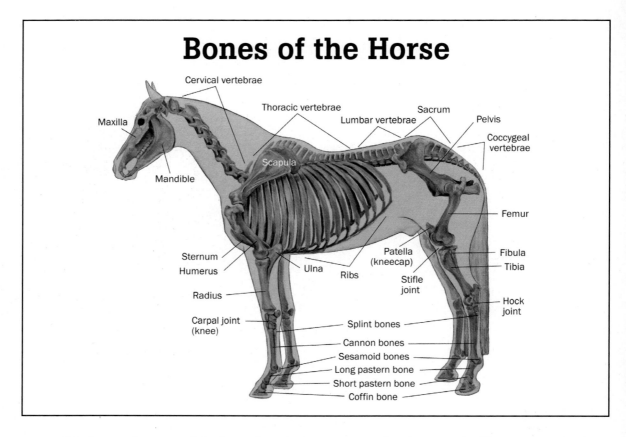

Bones of the Horse

Cervical vertebrae

Thoracic vertebrae

Lumbar vertebrae

Sacrum

Pelvis

Coccygeal vertebrae

Maxilla

Scapula

Mandible

Femur

Sternum

Patella (kneecap)

Fibula

Humerus

Ulna

Ribs

Tibia

Radius

Stifle joint

Hock joint

Carpal joint (knee)

Splint bones

Cannon bones

Sesamoid bones

Long pastern bone

Short pastern bone

Coffin bone

responsible for creating most of the forward motion. The joints of the horse are designed to effectively absorb shock, allow fluid movement, and carry the horse's weight of 1,200 to 1,500 pounds (550 to 650 kilograms) or greater.

Muscles and Tissue

The skeletal framework is held securely together by connective tissue (ligaments and tendon structures), which is covered by muscles that contract and put the body into motion. There are three classes of muscles: smooth, cardiac, and skeletal. The smooth and cardiac muscles operate on an involuntary basis and aid the equine's body in digestion, respiration, circulation, and other automatic systems. Skeletal muscles move voluntarily, meaning that the brain consciously instructs them to move.

Horses have two types of muscle fibers: slow twitch and fast twitch. All horses have fast-twitch and slow-twitch muscle fibers, but genetics determine how those muscles are distributed within a particular horse. Slow-twitch (red) fibers are aerobic, which means they need

oxygen to function properly. They are more dominant and well developed in horses that have greater strength and endurance, such as steeplechasers and endurance horses. In contrast, fast-twitch (white) fibers are anaerobic, which means they need less oxygen to function than slow-twitch fibers do. They are more dominant in horses that have bursts of speed, such as Quarter Horse racers and cutting horses. They are, however, only able to perform for short periods of time.

Digestive System

The horse is unique in the world of domesticated livestock. It is considered a nonruminant herbivore (plant eater), meaning the horses does not have a multicompartmented stomach, as a cow does. Horses spend nine to eleven hours a day grazing—longer than other grass-eating ruminants do—because they do not chew cud. The horse's constantly working digestive system also generates heat, which keeps the horse warm in winter. Yet the unique digestive system of the horse is delicate. Horses are unable to vomit and can die of extreme digestive upset (colic) if left untreated (*see appendix*).

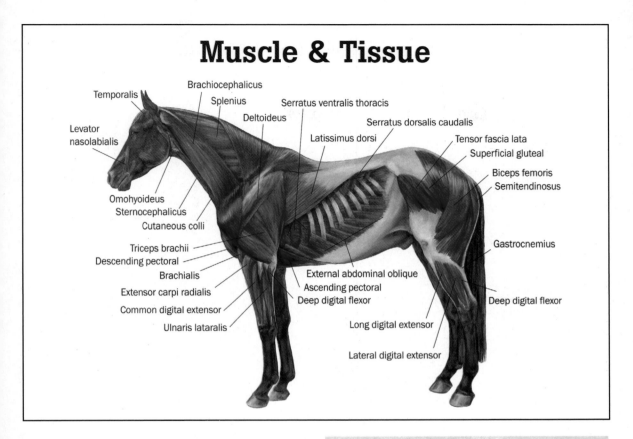

Muscle & Tissue

- Temporalis
- Levator nasolabialis
- Brachiocephalicus
- Splenius
- Deltoideus
- Serratus ventralis thoracis
- Latissimus dorsi
- Serratus dorsalis caudalis
- Tensor fascia lata
- Superficial gluteal
- Biceps femoris
- Semitendinosus
- Omohyoideus
- Sternocephalicus
- Cutaneous colli
- Gastrocnemius
- Triceps brachii
- Descending pectoral
- Brachialis
- Extensor carpi radialis
- Common digital extensor
- Ulnaris lataralis
- External abdominal oblique
- Ascending pectoral
- Deep digital flexor
- Deep digital flexor
- Long digital extensor
- Lateral digital extensor

Nervous System

The nervous system is the most important—and the most intricate—system in the horse, controlling everything within the horse's body. *(See illustration on page 36.)* It is composed of three main groups: the central nervous system (CNS), the peripheral nervous system (PNS), and the autonomic nervous system (ANS). The CNS is the center of all nervous control. It encompasses the brain and the spinal cord. From here, the CNS works as a sort of mission control for the horse. The PNS, which consists of sensory nerves and motor nerves, operates the nerves that are located within the periphery of the CNS.

The ANS, part of the PNS, governs the involuntary systems, such as the circulatory system (heart) and respiratory system (lungs). Within the ANS, the sympathetic nervous system (SNS) directs the horse's instincts. This keeps the horse safe—for instance, alerting him to danger and invoking the fight or flight reaction. Lastly, the ANS's parasympathetic nervous system controls more of the horse's relaxed systems, including those of sleep, digestion, and relaxation.

BUILT FOR SPEED

The horse is one of the fastest land mammals. The Quarter Horse, for instance, can reach 47.5 mph in a quarter-mile sprint. While some horses are built for great speed, others are built for great power. One team of draft horses was recorded pulling a load weighing 5 tons. Well-conditioned horses have large hearts and lungs, not only for speed but also for stamina.

Senses

The senses of a horse work together not only to protect him but also to ensure that he thrives. Eyes are positioned on each side of the head, providing almost 360-degree vision. Acute hearing allows the horse to react to danger before it actually presents itself. The horse's sense of smell, which is his most developed sense, not only serves as a way to recognize friend or foe but also aids in survival by helping him locate food and water.

Hearing: Hearing is one of the horse's most acute senses, one that the horse relies on to keep him safe. The

BLANKET

BRINDLE

DAPPLE GRAY

LEOPARD

STRAWBERRY ROAN

TOBIANO

Coat Colors and Patterns

Horses comes in many different coat variations; from dark to light and spots to splashes, there's plenty of variety. Here are some common and not-so-common coat colors and patterns.

BAY: Bay horses sport various shades of brown coats (from red brown to seal brown) with black points on the legs, the mane, and the tail.

BLACK: Black horses have black coats (no trace of brown hair), with black tails and manes. They may have some white markings on their faces or on their legs, as can horses of other coat colors.

BLANKET: Horses with blanket coats have white on their hips as well as on their loins and may or may not have some spots.

BLUE ROAN: Horses categorized as having blue-roan coats, have black or black-brown coats with intermingling white hairs.

BRINDLE: The horse with a brindle coat has a rare striped pattern that is found along his neck, back, hind-quarters, and upper legs.

CHESTNUT: Chestnut horses have coats of various shades of red, which range from golden copper to dark liver.

DAPPLE GRAY: The dapple gray horse has a coat of basic gray with darker grays hairs that form rings.

DUN: Dun horses are sandy yellow or reddish brown and sometimes have a dorsal stripe and other dark marks.

GRAY: A gray horse has a white coat or a gray coat with dark skin underneath.

GRULLA: Grulla horses have a type of dun coloration with smoky or mouse-colored coats and high black points and primitive markings, such as stripes.

LEOPARD: Horses with leopard markings have light-colored coats and have Dalmatian-like spots over their entire bodies.

OVERO: The overo pinto has uneven splashes of white that are to be found on the belly (but do not cross over the top line), the legs, the neck, and the head.

PALOMINO: Palomino horses have golden coats, with white or light-colored manes and tails, and a minimal amount of black.

PIEBALD: The piebald is a pinto horse sporting large patches of black and white. This is a term commonly used in the United Kingdom to describe pinto coloring.

SKEWBALD: *Skewbald* is a term commonly used in the United Kingdom for a pinto horse with large patches of brown and white.

SNOWFLAKE: A horse with snowflake markings has white spots or flecks on a dark body.

STRAWBERRY ROAN: The coat of a strawberry roan horse is chestnut in color with intermingling white hairs.

TOBIANO: A pinto horse with this pattern has white across the topline (which extends downward), white legs, and a dark head.

TOVERO: A pinto with this rare coat pattern may have one or two blue eyes and a mostly white body, except for the ears, poll, and neck.

WHITE: A white horse must have a white coat with pink skin and hazel, blue, or brown eyes.

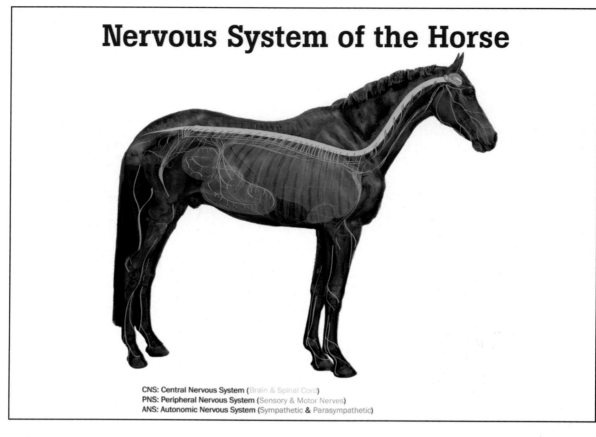

Nervous System of the Horse

CNS: Central Nervous System (Brain & Spinal Cord)
PNS: Peripheral Nervous System (Sensory & Motor Nerves)
ANS: Autonomic Nervous System (Sympathetic & Parasympathetic)

ears of a horse rotate nearly 180 degrees, which allows him to detect and identify sounds, determine where they are coming from and whether they mean danger, and decide what direction he can run in for safety.

Touch: Receptive nerves under the coat allow the horse to feel the tiniest sensations, such as the presence of a fly on his back. His sensitive muzzle acts like a hand in some cases. It processes information that the horse cannot see in front of his face. A horse's sensitive lips can actually sort out bitter medicine from regular feed.

Taste: Horses take pleasure in eating and, when presented with a choice, will select food that is flavorful. They have a palate that prefers sweet to bitter. They often reject water with an unfamiliar taste.

Smell: This is the horse's keenest sense. A mare can identify her foal, a stallion can detect a mare in season, and any horse can locate water, even underground—all by scent.

Sight: Horses can see with each eye separately, an ability called monocular vision. This means the horse can process information in front of and behind him simultaneously. Although they are not color-blind, horses do process colors differently than humans do.

Body Language

Horses display how they are feeling with various types of body language. Although horses make a host of different sounds, most of their communication is nonverbal. Mouthing, chewing, or licking the lips depicts submission, usually by a young horse or a horse willing to take a lower position within the hierarchy. A horse might display this to a human when he is being compliant during training. It can also be a sign that the horse is experiencing anxiety or confusion.

Pawing: Pawing indicates impatience, nervousness, or irritation; a horse tied to a post, for example, will paw to show that he is bored and wants to be released. A horse will also paw to show an eagerness for a reward, such as a treat or some attention. When a horse paws the ground during turnout, he's checking out the ground quality before rolling around.

Points of the Horse

Ear position: The horse's ear position can depict many different emotions. Ears held upright indicate interest or curiosity, those laid back flat show displeasure, and those pricked well forward point to a feared item. One or both ears swiveling back show the horse is listening to his rider—meaning his attention is on the rider's cues.

Tail position: The horse's tail is quite expressive. A tail held upright can reveal that the horse is feeling spirited or fresh, while a clamped tail shows fear. A lashing tail indicates displeasure. A softly swinging tail represents contentment.

Gaits and Movement

The horse has four main gaits, or paces: walk, trot, canter, and gallop. In addition, some horses have a unique way of going—outside of these four gaits—that comes naturally to them (although many are enhanced with training). These are ambling gaits such as the running walk, the foxtrot, the rack, and the single-foot.

The walk: This is the slowest pace and is known as a four-beat gait because each footfall is separate. It averages about 4 miles per hour.

The trot: This is a two-beat gait, in which the legs move in diagonal pairs (front right with the back left, and front left with the back right). While the average speed of a saddle horse is about 8 miles an hour (similar to the pace of a running human), there are breeds of horses that are trained to trot as fast as a Thoroughbred can gallop, such as the Standardbred. A two-beat gait called the pace also exists, with the horse's legs moving in lateral pairs instead of in diagonal pairs.

The canter: This is a three-beat gait that is faster than the trot but slower than the gallop. It is a natural gait for all breeds, even for gaited horses. The direction is determined by which leg the horse leads with. In the canter, one of the horse's rear legs—the right rear leg, for example—sends the horse forward. At this time, the horse is supported only on that single rear leg, while the remaining three legs are in the air. On the next beat, the horse lands on the left

A western rider lopes his Appaloosa. The lope, a slower version of the canter, was used to cover long distances over the range.

rear and right front legs, while the other hind leg is still momentarily on the ground. On the third beat, the horse lands on the left front leg, while the diagonal pair is momentarily still in contact with the ground. The process then repeats itself.

The gallop: This is like a very fast canter, except that there are four footfalls to the beat of the gallop and the horse's stride lengthens considerably to cover a greater amount of ground. At one point during the gallop, all feet will be off the ground. The gallop averages 30 miles an hour.

In addition, there are two more gaits western riders train into their horses: the *jog* and the *lope*. The jog is a very slow trot. It was originally developed so that the western rider would be able to sit the gait comfortably. The lope is a slower version of the canter. Like the jog, the lope was created for the rider's comfort and control of the western horse.

LIFE CYCLE OF THE HORSE

The gestation period for a horse is eleven months. As with most prey animals, a foal is born quickly so his mother can recover and rejoin her herd in a short time. Foals will rise to their feet usually within an hour of birth and will be nursing within about two hours. They must nurse right away to get the all-important first milk, or colostrum, into their systems. This milk, produced by the mother for only a short time, contains antibodies that help stave off infection and disease.

A foal that does not receive this milk from his mother may not survive.

After feeding, the foal gains strength and can walk and trot easily. After his first day, the foal grows stronger and more agile—being able to gallop, buck, and play. The milk of the mother helps her foal grow quickly, packing on the pounds. He will begin nibbling on grass or feeding within about two weeks of birth, trying out different flavors and mimicking the older horses. At about eight weeks of age, he will start to need nutrients added to his diet, as his mother will not be able to supply all the nutritional requirements of her growing baby.

Foals are usually weaned from their mothers between four and six months (although some handlers may choose to wean later, as it is more natural). These animals are now called weanlings. At the age of one year, they are called yearlings, which is the puberty stage of a horse's life. Yearlings are awkward and gangly, as they experience a rapid growth between the ages of one and three. While some horses, particularly racehorses and western stock horses, are trained and ridden as two-year-olds, the general school of thought is that horses are not fully grown until they reach their third year of life.

The bodies of most horse breeds are mature at this stage, but three-year-olds still have the desire to play and can be clumsy. They have the capacity to learn, but their attention spans are short. As they grow older, their mental maturity allows them to retain information from their lessons, focus, and respond more quickly.

Horses from the ages of five to seven are often considered green, meaning that they are still in the learning phase of their training. After this age, they are in their prime of adulthood, from eight to twelve years, with many competition horses considered to be at their peak performance. Although horses age twelve to fifteen are still competitive, they are considered more seasoned and settled into their roles.

After age fifteen, many horses begin to develop soundness or health issues, such as degenerative joint disease. Although the horses still have many years of life ahead, some people consider them aged horses. A horse in his late teens is a senior horse. A horse is considered to be old or geriatric when he has reached twenty years of age, but many horses this age are still physically fit if they have received proper care and use throughout their lives. Others, however, are only suitable for light work or must be retired to pasture.

COMMON CONFORMATION DEFECTS

It's important to be aware of conformation defects and the right terms for them when you're buying a horse.

■ **Back at the knee:** The knee is set behind the front lateral leg plumb line.

■ **Base narrow:** The legs are closer together at the hoof than they are at the chest.

■ **Base wide:** The legs are farther apart at the hoof than they are at the chest.

■ **Bench knees:** The cannons are set to the outside of the knee so an imaginary plumb line does not fall through the middle.

■ **Coon-footed:** The slope of hoof wall is steeper than the pastern, often associated with long, sloping pasterns tending to the horizontal.

■ **Cow-hocked:** The hocks are close together.

■ **Ewe/upside-down neck:** A neck with internal structures that causes it to invert.

■ **Jumper's bump (hunter's bump):** A protruding bump at the top of the croup, or a misalignment.

of the croup with the pelvis and lumbar vertebrae, caused by the tearing of a ligament at the top.

■ **Knife-necked:** A long, skinny neck, with poor muscular development along the crest and throat. It looks like a straight crest with little substance below.

■ **Mutton withers:** Flat and wide withers, from short spines off the eighth through twelfth vertebrae, resulting in a rounded look at the front of the spine.

■ **Over at the knee:** The knee is set in front of the front lateral leg plumb line.

■ **Overshot or undershot jaw:** The upper jaw juts out farther than the lower (overbite), or the lower jaw sticks out farther than the upper (underbite).

■ **Pigeon-breasted:** The front legs are situated too far back under the body, giving a bulky, puffy look to the chest area when viewed from the side.

■ **Sickle-hocked:** The hind hocks are in front of the rear plumb line and set too far under the body.

■ **Toed-in:** The horse stands in a pigeon-toed stance.

■ **Toed-out:** The horse stands in a duck-footed stance.

REPRODUCTION

In the natural world, horses breed in mid- to late spring, so that means the subsequent early spring would be the ideal time for mares to foal. When a mare is ready to breed, she is said to be in season. Depending upon the mare, she may ovulate (go into estrus) several times a year. A mare is a seasonally polyestrous animal, meaning that she has regular estrous cycles during a certain time of the year (usually late spring through early fall) and none during others (winter). This prevents the arrival of a foal during bad weather. The entire estrous cycle from ovulation to ovulation takes approximately twenty-one days.

The mare will display receptive behavior (estrus) toward the stallion for five to seven days and will ovulate in the last twenty-four to forty-eight hours of that period.

Although the natural breeding process for horses is to mate in mid- to late spring, people often alter the schedule. Thoroughbreds bred for racing, for example, are all given an automatic birthday on January 1; therefore mares are bred so that their foals will be born as close to January 1 as possible. That way the horses will have as an advantage competing against others in their age group born later in the year. Sometimes this backfires, however; a foal born on December 31 and

A Quarter Horse mare watches over her foal. Breeders often use various techniques to affect the timing of a foal's birth.

Desensitization: Although their flight instincts are strong, horses can overcome those instincts. Police horses are prime examples of desensitized animals. The key is to introduce a horse to a stimulus in small doses. A trainer, for example, can work on desensitizing a horse to a fear of a large truck by allowing the horse to slowly approach the truck and stop when he begins to show a frightened response. When the horse has lost his trepidation at that range, the trainer can move a step forward, and then another until the horse calmly accepts the truck in his path.

Classical conditioning: Like Pavlov's dog, which learned to associate the sound of a bell with food, a horse can be conditioned to give a specific response. Through conditioning, a horse can learn to respond to verbal cues as well as physical ones. If a reining rider tells a horse to "whoa" and gives all the necessary cues to pull that horse into a sliding stop, the horse will eventually slide to a stop hearing just the word.

Operant conditioning: Most horses are trained with this method. When a horse is being taught a new task, he usually doesn't know quite how to respond to a cue. He will give a random response. Eventually, the

registered with The Jockey Club automatically is one year old the next day. When an early foal is desired, breeding barns will keep mares under lights during the winter in order to simulate lengthening days, which triggers the mares to go into estrus sooner.

INTELLIGENCE

Horses can be very clever; they can learn, and they can reason. Their reasoning, coupled with their instinct, helps them quickly read the body language of other horses and humans. They respond to non-verbal cues. In the natural world, horses need their intelligence to deal with many challenges, such as inconsistent meals and water, predators that can change tactics and location, and an important pecking order that is in a constant state of flux.

Domesticated horses have even greater challenges because of the largely artificial environment that they exist in. Humans ask them to suppress their natural instincts, while responding to cues and training that is very foreign in the natural horse world. Horses learn in a many ways, including desensitization, classical conditioning, operant conditioning, and habituation.

A trainer rewards a well-behaved horse with a treat. Horses are smart and learn from various training techniques.

response will be correct. When that happens, the horse is given positive reinforcement immediately. This tells the horse that he performed the right behavior. With this conditioning, the horse learns how to respond appropriately to the stimulus.

Habituation: This simple form of learning basically states that if the horse is repeatedly exposed to a stimulus, the fear reaction is reduced or disappears. For example, a horse may be frightened of a tarp flapping on a roof. Extended hours in the presence of the flapping tarp will eventually make the horse overcome his trepidation.

BEHAVIOR

Horses behave in myriad ways that may seem baffling to humans. Of course, the behavior makes perfect sense to the horse. So it serves any equestrian well to learn what's behind a horse's action. Understanding the complex world of equine behavior is key to a successful human/horse relationship.

Herd Hierarchy

We use the term *herd* to refer to groups of horses, but the grouping is more complex than meets the eye. In the wild, herds comprise tightly knit groups called *bands*. A band may consist of one stallion, a mare or harem of mares, and their foals. The lead mare (sometimes called the *boss mare* or *alpha mare*) runs the herd, directing everyone to safety, locating food and water, and controlling the group when the stallion takes off on his own. The stallion in the herd is in charge of keeping everyone together as a unit and protecting them from predators and from being taken over by another stallion.

The young fillies and colts between two and four years of age usually drift in and out of the band until they leave for good to form their own bands. When young males reach sexual maturity, the lead stallion drives them out of the herd. Because horses are social creatures, older colts and young stallions must have company to survive. They form bachelor bands until they gather mares of their own.

Chincoteague horses off the Virginia coast cross the water in an orderly procession, displaying typical herd behavior.

Quarter Horses catch a light nap in the snow on a horse ranch in Colorado. For deeper REM sleep, horses have to lie down.

A social order within herds allows for a peaceful, harmonious life. The hierarchy allows orderly access to limited resources, such as the best forage, water, shade, and shelter. Hierarchy is established among the mares and their foals, and the pecking order changes whenever members leave or join the herd.

Sleep Positions and Patterns

Horses can enter a light sleep while standing up, though they will lie down for short periods of time to allow for deeper, REM sleep. Modifications in their muscles, ligaments, and tendons allow them to rest or doze while standing. Without any active effort or energy, horses can use these structures to hold their joints in place.

Horses are believed to have evolved in three ways to allow them to stand while sleeping: the reciprocal mechanism, the stay apparatus, and the locking mechanism in the stifle joint. The horse's neck lowers during sleep and is supported by the suspensory ligament of the neck. Experts think horses developed the

ability to sleep while standing because they were prey animals on the open plains. Equids could flee faster if they napped on their feet—it takes much longer to wake up, get up, and run than to just wake up and go. That head start can mean the difference between staying alive and becoming a tasty meal.

There are other theories about why horses developed the standing-sleep ability: the longer an animal could stand, the more grass he could eat; and lying down to snooze was too hard on the horse's ever-increasing size.

Horses can and do lie down, though not for the eight or so hours that most people do. Their weight affects the position in which they lie down and the length of time they do it. Horses usually sleep upright on their chests rather than flat out. Sometimes horses do sleep flat out on their sides, but it's usually only for fifteen or twenty minutes at a time. If a horse does lie down for a long period of time, there can be serious consequences. Doing so can interfere with blood flow. (That's one reason why horses under anesthesia can sometimes have health problems.) Young foals will sleep on their sides for longer, but only because they don't have the same body-weight dilemma.

Horses get most of their rest in short naps and light sleeps. Because little research has been done in this field, we don't know how much REM sleep horses actually need each night; it's probably minutes, less than half an hour. The slow-wave sleep is certainly needed for hours, about three or four. This sleep is apportioned out in short naps and mostly during the night, between midnight and 4 a.m.

Stereotypies

Horses, like other species, get stressed out. Horses want routine, adherence to a schedule, and mental stimulation. When these needs aren't met, horses may show stress in a variety of ways. Stereotypies, or vices as they are commonly called, are behaviors that some horses display when they are stressed or bored. These behaviors occur in domestic horses, not feral ones, and usually in horses that are kept in confined spaces.

Some people theorize that diet plays a role in vices, that feeds are too different from the forage horses would normally select while grazing. Whether this is true or not, horses are grazers, and they can be stressed if they aren't allowed to graze. Here are some common stereotypies.

Cribbing or wind sucking (aerophagia): Bored horses may develop the habit of cribbing, a vice that is extremely difficult to break. A cribbing horse will place his upper front teeth on a stall door, rail, or post, then tense his neck muscles and gulp air, emitting a grunting sound. This is believed to release endorphins in the brain, which the horse becomes addicted to. Studies have also shown that ulcers are a primary cause of cribbing.

Stall walking: like a caged animal in a zoo, a horse left in his stall for extended periods of time with no stimulation and no exercise will begin to pace or walk circles around his living area.

Weaving: A weaving horse will shift rhythmically from forefoot to forefoot, back and forth, sometimes swinging his head as well. Stall weaving, which is caused by stall confinement and lack of exercise, can lead to weight loss, poor performance, and lameness.

Wood chewing: This term refers to when a horse actually bites and chews wood, destroying fences and barns—as well as his teeth—in the process.

To avoid these vices and others, allow your horses to live in groups in large paddocks. If horses must be confined in smaller spaces, make sure that the stalls allow the horses to see and even interact with their neighbors. Additionally, horse owners or handlers should provide an adequate exercise regimen and enough forage to satisfy the stalled horse's need to graze.

To relieve boredom, horses often crib, as this one is doing. This stereotypy, a compulsive behavior, can be difficult to break.

Breeds and Types

Section 2

Horses evolved over the course of 60 million years, largely as a result of natural selection. When the horse was domesticated, humans began to breed equines with the characteristics various groups of people preferred, such as strength and size for working the land or speed and stamina for racing. Breeders drew from existing bloodlines and perfected characteristics over generations until a breed was fixed. Through careful manipulation, humankind has created a variety of breeds and types of horses for all needs, including work, competition, and pleasure. Breeding is as much a science as it is an art, and the resulting horses come in a host of different sizes, colors, and capabilities.

Chapter 3

Type versus Breed

Some horses fall into the category of type, in particular the mixed breeds or unknown breeds that don't fit into a specific breed registry. Horses within a certain breed share distinctive inherited characteristics (such as the uniquely shaped ears of the Marwari), as well as common ancestors. The characteristics shared by horses of a certain type are more general (such as strength or agility), and the horses don't necessarily have common ancestors.

All modern horses are descended from three ancient foundation wild horses; from those foundation horses came four basic types of ponies and horses.

ANCIENT AND BASIC TYPES

The modern horse originated from three foundation wild horses: the Forest Horse, the Tarpan, and the Asiatic wild horse. It is thought that four basic types of horses and ponies, derived from the three ancient types, created a secondary foundation line.

Three Ancient Ancestors

The Forest Horse (*Equus caballus silvaticus*) lived through the postglacial times, inhabiting the forests and marshes of northern Europe. It stood 15 hands high and was a sturdy animal with coarse features. It had a substantially thick coat, mane, and tail, as well as the large hooves appropriate for a swampy environment.

The true wild Tarpan (*Equus caballus gmelini*) was basically hunted to extinction during the eighteenth

and nineteenth centuries; the last one died in 1887, in the Munich Zoo. At one time, Tarpans roamed over all of eastern Europe and the western part of Russia. This sturdy horse had a large head, a thick neck, an upright mane, and strong jaws capable of chewing woody forage. The breed was extremely hardy and very resistant to disease. Tarpans were small horses, standing 12 to 13 hands high and were grulla (a mouse dun) in coloring, with primitive markings, such as a dorsal stripe.

The Asiatic wild horse (*Equus przewalskii* or *Equus caballus przewalskii,* depending on whether the horse is viewed as a subspecies of *Equus caballus*) is the only one of the three still in existence and in its natural environment. It is most commonly known as the Przewalski horse. It was thought to be extinct until Russian cartographer Colonel Nikolai Przewalski saw a herd of dun-colored horses while he was in Mongolia in 1879. These horses evolved on the steppe—a large, semiarid, grass-covered flat land sprinkled with woods.

Four Basic Types

These postglacial types developed 5,000 to 6,000 years ago, before horses were domesticated. The conformation, hardiness, size, and movement of each breed that originated from these four types emerged mostly because of its environment. Later, humans crossbred horses or selected for certain characteristics, such as movement or strength, to create the breeds we know today.

A pair of Highland Ponies cross a snowy field in Britain, inured to the cold like their hardy Eurasian ancestors.

Competitors show their Caspians at the Aspahan Horse Festival in Iran, where this small elegant breed originated.

Pony Type 1: This was a small (around 12 hands high) bay or brown equine from northwest Europe that was nearly impervious to cold and rain. It is thought to have derived from Tarpan stock. The closest equivalent of this type is the Exmoor Pony (*see page 46*).

Pony Type 2: This was a larger (around 14.2 hands high), heavier dun pony from northern Eurasia that was resistant to cold and snow. It is thought to have derived from the Asiatic wild horse. The equivalent of this type is the Highland Pony (*above*).

Horse Type 3: This was a desert horse that lived comfortably in harsh, hot climes. It is thought to have originated from the Tarpan and to a lesser extent from the Asiatic wild horse. Type 3 developed in Central Asia and migrated into areas of Spain. It was a thin-skinned, narrowly built horse with a long head and ears that stood around 14.2 hands high. The closest equivalents are the Akhal-Teke and the Sorraia (*see chapter 4*).

TUNDRA HORSE

The Tundra Horse is another extinct wild horse. However, because it had little bearing on the breeds today, it is not considered a foundation horse. The Tundra, which developed in the Arctic Circle, was pure white with very long hair. It could withstand severe weather conditions. It is thought that a small pony found in Siberia called the Yakut is a direct descendant of the Tundra. Some argue that the pony comes from the Asiatic wild horse.

Horse Type 4: This type developed in western Asia and was resistant to heat and drought. It stood only 12 hands high. This horse was more elegantly built than the other types, with a dished face and a high set on neck. It is thought to have come from Tarpan stock. It's close modern equivalent is the Caspian (*above*).

MODERN TYPES

The modern types of horse have particular looks or specific uses. Cob types, for example, are compact and powerful, as well as smooth-gaited.

Baroque: The Baroque breeds descended from horses of the Middle Ages (ca. AD 500 to 1500), such as the Neapolitan Horse, the Barb, and the Iberian Horse, represented today as the Sorraia. Modern breeds in the Baroque category include the Andalusian/Pura Raza Española, Friesian, Lipizzan, and Lusitano. Lesser-known Baroque breeds include the Czechoslovakian Kladruber, the Danish spotted Knabstrupper, and the Fredricksborg. (*See these breeds in chapter 4.*) They are usually used in dressage. The name Baroque is derived from the Baroque period (ca. 1600 to ca. 1750), when riding was elevated to an art form.

Cob: With the exception of the Welsh Cob, which is a true breed, the term *cob* denotes a type of horse, not a breed. A cob is usually 14 to 15 hands high and chunky in build, yet still very smooth in his gaits. The steady cob is easy to ride, keep, and train. He is seen ridden, driven, and shown. His mane is traditionally shaved off, or hogged.

These two sturdy cobs demonstrate the steady trot that makes them popular riding horses in the British Isles.

Hack: The hack is an animal chosen for its comfortable gaits and rideability. The hack type is common in the United Kingdom. The terms *hack* and *hackney* come from the Norman word *haquenné*, which means a "riding horse." Today's modern hack is called a show hack and is ridden English: sidesaddle and astride. Show hacks must have impeccable manners and flowing, easy-to-ride strides. They are generally a fuller-bodied animal. The hack stands from 14.2 to 16 hands high and is found in all solid colors.

Hunter: The brave hunter is a sturdy and well-muscled horse or pony that is confident moving over open country and jumping over natural obstacles at an even pace. He's bold, sure-footed, and easy to handle. Top hunters are often Thoroughbred, draft horse, and pony crosses. Today hunters are found in the field and in the show ring.

Polo Pony: Nimble and fast, the polo pony is not exactly a pony; in fact, there is no height limitation, although heights average around 15 hands high. Polo ponies have to be able to spin on a dime and accelerate at a rider's whim, while also being calm enough to handle a mallet swinging around their bodies. Many polo ponies are Thoroughbred or Quarter Horse crosses.

Sport Horse/Sport Pony: Sport horses and sport ponies are bred for the English sports of dressage, jumping, and eventing. These horses come in all sizes,

This hack performs at the famous 2010 English Hack Championship at Hickstead, in England.

These two foxhunters take the fences in their stride while out hunting.

Archaic Types and Extinct Breeds

Although breeds did exist as far back as the medieval period, few pedigrees were kept, and horses were usually referred to by the work they did. Archaic types include:

Courser/Charger: A light, swift horse, the Courser was used in battle.

Destrier: This compact, strong battle horse was trained primarily for jousting.

Hobby: This small riding horse was usually found in the British Isles.

Palfrey: The Palfrey was a well-bred, docile riding horse, usually gaited.

Rouncey: Rouncey referred to an ordinary, all-around riding horse or packhorse.

Certain breeds became extinct because they were amalgamated into other breeds and were no longer purely bred. They include:

Chapman Horse: The Chapman was an ancestor of the Cleveland Bay.

Galloway Pony: Native to England, the Galloway Pony was an ancestor of the Fell Pony and the Highland Pony.

Irish Hobby: This small, handy horse was developed in Ireland from the Barb.

Narragansett Pacer: This gaited horse was developed in the United States in the eighteenth century from Spanish and British bloodlines.

Neapolitan Horse: The Neapolitan was bred in Naples, Italy, from the fifteenth to the eighteenth centuries, for cavalry use and riding. Two purebred Neapolitans, Neapolitano and Conversano, served as foundation stock for the Lipizzan. Another Lipizzan foundation stallion, Maestoso, was half Spanish and half Neapolitan.

Norfolk/Yorkshire Trotter/Roadster: Developed in England during the 1500s, this breed (known by several names) influenced the Hackney and the American Standardbred.

Old English Black: The Old English Black developed in Britain in the eleventh century from heavy horses brought to the country by Norman invaders. This breed influenced the Clydesdale and the Shire.

Spanish Jennet: This was a small Spanish horse with an ambling gait. Many Spanish Jennets were exported to America, and many American breeds have the Jennet blood. The Paso Fino and the Peruvian Horse are the closest remaining relatives.

Turkoman: The Turkoman breed developed in the deserts of Central Asia. One of the Thoroughbred's foundation sires, the Byerley Turk, is believed to have been a Turkoman. The Akhal-Teke is the closest living relative of this now-extinct breed.

Yorkshire Coach Horse: A flashy coach horse created by crossing the Cleveland Bay with the Thoroughbred. Breeding came to an end in the 1930s, when carriage horses were no longer needed.

---

shapes, colors, and heights and are of various breeding. Sport horse breeders often draw from the best athletic breeds, such as the Thoroughbred, the Irish Draft, and the Connemara. The most important aspect to look for when choosing breeding stock is the horse's athleticism. Sport horses and ponies from New Zealand, Australia, and Ireland are in great demand.

Stock horse: The stock horse is a type or breed of horse skilled at ranch work, in particular working with cattle. In fact, the name comes from the fact that they work "stock." Although heavily muscled horses, they are extremely agile, with the ability to move quickly in any direction, which is important when working with livestock. Stock horse breeds include the Quarter Horse, Australian Stock Horse, and Paint (*see chapter 4*).

As players battle in close quarters for the ball, their horses display the calmness and agility requisite in polo mounts.

Irish Sport Horse Tankers Town and Sharon Hunt compete at the 2008 Olympics. Sport horses are bred for athletic ability.

A Suffolk Punch pulls a plow through a field. Draft horses are known for their great strength and stamina.

WHAT IS A BREED?

Because of their foundation stallions, some breeds, such as the Morgan and the Suffolk Punch, are very uniform in appearance. Others have variations within their breeds, such as the stocky working cow horse and the refined hunter in the Quarter Horse breed. Each variety holds the breed's essence but differs from others according to its use. Breeds fall into categories: draft, light, hot/cold/warmblood, gaited, pony, and color.

Draft Horse/Heavy and Light

The term *draft* (also spelled *draught*, particularly in Great Britain) was initially used to describe any horse that drew a vehicle. Originally bred for farmwork, draft horses are the largest breeds of horses and are incredibly powerful. The larger draft horses, such as the Shire and the Suffolk, are called heavy horses. The Irish Draught is considered a light draft horse. Smaller than other draft horses, the Irish Draught is used increasingly for riding these days. Other examples of draft horses include the Clydesdale, the Belgian, and the Percheron. The body of a draft horse is large and dense, with a muscular neck and broad, deep chest. Proportionally, the legs are usually short.

Light Horse

The term *light horse* is used to describe any breed used primarily as a riding horse. Although light horses can be any size (the diminutive Caspian and the Miniature Horse are light horses), they usually stand 15 to 17.2 hands high, with a proportionate body. That means, viewed from the side, the horse can be visually divided into three equal parts: neck, shoulders, and forelegs; back and rib cage; and hips, hindquarters, and hind legs. The light horse is well muscled and agile, with easy gaits. Some, such as the Arabian, have great stamina, while others, including the Quarter Horse—also categorized as a stock horse, which denotes the traditional western breeds—have brilliant cow sense. Still others, such as the Thoroughbred, are fast, and others are high stepping, like the Saddlebred. Light horses are found in western and English riding, as well as under harness.

Hot/Cold/Warmblood

Within breeds there are blood designations. Coldbloods are draft horses. Hotblood refers to Arabians, Barbs, and Thoroughbreds and their unique bloodline. Light

Arabians thunder across the Royal Stable grounds of Abu Dhabi, displaying the fierce spirit of the hotblooded desert breeds.

horses are often called warmbloods because they are a mixture of cold- and hotblooded breeds. *Warmblood* today denotes a sport horse of European ancestry.

Pony

Ponies stand below 14.2 hands high, but all short equines are not ponies. Ponies and horses share common ancestry, but ponies have different body proportions. The body length is bigger than the height at the withers, the legs are shorter, and the head length generally equals the length of the shoulder. Ponies also tend to have more primitive features, such as long, thick manes and tails, heavy coats, and hard hooves. They are hardy, easy keepers. Ponies haven't been crossbred as much as horses have and are therefore closer to the natural instincts. As a result, ponies can be more independent thinkers. Ancient pony breeds include the Shetland and the Exmoor. More modern breeds include the Pony of the Americas and the Hackney Pony (*see chapter 4*).

Gaited Horse

There are many breeds of gaited horses and many types of gaits. The most popular breeds with natural gaits include the Tennessee Walking Horse (aka Walking Horse or Walker) and the Missouri Fox Trotter. The Walker has

A regal Paso Fino performs the classic fino gait, which includes rapid, deliberate footfalls.

a running walk, a four-beat gait in which each hoof strikes the ground independently. The Fox Trotter's hallmark is a broken gait, in which the front legs walk and the back legs trot. The Racking Horse and the Rocky Mountain Horse have gaits that are similar to those of the Walker.

Other naturally gaited horses include the Icelandic Horse, which has five gaits, including the *tölt* (a four-beat walk) and the *pace* (a speedy, lateral trot).

A golden coat and white tail and mane are hallmarks of a palomino, one of the four color breeds.

Some Spanish horses, such as the Paso Fino and the Peruvian Horse, walk and canter but do not trot. The Paso Fino's gaits include the *classic fino* (the slowest gait, covering as much ground as possible using rapid footfalls), the *paso corto* (smoother than the trot but comparable in speed), and the *paso largo* (the fastest gait, similar to a canter or slow gallop). Each foot in these gaits strikes the ground independently. The Peruvian Horse has the same four-beat gait as the Paso Fino, but the slower gait is the *paso llano* and the faster the *sobreandando*. Enthusiasts of gaited horses prize the mounts for their easy-to-sit, comfortable strides.

Color

In addition to horses bred for their conformation and ability, some horses are prized specifically for coat color. These horses generally can be of any height, breed, or conformation; here the coat color is what's desired. Horses may be registered in a breed association and be registered in an additional color-breed association. Color breeds include American White and American Creme, buckskin, palomino, and pinto.

AMERICAN WHITE AND AMERICAN CREME

White horses are different from Cremes; they can vary from off-white to an almost palomino color. They often are born slightly deeper in color and lighten within their first year. A Creme can have white markings—such as stockings, a star, or a blaze—but these generally will only be visible when the coat is wet. Their manes can be the same color or slightly darker than their coats. The skin must be pink or shades of tan, never gray. Cremes

Horse talk

TYPES OF HORSES

Knowing the following terms will help you to talk horse types.

■ **Albino:** A misnomer for horses, because an albino technically must be without any pigmentation, including eye color.

■ **Cob:** A medium-size type of horse, very chunky in build yet still very smooth in his gaits.

■ **Coldblood:** A draft horse.

■ **Draft (draught):** A strong horse used to pull equipment.

■ **Hogged (roached) mane:** Mane clipped completely off.

■ **Hotblood:** Arabian, Barb, or Thoroughbred.

■ **Hunter:** A horse that is used in the sport of foxhunting.

■ **Light horse:** Finely built breeds used for riding.

■ **Sport horse:** A type of horse that excels in English sports such as dressage and jumping.

■ **Sport pony:** A type of pony that excels in English sports such as dressage and jumping.

■ **Warmblood:** An amalgamation of hot and cold breeds; today it denotes European sport horses.

differ genetically from Whites. The genes of Cremes are diluted from darker horses such as palominos and sorrels, and their eyes are most often lighter than those of Whites, in shades such as amber and light blue.

Even if a horse looks white or cream, he still must meet certain genetic criteria: he must be born white

This Quarter Horse gets the buckskin color from the Spanish bloodlines, which carry this particular color gene.

a Spanish breed, that with the color comes inherited hardiness. Breeds with buckskin color include the Lusitano, the Quarter Horse, and the New Forest Pony. There is a range of shades in the buckskin category.

Buckskin: This is a deer hide color with black or brown mane, legs, and tail. Coat colors range from yellow to dark brown.

Dun: This is a duller buckskin with smuttiness (buckskin guard hairs flecked through the coat) and the dun factor, which includes dorsal (down the back) and leg stripes and shoulder bars. Dun horses also have black or brown manes, legs, and tails.

Red Dun: This one has the same markings as the dun but with colors ranging from peach to a deep red. The mane, tail, and legs are chestnut or dark red.

Grulla: This coat is dove, mouse gray, slate, or blue. It has dark sepia or black mane, tail, and legs. It also has dorsal and leg stripes and shoulder bars.

Brindle Dun: The brindle dun is a rare and ancient coat color with zigzag striping throughout the coat.

PALOMINO

The palomino coloring has always been prized. Horses with golden coats were sought after in sixteenth-century Spain; horses that had such coats were called Isabellas, after the queen. A true palomino has dark skin, a cream to dark-gold body, a light mane and tail, and matching dark or hazel eyes, never blue. Breeds with palomino coloring include the American Quarter Horse, the American Saddlebred, the Morgan, and the Lusitano. Palominos can't be specifically bred because the coat color is an incomplete dominant gene and as such does not breed true.

PINTO

The pinto's two-toned coat pattern probably came to North America in the fifteenth century through Arabian and Spanish stock accompanying the explorers. Native Americans preferred the spotted color and bred horses for this characteristic. Western settlers later bred their own horses to Indian ponies out of necessity, which also perpetuated the coloring. The pinto and Paint are often lumped together as the same breed; but the Paint horse must have Quarter Horse or Thoroughbred parentage; the pinto is a color breed and can be of any parentage as long as it fulfills coat requirements. There are four pinto types: **hunter** (Thoroughbred-, racing Quarter Horse–, or warmblood-type conformation); **pleasure** (Arabian- or Morgan-type conformation); **saddle** (Saddlebred-,

or cream colored, which is different from gray horses that are commonly born chestnut or even near black, and then whiten with age. The skin underneath must be pink—also different from gray horses, which have dark skin under their white coats. A White horse also has a white mane and tail, although his eyes can be any color. A White or Creme horse can be of any breed and can come in any size or conformation.

BUCKSKIN

The buckskin color is thought to have originated with Spain's Sorraia. Because the Sorraia blood flows in many breeds, buckskin is a common coat color. It is thought that because the buckskin color comes from

Fox Trotter–, or Walker-type conformation), and **stock** (Quarter Horse–type conformation). There are various colors and two main coat patterns: tobiano and overo.

Tobiano: This horse has vertical color patches across his back and often white legs and a dark head.

Overo: The white in the overo horse will generally not meet over the back. The overo usually has large white markings on the face. Overos come in three types: frame overo (a frame of color around the white markings on the body), sabino (roan at the edges of white markings, and white that extends past the face and chin; some maximum-white sabinos are nearly all white), and splashed white (blue-eyed horses with long white socks, white or blazed faces, and/or white on the bellies).

Tovero: A rare pattern, the tovero is a combination of the tobiano and overo. Characteristics of the tovero include one or both eyes blue; a mostly white body but with ears, poll, and sometimes the top of the neck another color. This feature is called the medicine hat or war bonnet (the war bonnet has less color than the medicine hat; sometimes the war bonnet is the only color) and was greatly prized by the Native American chiefs and medicine men.

Color patterns can also occur on the tovero's hindquarters, chest, or belly. Dark markings surrounded by a large patch of white, such as on the chest or face, is called a shield.

Overos and toveros may carry the "lethal white gene," which means there is a chance the mare will give birth to a lethal white foal that will die or have to be put down shortly after birth due to a defect in the digestive tract.

Native American chiefs and medicine men highly prized the rare tovero medicine hat markings found on pintos like this one.

Chapter 4

The Breeds

There are some 200 recognized horse breeds in the world, developed on every continent except inhospitable Antarctica. Some breeds have been with us for millennia, such as the Barb and the Akhal-Teke; others are only a couple of centuries or a few decades old, including the Georgia Grande and the Missouri Fox Trotter. With so many breeds to choose from, horsemen can find a horse to fit any need, whether it be for working cattle, hauling logs from forest and mountain, jumping cross country, or riding over trails.

North America gave birth to *Equus cabbalas* about 2 million years ago. Over the following millennia, horses migrated from North America across land bridges leading to South America, Asia, and Europe. Some 10,000 years ago, however, horses became extinct in the Americas. They did not return until European colonists brought them over by boat in the 1500s. The New World's inhabitants then began breeding for specific traits and eventually developed several uniquely American equine breeds.

ABACO BARB

Alternative Name(s): Abaco Colonial Horse
Region of Origin: Europe, Spain

History: The history of this small tribe of feral horses, residing on the island of Abaco in the Bahamas, dates back to the time of the Spanish exploration of the New World. Colonists set up areas within the Caribbean for horse breeding, producing cavalry, work, and saddle horses. Many ships carrying horses to these places never made it, and some researchers theorize that horses from shipwrecks near Abaco swam to the island. Others think the horses were taken to the island and later abandoned there.

The island proved to be a suitable home for the animals, which found protection and forage in the lush pine forests. Left without human intervention, the herd's bloodlines remained pure and their numbers grew. That all changed in the twentieth century, when people settled on the main island of Great Abaco and began to take advantage of all the natural resources. They hunted the island's wild game, which included the feral equines. The horses suffered a further blow in the 1960s, when an unattended child climbed onto one of them and was killed as the horse bolted. Villagers slaughtered all but three of the island horses, who were taken to a safe area by some residents. Within three decades, the horses had increased to a band of thirty-five; however, in the early 1990s, that number dropped to half. Horses died from lack of medical attention, pesticide poisoning, and mortal wounds from dogs and people.

In 2002, after extensive DNA testing, investigators determined that the Abaco Barbs were a strain of Spanish Barb horses, dating back to the original stock brought over from the Barbary Coast. The breed was added to the Horses of the Americas registry. Although the herd now has a preserve set aside so that the animals can live once

Abaco Barb

again in their original settlement, funding for the care and protection of the horses has been scarce.

Physical Description: Like its Barb brethren, the Abaco wild horse is a small, sturdy, and compact animal with light but extremely strong legs. Pesticide poisoning has weakened the breed's once-tough hooves, but the surviving horses remain nimble and sure-footed, with a great deal of stamina and endurance. Many of the horses bear splash markings (white patches usually along the lower half of the horse's body), a characteristic of the true Spanish Barb. Their height ranges from 13.2 to 14.2 hands high (hh).

BREED IN BRIEF

ALTÈR-REAL: PORTUGAL

The Altèr-Real comes from Lusitano and Andalusian origins. The breed's name derives from the horse's native village, Alter do Choã, as well as the Portuguese word for royal (meaning real). The breed was established by the Portuguese royal family from Andalusian mares imported from the Jerez stud in Spain in 1748. The Altèr-Real has weathered a difficult history, first nearly destroyed by Napoleon and his invading troops, then later by incorrect breeding. The breed was reestablished in 1932 by Dr. Ruy d'Andrade, using three remaining stallions. The Altèr-Real is an extravagant mover and well-suited to high-level dressage. This breed can be found in all solid colors, usually bay or brown; the horses are 15 to 16 hh.

ANGLO ARAB: FRANCE

The Anglo Arab gets his name from his two foundation breeds: the English Thoroughbred and the Arabian. Although the breed is found across the globe, the French have developed the Anglo Arab into the sporting saddle horse that we know today. The Anglo Arab was developed in the French studs of Pompadour and Tarbes. This elegant breed is raced, driven, and ridden in all English pursuits, particularly jumping and eventing. The Anglo Arab appears in all solid colors and stands 16 to 16.2 hh.

Uses: The Spanish colonists who were responsible for introducing the horses to the New World intended to use them as cavalry, work, and saddle horses.

Of Note: The endangered Abaco Barb is a fragile branch on the Barb's family tree. No foals have been born since 1998. As of January 2011, only five horses remain.

TEKE'S NECK

The Teke's unusually thin, yet flexible neck makes him seem "above the bit" by modern standards. Although the picture may look wrong, the Teke's mouth remains even with a rider's hands.

AKHAL-TEKE

Alternative Name(s): Teke
Region of Origin: Central Asia, Turkmenistan

History: Through archaeological evidence, experts say the Akhal-Teke breed is at least 3,000 years old. They surmise that the Teke is the last remaining strain of the Turkmene (a horse that has existed since 2400 BC). Beginning around the sixth century AD, Teke horses found human companionship in the Turkmen, a nomadic tribe who lived around an oasis near the Kopet Dag Mountains in Turkmenistan. The horses thrived where food was scarce and weather fluctuated between extreme heat and bitter cold. The Tekes survived on the food the nomads provided, a pemmican-type cake of grains and mutton fat fed by hand. The nomads cherished their horses and treated them as part of the family, tethering them individually near their shelters.

In 1881, Turkmenistan became part of the Russian Empire. The Russians originally called the nomad's horses Argamaks, which means "cherished Asian horses." A Russian general founded a breeding farm in Turkmenistan and named the horses Akhal-Tekes, which referred to their connection with the Teke Turkmen tribe who lived near the Akhal oasis. The Russians tried to improve the breed by crossing Tekes with English Thoroughbreds to make the horses taller and stronger. Unfortunately, the resulting Anglo-Akhal-Tekes were so weak that they couldn't withstand the harsh conditions of Central Asia.

In 1973, to protect the purity of the breed, authorities decided all foals would be blood-typed before they were included in the studbook. Today, all new breeding stock is inspected by the Russian Institute of Horse Breeding, and any stallion that is not producing the right type of offspring is stricken from the studbook.

Physical Description: The Akhal-Teke is a standout in form and figure—more than any other horse breed in existence. The breed's odd, gazelle-like appearance gives the impression that this animal is another species. For thousands of years, the Teke developed without any other breed influence. The Teke evolved into a long narrow frame, which created a flat, gliding gait perfect for striding over the stony plains of Turkmenistan's Karakum Desert. The Teke has a small head and a high-set neck, almond-shaped eyes, and a shimmering coat. Tekes stand 15 to 16.2 hh and are found in bay, grey, black, dun, chestnut, and gold.

Akhal-Teke

American Indian Horse

The American Indian Horse Registry (AIHR), established in 1961 "for the purpose of collecting, recording and preserving the pedigrees of American Indian Horses," considers all horses of Native legacy as American Indian Horses regardless of their breed. To the AIHR, the history and origins of all these breeds link them all.

There are different strains of the American Indian horse around the country, such as the Marsh Tacky from South Carolina, the Nokota from North Dakota, and the Florida Cracker. American light horse breeds, such as the Appaloosa and Quarter Horse, are all eligible. Because this is a large group of breeds, the registry distinguishes horses through five classes of registration, determined through photographic inspections. The AIHR also holds shows that include traditional Native American games such as "backfiring the prairie," in which a rider carrying a torch jumps over a brush pile, drops the torch to ignite the pile, then races out of the arena. The fastest time wins.

Despite the variation of breeds, American Indian Horses share similar characteristics. The horses have long, not bunching, muscling, and they usually are not heavy, although the northern-bred horses have developed to be somewhat heavier than southern-bred horses because of the climate. The heads usually have a straight profile although Roman noses are sometimes found. Horses should be hardy and sure-footed. The American Indian Horse is a versatile horse in both English and western pursuits but shines on the trail. All colors are found and heights range according to breed and classification.

American Cream Draft

Uses: The Akhal-Teke excels in flat racing, show jumping, dressage, and endurance racing.

Of Note: Two of the Teke's most striking features are extremely rate. The startling almond-shaped eyes look more like a cat's than a horse's. And the coat is burnished with a shimmering gold polish. Although some shimmer more than others, all Tekes have a gold, metallic polish over the top of their base colors. Other breeds that have this coat feature are thought to have Teke blood.

AMERICAN CREAM DRAFT

Alternative Name(s): Cream
Region of Origin: United States

History: The American Cream Draft is the sole draft breed originating in America, springing from one large, coldblood-type mare whose coat was a rich, milky-yellow color. Sold at a livestock and farm auction in 1911 in Story County, Iowa, to a horse dealer named Harry Lakin, Old Granny, as she came to be called, later became the first registered American Cream. Lakin used Old Granny in his breeding program and was delighted to see her produce the same light-colored foals for him, all of which sold handsomely.

In the 1940s, C. T. Rierson of Hardin County, Iowa, further grew the breed by buying available colts sired by the foundation stallion, Silver Lace, and building his own herd. With the help of the horses' former owners, Rierson painstakingly recorded the pedigree of each

ARIEGÈOIS: EASTERN PYRENEES, SPAIN, AND SOUTHERN FRANCE

This hardy pony is also known as the Cheval de Mérens. The Ariegèois breed developed in the Pyrenees and Ariegèois mountains under harsh conditions. Although this pony is thought to be ancient breed, it also has Arabian and Barb blood from later crossbreeding. This nimble breed is prized for trekking mountains and for farming. Ariegèois are always black; white markings are rare; in the winter, the pony's coat can develop a reddish sheen. The breed stands 13.1 to 14.3 hh.

---

horse. In 1944, Iowa granted a charter to the American Cream Horse Association of America. Six years later, the Iowa Department of Agriculture and Land Stewardship officially recognized the association's breed standard. Genetic testing shows that Creams are a distinct breed among drafts.

Physical Description: The American Cream Draft has a medium draft-horse build, standing 15.1 to 16.3 hh. The ideal horse is a medium cream color with a white mane and tail, pink skin, and amber eyes. The Cream is lighter than a palomino, but darker than a gray. White markings, such as blazes and stockings, are desirable. Pink skin, as opposed to dark or gray, is the determining factor in securing this rich cream color. Therefore, the most desirable American Creams have pink skin, correct color, and white markings that contrast beautifully with their coats. Unlike true white horses, Creams feature eyes the color of amber. Foals are usually born with eyes so light they are nearly white, but they darken with time; when the horses reach maturity, their eyes are topaz. They have well-proportioned heads with straight profiles, arched necks, and muscular bodies. They have nice activity at all gaits.

Uses: American Cream Drafts are small enough to ride under saddle for pleasure riding, but they are still used in harness.

Of Note: After the market for draft horses collapsed in the mid-twentieth century, the numbers of American Cream Drafts dropped drastically. Through the efforts of individual owners and the American Cream Draft Horse Association, the breed has survived and the horses' numbers have slowly increased.

AMERICAN PAINT HORSE

Alternative Name(s): Paint
Region of Origin: United States

History: The American Paint can trace his ancestry to the horses of the Spanish explorers who landed in Central America during the sixteenth century. Later, the native peoples quickly and skillfully developed their horsemanship skills using horses bartered or stolen from the explorers or recovered when the explorers abandoned them; often, the splashy colored horses were tribe favorites. Pinto-marked horses migrated into North America as well, and since Colonial times, these horses of Spanish ancestry have worked on cattle ranches and American farms.

But not all spotted horses are necessarily American Paint Horses. Paint Horses' official origins are a result of the development of the American Quarter Horse. When the American Quarter Horse Association formed in 1940 to preserve its stock horse breed, the organization excluded pinto and "crop out" horses (those born with white body spots or white above the knees and hocks). Pinto stock horse enthusiasts formed a variety of organizations to preserve and promote their animals, and eventually these groups merged in 1965 to form the American Paint Horse Association.

Physical Description: The American Paint Horse is a breed that bears stock horse conformation and a pinto coloration. (Offspring of Paint horses that do not have spots are referred to as breeding stock paints.) The Paints usually range in height from 14.2 to 16.2 hh and come in a variety of colors and in two main

Colors Wanted

Too much white, such as big splashes of white on the body or white splashes above the knees, mark a Quarter Horse as a crop-out. Crop-outs started the American Paint Horse breed, so a Quarter Horse that is considered "colorful" can be registered with the American Quarter Horse Association, and ones that are gelded or spayed can be registered under a hardship clause.

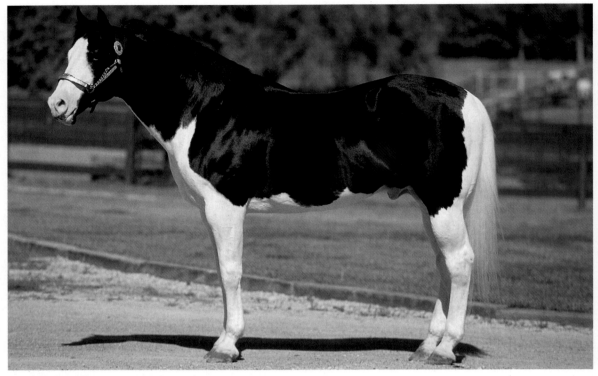

American Paint Horse

coat patterns: tobiano and overo. Tobianos have well-defined vertical patches of color across their backs, and they often have white legs and dark heads. Overos come in three different types: frame overo, sabino, and splashed white.

Frame overos usually have a lot of white on their heads, and they often have one or two blue eyes. They have white horizontal patches on their necks and sides. Frame overos can have white feet or socks, but their legs are usually dark. Sabinos have white on their legs and heads, as well as a bit of white on their bellies.

Roan colorations are typical in sabinos, and some have blue eyes. Splashed-patterned horses have a lot of white on their heads, as well as some on their legs and bellies. Blue eyes are common in this type of overo.

Uses: Paint horses excel in western and performance events; however, those with Thoroughbred bloodlines also do well in English classes.

Of Note: To qualify as a Paint, a horse must have a minimum of natural paint markings. These include: white leg markings that extend above the knees and/or hocks; a blue eye; an apron or bald face; white on the

BREED IN BRIEF

AUSTRALIAN PONY: AUSTRALIA

There are no native horses in Australia, so any Aussie breed has originated from animals brought by early European settlers. The Australian Pony's characteristics reflect his foundation breed, the Welsh Mountain Pony. Other breeds that have contributed to the genetic mix include the Exmoor, the Shetland, the Timor, the Thoroughbred, and the Arabian. The official studbook for the Australian Pony was established in 1931.

This sweet pony is ridden by children and by small adults in gymkhana and in all English pursuits. Although Australian Ponies can be found in all colors, gray is the one that is most often seen. Australian Ponies range from 12 to 14 hh.

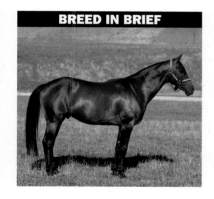

AUSTRALIAN STOCK HORSE: AUSTRALIA

The Australian Stock Horse was originally named the Waler (after New South Wales, where the breed developed). The horse's original bloodlines come from the Cape Horse of South Africa. Because Australian Stock Horses were so strong and versatile, they were used widely as cavalry horses; so many were exported to Europe during WWI that few were left in Australia. Breeders introduced Arabian and Thoroughbred blood after the war to increase the breed's numbers. Today, the Australian Stock Horse looks much like a Thoroughbred, but the breed has a hardy countenance. As this breed is the most plentiful in Australia and is very versatile in use, the horse is used for all English and western pursuits. The breed stands 15 to 16 hh in all solid colors, with bay the most common.

jaw or lower lip; a blue zone around the marking; a two-color mane, one color being natural white; dark spots or freckles in white hair on the face or legs; white areas in the nonvisible zone, excluding the head, completely surrounded by a contrasting color; or the reverse.

AMERICAN QUARTER HORSE

Alternative Name(s): Quarter Horse
Region of Origin: United States

History: During the colonial era, horse breeders worked to develop horses that could work during the week and still be able to run in weekend races, as well as other leisure pastimes. Match races held in fields and along country roads were about a quarter-mile long and required sprinting horses. The colonists began blending Chickasaws, Thoroughbreds, English Pacers, and even some French stock.

Of the Thoroughbred stallions used in breeding from 1746 to 1800, the one with the most influence was Janus, who was directly related to the Godolphin Barb that helped create the Thoroughbred breed. Janus was built more muscled and close-coupled than long-legged Thoroughbreds, and his foals carried the same traits of ruggedness, strength, and athletic ability. The fastest resulting offspring were called Celebrated American Quarter Running Horses. Janus put his stamp on the breed, and even several generations later, his impact on the horses' conformation remains indisputable. Today, all but two of the original eleven foundation lines trace to Janus.

When Americans began adding furlongs to races, the Quarter Horse fell out of favor in the sport, and the Thoroughbred took the racing spotlight. The Quarter Horse went West with pioneers headed to new frontiers in the early 1800s, serving as their primary mode

American Quarter Horse

of transportation, both for riding and pulling wagons. Throughout the new lands, the breed worked on cattle ranches, demonstrating an innate ability to understand a cow's body language.

Physical Description: The average height of the breed is between 14.2 and 15.3 hh. The horse has a compact head with wide-set eyes, small, alert ears, a delicate muzzle, and large nostrils to enhance breathing. The jaws are trademark on some bloodlines: many sport large, powerful-looking cheekbones that have earned them the nickname "bulldog." Quarter Horses have compact frames, sloping shoulders, deep barrels, broad chests, and clean legs. The hindquarters are what make this horse legendary. They're broad, well muscled, and heavy, giving them the ability to propel themselves forward without hesitation.

The American Quarter Horse Association recognizes thirteen color variations, including bay, gray, palomino, and black. The most common is sorrel, a burnished chestnut. The Quarter Horse can have some white on the face and white below the knees. There is also a registry for the Foundation Quarter Horse, which is small, compact, and upright.

Uses: The American Quarter Horse Association formed in 1940 to preserve and promote the breed as a horse for all reasons: a ranch horse, a performance horse, a pleasure horse, and a racehorse.

Of Note: The American Quarter Horse is the most popular and populous breed in the world, with more than 4 million registered horses.

AMERICAN SADDLEBRED

Alternative Name(s): Saddlebred, Kentucky Saddler
Region of Origin: United States

History: In eighteenth-century New England, the British colonists needed to produce a horse suitable for their new environment. Crossing imported stock with sturdy Canadian horses, they created a horse

American Saddlebred

that ambled, which they called the Narragansett Pacer. Although the Narragansett was the foundation breed for many American horses, crossbreeding and exporting led to the breed's extinction. By crossing English Thoroughbreds with the Narragansett around the time of the American Revolution, colonists created a lean, elegant mount that possessed smooth, floating gaits; they named the new breed the American Horse.

These horses possessed flowing gaits that enabled farmers to ride from field to field in comfort, yet the breed was solid and hardy enough for moderate farm work. Thoroughbred blood continued to infuse the breed, which soon became known as the American Saddlebred Horse.

In 1891, in Louisville, Kentucky, owners and breeders founded the American Saddlebred Horse Association—the first-ever breed association designated for an American horse.

Physical Description: The American Saddlebred's high-stepping, graceful action comes naturally. He is an elegant horse with a commanding presence in the show arena. His long and lean appearance is the

BREED IN BRIEF

AVELIGNESE: NORTHERN ITALY

Bred in the mountainous areas of Italy during the late nineteenth century, the Avelignese is closely related to the Haflinger as well as to the Bardigiano, but the Avelignese is a slightly taller and heavier equine than are the other two breeds.

This unflappable, sure-footed, and hardy pony is used on mountain and other trails and in light draft work, such as packing and farming. In appearance the Avelignese is always chestnut in color, with a flaxen mane and tail. Avelignese ponies range in height from 13.3 to 14.3 hh.

BARDIGIANO: NORTHERN ITALY

This powerful, sure-footed pony developed in the mountains of northern Italy sometime in the Middle Ages. Closely related to the Haflinger and the Avelignese, the Bardigiano has a small and pretty head and a compact body.

Bardigiano ponies are prized trekking ponies. And because they are docile in nature, these ponies are also used as therapy animals, working with physically disabled children and adults. Bardigiano ponies can be found in all solid equine colors, and they range in height from 12 to 13 hh.

---

result of careful breeding that strives for a refined, stylish animal. The ideal American Saddlebred is well proportioned and presents an exquisite overall portrait. He is slender, but not thin, with good muscle tone and a smooth, shiny coat. The horse's average height is 15 to 16 hh. All colors are acceptable; the most prominent colors are chestnut, bay, brown and black, with some gray, roan, palomino, and pinto. For the gaited classes in the show ring, the horse's tail is set artificially high by a device known as a tail set. This device props the horse's tail up when he is not in the show ring. Other methods of setting tails include surgery.

Uses: The Saddlebred is known for being the epitome of a flashy show horse, one that demonstrates outstanding refinement and elegance, manners, expression, and brilliant gaits. But the horses also excel as dressage horses, as jumpers, and as western mounts.

The Saddlebred is shown in either three- or five-gaited classed under saddle, and in fine harness classes. Three-gaited horses are judged at the walk, trot, and canter. The five-gaited Saddlebred is shown in the same three natural gaits, but he also has two man-made gaits: the slow gait and the rack. Both of these are four-beat gaits, with the slow gait performed by moving in an almost prancing motion, lifting the legs very high and elevating the forehand. When the Saddlebred racks, he moves faster in a ground-covering stride, snapping up his knees and hocks quickly.

Of Note: During the Civil War, Saddlebreds could be found on both sides of the battlefield. Confederate general Robert E. Lee's brave partner was known as Traveler; Union Army general Ulysses S. Grant rode Cincinnati (*see page 148*); and even Confederate general Thomas Jonathan "Stonewall" Jackson rode on Little Sorrel—all of them Saddlebred-type horses.

Andalusian

ANDALUSIAN

Alternative Name(s): Pure Spanish Horse, Pura Raza Española (P.R.E. Horse)
Region of Origin: Spain

History: Hailing from the Iberian Peninsula, the Andalusian takes his name from the Spanish province of Andalusia, where the horse was most famous. Historically, however, horses on the Iberian Peninsula (which today includes the two countries of Spain and Portugal) were simply called Iberian horses from 1578 when Portugal was annexed to Spain. The Iberian Horse (also called a Jennet) was an ancient breed; 20,000-year-old cave drawings throughout the peninsula show a similar type of horse. During the Moorish invasion in AD 711, Iberian horses were bred to the invaders' horses, the Barbs.

Although Andalusians and Lusitanos are historically the same breed, disagreements between Spain and Portugal have separated what was once one breed into two separate breeds. In 1954, breeders in Spain and Portugal attempted to combine the breeds into one studbook, but a common name was never agreed upon and Iberian studbooks were never united. Thus the breeds share many characteristics, with differences only stemming from lineage or usage. Although each country may have appreciated and bred horses with certain attributes and bloodlines, Iberian horses have always interbred between the two countries, often making it difficult to tell by sight whether a horse is of Portuguese or Spanish descent.

Andalusians are more commonly called Pura Raza Española or P.R.E. Horses, particularly in Spain. The Pure Spanish Horse breed was formalized in 1567 by King Felipe of Spain and has been in existence ever since that time. Originally, the major breeders in Spain came from Andalusia. However, in 1911, when the Cría Caballar formed the national studbook, it was decided that the correct name for the breed was Pura Raza Española.

Physical Description: Andalusians are renowned for their athleticism and dramatic movement. The breed features a short back, high-arched neck, and rounded frontal profile with a natural ability for collection. The Andalusian has a confident attitude and a presence that makes the breed stand out in a crowd. Long, flowing manes and tails, deep barrels, and substantial bone contribute to the picture of strength and nobility.

The Andalusian has a straighter profile and finer hindquarter than the Lusitano, and he moves with more action in the trot. Most Andalusians are grey; less common color occurrences include bay, black, dun, and palomino.

Uses: The Andalusian excels in dressage and cavalry displays, as well as traditional Spanish equestrian pursuits, such as bullfighting and ranch work. The breed makes a very good saddle seat horse, jumper, and carriage horse.

Of Note: In North America, the International Andalusian and Lusitano Horse Association registers horses from either country's line as purebred Andalusians, noting the specific country of lineage on the animal's registration papers.

APPALOOSA

Alternative Name(s): Palouse Horse
Region of Origin: United States

History: Spotted horses have existed since prehistoric times. But centuries later in Denmark and Austria, horses were specifically bred with small, circular patches. Most likely, these were the horses that accompanied New World explorers on their travels. Like many of the European horses that survived the journey to the American shores, some of these spotted horses escaped, others were lost, and still others were stolen by native peoples captivated by their beautiful coat patterns.

Most American Indian tribes in the Pacific Northwest had experience working horses by the turn of the eighteenth century, but one tribe emerged as expert horsemen—the Nez Perce, in Oregon and Washington. They were adept at bringing out the best in horses by practicing selective breeding. Chosen horses strengthened the Nez Perce's herd, producing even hardier, more athletic equines. Their legs became stronger and hooves grew harder. Their tails and manes became sparse, so they wouldn't catch on brush and brambles. The horses' spots not only served as coveted markings but also worked as intricate camouflage, to better hide from prey or foe. Over time, the number of spotted horses soared.

In the 1800s, European settlers began their trek westward. They saw the Nez Perce's spotted mounts and referred to them as "Palouse" horses, named for the Palouse River, which ran through the area. By the mid-nineteenth century, the US government took the western lands for its citizens by displacing the native peoples and housing them on reservations, leading to wars between the US government and various tribes. In 1877, the US Cavalry killed the Palouse horses, or Appaloosas, as they had been renamed, to ensure that the Nez Perce would never rise again. Some horses managed to escape to the hills, where settlers and ranchers secretly retained a handful of the horses.

BREED IN BRIEF

BASUTO: SOUTH AFRICA

The Basuto (also spelled Basotho) is descended from four individual horses that were brought to South Africa by the Dutch East India Trading Company in the 1600s. The breed looks very straight-backed and angular. Although the Basuto breed was nearly destroyed during the Boer War in the nineteenth century, enthusiasts are working to bring the breed back. Today, the Basuto is used for trekking and polo. Basutos are 14.2 hh, in chestnut, brown, bay, or gray coloring.

BATAK: INDONESIA

The fast and nimble Batak is one of Indonesia's eight native ponies and comes from central Sumatra, where they were highly revered. Throughout the history of Sumatra, every clan of the native Toba tribe kept three sacred Batak horses, which were dedicated as the tribe's trinity of gods; when the horses grew old, they were sacrificed. The Batak's heritage includes Mongolian and Arabian bloodlines. With fine Arabian-like features, the Batak is the most refined of the native Indonesian breeds. The breed is a popular saddle horse and is often used to improve the other native breeds. The Batak is found in all colors and stands 13 hh.

These Appaloosas were crossed with Spanish mares to produce stock to work the farms.

Dedicated breeders formed the Appaloosa Horse Club in 1938, and members found the remaining horses and attempted to refine the draft/farm Appaloosa crosses with Arabian and Quarter Horse blood. Their numbers increased, and today the Appaloosa is counted among America's favorite stock breeds.

Physical Description: Depending on the genetic background of the horse, the breed can have a somewhat muscular build and more upright carriage, or the horse can be bulkier. Appaloosas range anywhere from 14.2 to 16 hh, with a small, well-formed profile and pointed ears. The horse has expressive eyes, with a white sclera that rings the eye. The Appaloosa has a slope to his shoulder, a deep chest, and good bone underneath him. Another characteristic of the breed is mottled, freckly skin around the muzzle and genitals.

The Appaloosa has several color patterns: a snowflake pattern is a solid coat sprinkled with a dusting of white; a leopard pattern is spotted all over; a blanket marking has a solid white area over the horse's hips or up to his shoulder, and sometimes there are spots found within the blanket. Some Appaloosas are even roan, which is a distinctive pattern, complete with dark strips along the face bones, known as varnish.

Uses: The Appaloosa is used in many events and riding styles such as combined training, endurance racing, western riding, dressage, and show jumping. Appaloosa racing is common along the American West

Appaloosa

Arabian

Coast, where horses that are as lean and nearly as swift as racing Thoroughbreds pound the racetracks.

Of Note: The flashy appearance of the Appaloosas has made them popular in entertainment circles.

ARABIAN

Alternative Name(s): Arab
Region of Origin: Middle East, Arabia

History: For thousands of years, Arabian horses have been companions to the desert tribes of the Arabian Peninsula. Known as Bedouins, these nomadic tribesmen bred only the finest Arabians, cherishing them so much that they shared their own tents with them. Because of the harsh desert conditions, the Arabian developed a large lung capacity and incredible endurance.

One of the breed's greatest fans was the Islamic prophet Mohammed (seventh century AD). He needed many men on horses to spread the word of Islam, and in this way the breed grew from a Bedouin warhorse to one of worldwide fame. Stories of the Arabian horses, which spoke of their amazing stamina, bravery and loyalty, traveled to the Western world. During the Crusades, the horses were taken back to Europe as war spoils.

Arabian horses came to the American colonies in the seventeenth and eighteenth centuries, but the breed didn't gain popularity until the Chicago World's Fair

THE BEDOUINS

The Bedouins have roamed the Middle Eastern deserts for millennia, as herders of camels, horses (particularly Arabians), sheep, and goats. The Bedouins were fanatical about maintaining the purity of their horses' bloodlines, which they traced primarily through the mares, rather than the stallions. This mid-1930s photograph shows a Bedouin tribesman holding the Sheik Majid's mare at an encampment in the Jordanian desert.

of 1893, when the Turkish government sent forty-five horses for the event. The horses were later auctioned to Americans, and industrialist businessman Peter B. Bradley bought many. In 1906, Bradley funded an expedition for American cartoonist Homer Davenport to the Middle East to import more horses to the United States. Two years later, Davenport and other Arabian enthusiasts founded the Arabian Horse Registry of America.

The International Arabian Horse Association was founded in 1950, offering horse shows specifically for Arabians and registering half-Arab and Anglo-Arabs. In 2003, the International Arabian Horse Association and the Arabian Horse Registry merged to form the Arabian Horse Association. Today, Arabian blood is found in almost every breed of saddle horse.

Physical Description: The ancient Bedouin breeders were fanatical about the purity of their horses and as a result, a type was fixed. Today's Arabian is as distinctive in appearance as it was millennia ago. Its most discerning features include an elegant arched neck, a high-set tail, a dished profile with large eyes, and a small muzzle with large nostrils.

The Arabian is also unique in that it has five lumbar vertebrae, seventeen ribs, and sixteen tailbones whereas other breeds have six lumbar vertebrae, eighteen ribs, and eighteen tailbones. Arabians stand from 14.2 to 15.2 hh. Their colors are black, gray, chestnut, roan, and bay. All Arabians have black skin, except under the white markings. Black skin historically protected the horse from the desert sun.

Uses: The Arabian moves with brilliant action and grace, lending it's easy-to-ride gaits in all disciplines. Arabians are found in all equestrian disciplines from western and English pursuits: jumping, dressage, and eventing to driving and endurance racing.

Of Note: Arabians are bred to other breeds to create crosses such as the athletic Anglo-Arab (Thoroughbred x Arabian), the versatile Morab (Morgan x Arabian) and the flashy National Show Horse (Saddlebred x Arabian). The Arabian was one of the foundation breeds for the Thoroughbred in the early eighteenth century.

AZTECA

Alternative Name(s): None
Region of Origin: Mexico

History: The Azteca is a relatively new breed, less than four decades old. This breed is the result of Spanish Andalusian and American Quarter Horse breeding, with the ideal Azteca bearing five-eighths Spanish horse blood and three-eighths Quarter Horse. In Mexico, three generations of Azteca-to-Azteca breedings are common, while an Azteca in the United States can still be crossbred and registered.

A first-generation American Azteca registered in America is the result of crossing a registered Andalusian to a registered Quarter Horse. Subsequent generations may be bred back and forth, as long as neither parent exceeds three quarters of the whole.

The first generation foundation stock is half Quarter Horse and half Andalusian. Second generation foundation horses are usually a blend of three quarters Andalusian and one quarter Quarter Horse, although the reverse is often favored by cutting-horse enthusiasts. The breed association considers a pure Azteca one that is third generation: five-eighths Andalusian and three-eighths Quarter Horse or vice versa.

Physical Description: Aztecas are often born dark then lighten and turn gray when they reach adulthood. Gray is the dominant color, but since many Andalusians are now bred to be bay or black, we will also start to

Azteca

BELGIAN WARMBLOOD: BELGIUM

Like many warmbloods, the Belgian Warmblood was developed from a cavalry, farm, and carriage horse into a sport horse after World War II. The breed was first crossed with the Gelderlander. Later, Hanoverian blood, Thoroughbred blood, and Selle Français blood were included in the mix.

In 1955, the Belgian National Breeding Association of Agricultural-Riding Horses was founded. Fifteen years later, the name was changed to the National Breeding Association of Warmbloods. Today, the Belgian Warmblood is highly prized as a show jumper. The breed can be found in all solid colors and ranges in height from 15.2 to 16.2 hh.

---

see more colored Aztecas. The breed's legs are strong and dense, with hard, durable hooves that grow very slowly. The Azteca generally has an upright carriage and a close-coupled body, with full mane and tail. His profile is generally straight. The Azteca stands about 15 to 16 hh.

Uses: The Azteca is able to execute intricate classical dressage movements, as well as deal with errant cows on the ranch.

Of Note: This is the first horse breed to be developed on Mexican soil. The Azteca is considered the official horse of Mexico.

BANKER HORSE

Alternative Name(s): Shackleford, Corolla, Ocracoke, or Shackleford Banks

Region of Origin: United States

History: The wild Banker Horse makes his home on a group of islands called the Outer Banks, located off the coast of North Carolina. Genetic testing of these horses has shown the blood variant Q-ac (a rare and ancient Spanish marker) is present in these horses, proving they have existed on the island for centuries.

In 1492, Spanish explorers set up New World breeding stations on the Caribbean island of Hispaniola for saddle and workhorses. In 1526, eighty-nine of these horses and 500 people, led by Luis Vazquez de Ayllon, traveled up the coastline of present-day North and South Carolina and Virginia to create a colony. It wasn't an auspicious start, and the colony disbanded within the year. Many people died, including de Ayllon. The ragtag group that returned to the Caribbean Antilles left behind many surplus horses. The abandoned horses migrated to North Carolina's Shackleford Banks island and other nearby islands

where their numbers grew, helped along with more horses abandoned during shipwrecks in the late 1500s. With no further influx of horses, the bloodlines on the islands remained true.

In 1926, National Geographic magazine published an article on motor coaching through North Carolina, stating that there were between 5,000 and 6,000 horses on the islands. That number took a steep decline in the 1950s, when thousands were removed in a mistaken belief that the horses and other livestock would cause

Banker Horse

the Outer Banks to wash away. No one knows what happened to the horses, but many residents on Shackleford begged the state legislature to leave the horses until research could prove they were causing damage. Some horses remained.

Today, there are fewer than 350 Banker horses left in the world. The biggest herd is on Shackleford (about 117), which is also the largest, most genetically diverse herd. The legislation that protects them limits the number. Herds are kept in control using birth control approved by the Humane Society of the United States on select mares. Excess horses are adopted out. In cooperation with the Foundation, the US National Park Service at Cape Lookout National Seashore manages the Shackleford herd.

Physical Description: Although the Banker horse is a shorter breed, standing under 14.2 hh, he is well proportioned and compact, with strong haunches and slender legs. He has a long head with a straight profile. The Banker horse belongs to a small group of genetically pure Spanish horses that includes the Paso Fino. He is also typical of old-style Spanish horses, possessing inherited gaits such as the running walk, the single foot, the amble, and the pace. The Banker is seen in buckskin, dun, bay, chestnut, and brown coloring; in the Shackleford herd, some horses have pinto colorations.

Uses: Bankers are protected in their native habitat, but a number of people took these horses from the Outer Banks in the past years and trained them for personal use. Those Banker horses have been used for driving, for trail riding, and occasionally for mounted patrols.

Of Note: Banker Horses survive on the Outer Banks by eating the marsh grass and scratching through sand for water seepage.

BARB

Alternative Name(s): None
Region of Origin: North Africa

History: The Barb is an ancient breed of unknown origin that was established in the Fertile Crescent of North Africa, which today encompasses Morocco, Algeria, and Libya. The fast and agile Barb was a favored mount for the Berbers of North Africa, who formed a large part of the Muslim armies that invaded Spain in the eighth century. In fact, the animal draws its name from this group of barbarous people. Barbs

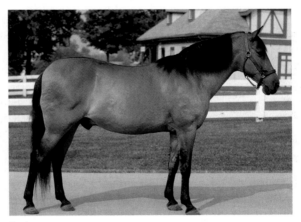

Barb

were originally prized warhorses, which explains their worldwide distribution. As the Berbers conquered new lands, the horses left behind were bred with native stock.

The Barb was a genetically dominant horse that had a large impact on the European and American breeds. It greatly influenced the Spanish breeds, which later influenced other breeds such as the Friesian and the Lipizzan. One of the Thoroughbred's foundation stallions, the Godolphin Arabian, was actually a Barb. Blood of the Barb flows through the veins of many American breeds, including the Quarter Horse, the Standardbred, and the Kiger Mustang.

Physical Description: The Barb is a stocky yet expressive horse, standing 14.2 to 15 hh, and is found in many colors, including dun and palomino. As he is a desert horse, the Barb is fine-skinned and resistant to heat. The neck is arched and well-muscled, and the quarters are rounded, sloping to a low-set tail. The head is straight and sometimes Roman (or convex), an attribute that is seen in many of the Iberian breeds. The Barb has seventeen ribs and fewer lumbar vertebrae than other horses do. Other horses have eighteen ribs. The Barb is a hardy horse and can exist on only a few rations.

Uses: Barbs are fast and agile, which makes them perfect endurance horses. Barbs are also used English and western riding.

Of Note: Few original Barbs remain in the world outside of North Africa. Most were gathered from the wild in the United States by Barb Horse Registry founder Richard Painter in the 1950s to re-create the original type.

THE BLACK FOREST DRAFT

This 600-year-old light draft breed is endangered, with only 1,000 horses worldwide. Called the Schwarzwälder Fuchs or the Schwarzwälder Kaltblut in its native Germany, this strong, sure-footed breed was used for forestry work. Like many drafts, it fell victim to mechanization, and by 1981 only 160 broodmares remained. Germany helped revive the breed and instilled testing so only the best animals were used. Today the horses are ridden and used with carriages. This is a compact horse with rounded, well-muscled quarters, and an arched neck that comes high out of the withers. The breeds stands 14.2 to 16 hh and is sorrel to dark chestnut with a flaxen mane and tail.

BELGIAN

Alternative Name(s): Brabant
Region of Origin: Belgium

History: The small country of Belgium on the west coast of Europe is only about one-fifth the size of Iowa, but it is here that one of the largest and most powerful, popular, and influential of all draft horses was bred. The forefather of all draft horses was first bred here—a heavy black horse called the Flemish, which was used as a knight's mount. It is from this horse that the Belgian draft horse, also called the Brabant, developed; it still carries the characteristics of the Flemish.

The Belgian government refined the breed in the early nineteenth century by offering breeders prizes called bounties for all well-bred mares and stallions. The gentle nature, willingness to work, and strength of the Belgian draft horses made them the perfect choice as the foundation breed for other draft horses in the world. During the nineteenth century, Belgians were exported to government stables in the Austro-Hungarian Empire, France, Germany, Italy, and Russia. In 1866, Dr. A. G. Van Hoorebeke from Monmouth, Illinois, imported the first Belgian horses to the United States. These horses caught the attention of several businessmen in Wabash,

Belgian

Indiana, and in 1885 they started importing Belgian stallions and selling them to horse breeders throughout the Midwest. Widespread recognition of the breed, however, wasn't established in the United States until 1904, when the government of Belgium sent several horses to the St. Louis World's Fair and the Chicago Livestock Exhibition.

Physical Description: Originally, Belgians were bay, roan, black, gray, and chestnut/sorrel. Breeders preferred the sorrel coloring to any other. Today, Belgian horses are instantly recognized by their color (usually sorrel with white manes and tails), white markings on their faces (usually a blaze), and four white socks or stockings. Roan also occurs, as well as the occasional throwback bay.

Uses: The Belgian can pull a wagonload of 6,000 to 8,000 pounds and work eight to ten hours a day. The breed was primarily used for farm work, but later they were sent to cities to work alongside other draft breeds in warehouses, freight stations, and fishing wharfs. The Belgian matures early and is ready to work at eighteen months. He is used for hobby farming, logging, and pleasure driving. Some are even ridden.

Of Note: The Société Royale Le Cheval de Trait Belge was one of the more famous horse shows, running for more than fifty years. Most of the champions from this prestigious show became the foundation stallions of today's Belgian draft.

BRUMBY

Alternative Name(s): None
Region of Origin: Australia

History: The Brumby is the free-roaming horse of the Australian continent, descended from imported horses that either broke free or were turned loose. The first horses to land on Australian shores in 1788 were of English descent, from South Africa. Twenty years

Brumby

later, horse racing gained popularity Down Under, and it became a recognized sport in 1810. As a result, English Thoroughbreds were sent to Australia. Horses from Chile also were exported, as were Timor Ponies from Indonesia, British native ponies, Arabians, and draft horses including the Clydesdale and the Suffolk. These horses helped settle the Australian colonies, on farms and ranches. Explorers and traders used their horses to cover the long distances. The finer-bred horses were used as carriage horses and saddle horses.

Because of the large open spaces in Australia, many horses of various breeds and backgrounds simply slipped away. As agriculture became more mechanized, some horses were abandoned by their owners to run with the wild mobs, as their herds were called. These horses became known as wild Brumbies.

As their numbers grew—particularly in national parkland—many were gathered in "musters" to be caught and tamed for use on cattle and sheep stations.

BREED IN BRIEF

BOULONNAIS: FRANCE

The French Boulonnais is considered to be one of the most beautiful draft horses in the world. This striking breed was developed in the first century in the northwest of France by the Romans, who bred local heavy breeds to their Arabians. The Boulonnais was used as a warhorse in the medieval period. In the sixteenth century, Spanish blood was added. The Boulonnais has a dished head on an arched neck. The breed is used in France as a farm horse and for meat. Most of the horses are colored gray, although bay and chestnut varieties can also be found. They range in height from 15.3 to 16.3 hh.

BRETON: FRANCE

This draft horse hails from Brittany and was first developed from the Ardennais, Boulonnais, and Belgian drafts. In the early 1900s, the Breton further developed into several types. Today only two remain: the Postier, which was historically used to pull cannons; and the Heavy Draft, which was used for meat and farm work. The Postier has additional Norfolk Roadster (now extinct) blood in his lineage, which created a finer, faster trotting horse. Both types are compact horses; the Postier is a refined version of the Heavy Draft. Bretons are roan, bay, chestnut, or gray; and they range in height from 14.3 to 16.3 hh.

Some people felt, however, that the horses were eating pasture intended for cattle, and fouling up watering holes. And so people began aerial culling operations, which included shooting, to reduce their numbers. A group called Save the Brumbies Inc. (STB) formed in 2002 to lobby for changes in management of the feral horse population. The STB seeks to abolish shooting Brumbies as population control; members want sanctuaries set up so that the horses can run free on their own land. Brumbies are found all over Australia, with the largest number living in the Northern Territory, and the second largest in Queensland.

Physical Description: There isn't really one type of Brumby, due to all the different breeds and types of horses that have been introduced into Australia's wild over the years, but all Brumbies share a good deal of cunning, sure-footedness, hardiness, and intelligence. They are found in all coat colors and patterns, including spotted Appaloosa patterns. While their size varies, most are found in the 14.2 to 15.2 hh range.

Uses: Sometimes Brumbies are captured and trained as stock and saddle horses. Otherwise, they roam freely.

Of Note: The origin of the name Brumby is uncertain, but there are plenty of theories. Some speculate that the horses were named after an eighteenth-century soldier turned farrier named James Brumby. Others, however, point to native peoples for the origin of the feral horse's name. Aboriginal language of the Pitjantjatjara translates the word wild as "baroomby." In the introduction of his poem "Brumby's Run," Poet Banjo Paterson noted that the word meant "free roaming horse." Still another explanation points to Barambah, a creek and a cattle operation in the Queensland district of Burnett, which had been established in the 1840s and later abandoned. Some even say that the name comes from the Gaelic word *bromach* (or plural *bromaigh*), meaning "colt."

BUDENNY

Alternative Name(s): Budyonny, Budonny, Budjonny, Budennovsky, and Budennovskaia (Russian)
Region of Origin: Russia, Don steppe and Northern Caucasus

History: Russia's elegant Budenny (pronounced bood-yo-nee) was created to replace the mass equine casualties in the Russian Revolution and World War I and to breed the ultimate mount for the Red Cavalry officers. Certain characteristics were important. Officers' horses worked hard as they often had to leave the line and ride up and down bringing commands throughout the group. And as an officer was often preoccupied with relaying orders, his horse needed be independent and move out of the way of danger without being told, yet turn to meet it when his rider asked. Such a horse had to be strong, sure-footed, controllable, maneuverable, and spirited. And at the end of a long march, the horse also had to get by on meager rations and be ready for action the following day.

The Budenny is the result of crossing the native Don breed with English Thoroughbred stallions, which had come to Russia as racehorses in 1885. (The Don owes his origins to several horses—including Arabians, Karabakh, Persian, and Turkmenic breeds, which Russian Cossacks took as war booty.) The best of these Anglo-Dons were then sent to the Budenny Higher Cavalry School for further training and put into service as cavalry horses. Of these, only 10 percent of the mares and 5 percent of the stallions were accepted as breeding stock. The ones that remained after World War II were officially recognized in 1948 as their own breed and given the name Budenny. In the early 1950s, the Red Army cavalry was disbanded; but the rise of sport has brought the breed back from near extinction.

Physical Description: Careful mixing of the powerful Cossack's Don with the stamina and the

Budenny

elegance of the English Thoroughbred has given the Budenny the best attributes of both breeds. The Thoroughbred ancestry gave the breed an flowing and agile movement. The Budenny's Don heritage made the breed strong, hardy, and calm. The horses look like larger-boned Thoroughbreds, with the same long neck, slender yet strong legs, and pretty head found on a Thoroughbred bred in the West. The Budenny stands around 16 hh and is primarily found in various shades of chestnut. Black and bay are not as common. Some white markings are apparent. Some Budennies have a golden or metallic sheen, which is attributed to the Don heritage, although many people suggest this could be a link to the Akhal-Teke.

Uses: The Budenny has the right temperament for dressage, jumping, and eventing.

Of Note: This horse was named after the Red Cavalry officer Marshal S.M. Budenny, who oversaw the breeding.

CAMARGUE

Alternative Name(s): Horse of the sea
Region of Origin: France

History: The mysterious gray Camargue lives in the Rhône delta in the South of France, a harsh boggy environment of vast brine lagoons, called étangs. The free-roaming Camargue horse, called the horse of the sea, has thrived in the delta since prehistoric times. The

BREED IN BRIEF

CANADIAN HORSE: CANADA

The Canadian Horse, also known as the Cheval Canadien, is a rare breed that arose from horses sent from France to North America by King Louis XIV in the 1600s. The Canadian Horse's ancestors are believed to be Barbs, Spanish horses, and Arabians. The hardiness and workability of the Canadian Horse drove the breed's popularity. Many were exported and crossed to other breeds. Several American breeds, such as the Morgan and the Saddlebred, count the Canadian in their bloodlines. These powerful horses are used for driving and all riding pursuits. They can be found in black, chestnut, bay, and dark brown and range in size from 14 to 16 hh.

COMTOIS: FRANCE

The Comtois is a close cousin to the Ardennais. The breed was developed in the sixth century in the Franche-Comté region of the Franco-Swiss border. A lighter draft horse, the Comtois is well-known for his hardiness and agility. The breed was used for farming in the mountains and forests and for pulling carts in vineyards. Despite the breed's short, stocky build, the horses have a lovely free style of movement. Today, the Comtois is used in France for logging and for horsemeat. Found in bay and chestnut colors, with light-colored manes and tails, they range from 14.3 to 15.2 hh.

Camargue's origins may date back to the long-extinct Soutre horse (from 17,000 years ago) or the Arabian or Saracen horses, that came to southern France with barbarian invaders in the eighth century. Other invaders from Celtic and Roman regions found the Camargues on the Iberian Peninsula appealing, and as a result the horse's bloodlines became entwined with some of the Spanish breeds that lived nearby, particularly those in the northern part of the peninsula.

Camargue horses adapted to difficult environmental conditions, including biting pests, unbearably humid summers, frigid winters, and scant feed. The terrain helped mold them into athletic, hardy creatures of the wetlands. Although the breed has largely developed through natural selection, over the years, soldiers passing through the area bred their own mounts to the Camargue.

The Camargue still runs wild in the marshes, and breeding is now overseen by the Biological Research Station of the Tour du Valat. The Camargue Regional Park serves as a preserve for the horses. In 1976, the French government set standards for the breed and started registering the main breeders of the Camargue horse. The breed's studbook was set up in 1978. To be registered, foals must be born outside (as opposed to in a stable) and must be seen to suckle from a registered mare as proof of parentage. Foals born within the boundaries of the Camargue region are registered *sous berceau* (in birthplace), while those born outside of the region are registered *hors berceau* (out of the birthplace).

Physical Description: The Camargue is stocky, with stout legs, hooves, and haunches. The breed has primitive features, with heavy manes and tails and large square heads with large eyes flush to the skull. The broad hooves are evolutionary adaptations to their wet environment, and the teeth are adapted for eating tough marsh grasses. Foals are born black or dark brown, and they turn light gray or nearly white as they age. The Camargue stands from 13 to 14 hh.

Uses: Because of their calm temperament and agility, the breed is used for gymkhana, dressage, and endurance racing. Traditionally, Camargues are ridden by the local cowboys who manage the region's feral bulls, which are used for bullfighting.

Of Note: The caretakers of the Camargues are called guardians. They oversee the horses as they are rounded up for annual inspections, branding, and gelding of any stallions deemed unsuitable for breeding. Fillies are usually caught and branded as yearlings.

CAROLINA MARSH TACKY

Alternative Name(s): Marsh Tacky, Tacky
Region of Origin: United States, South Carolina
History: The Carolina Marsh Tacky, a close cousin of the Banker Horse and the Florida Cracker, arrived in the coastal area of South Carolina in a similar way

Camargue

Carolina Marsh Tacky

to the Banker, through Luis Vazquez de Ayllon's failed efforts to further Spanish colonization. The breed had evolved enough to tolerate the harsh and buggy swamps of South Carolina, so it was a highly prized work and riding horse for the colonists, as well as the Choctaw, Chickasaw, Cherokee, and Seminole native tribes. The Marsh Tacky also played a part in the Revolutionary War, carrying colonists through the swamps to fight the Red Coats.

Tacky is a British word meaning "common." But although the Tackies were common during the colonial time, today there are fewer than 300 horses left in South Carolina. The numbers dwindled severely when cars arrived. As time went on, people lost track of remaining horses and weren't interested in the breed. The biggest herd today (more than 100 horses) belongs to Carolina Marsh Tacky Association president and lifelong breeder D.P. Lowther. The American Livestock Breed Conservancy (ALBC) began trying to save the Tacky in 2006, collecting DNA samples

and documenting the remaining horses. In 2009, the ALBC created a studbook to help owners of domesticated horses make the right breeding choices.

Physical Description: It is thought that the gentle Marsh Tacky has changed very little since his colonial days. The breed is refined yet well-muscled. The horses usually stand under 14.3 hh and are found in all colors, including a primitive dun. The Tacky has superb endurance and can often ride all day without tiring. Some are naturally gaited.

Uses: The Carolina Marsh Tacky is a popular trail horse because his widespread "pan" feet allow him to move through swampy terrain with ease. The Marsh Tacky is loved by hunters because he is sure-footed and unafraid of guns.

Of Note: The Marsh Tacky is known for a special ability to calmly get out of boggy areas—areas in which other breeds would flounder and panic. In June 2010, the Marsh Tacky became the official state heritage horse of South Carolina.

CASPIAN

Alternative Name(s): None
Region of Origin: Asia, Iran

History: In 1965, Louise Firouz, an American living in Tehran, Iran, was looking for ponies for her new school in Norouzabad. She had heard of small horses living in surrounding villages in the Elburz Mountains and went to investigate. What she found was a beautiful small stallion easily pulling an overloaded cart. Firouz purchased the stallion and gathered others like him from a feral herd of thirty horses. Because the breed had no name, she called the horses Caspians, after the nearby sea.

Firouz had a feeling she had rediscovered a lost breed so important that it was like a paleontologist finding a live specimen of the Tyrannosaurus rex. Through various bone, blood, and DNA tests, archeozoologists proved the Caspian was the direct descendant of Horse Type 4, thought to be extinct for the past 1,300 years and thought to be the forerunner of the Arabian. The Caspian had survived throughout time in small numbers because they were hemmed in by the mountains on one side and by the Caspian Sea on the other side.

The Shah Reza Pahlavi of Iran founded the Royal Horse Society in 1970 in order to help preserve the Caspian breed, along with other native Iranian breeds. In 1971, Great Britain's Prince Philip offered to help with the exportation of horses to the United Kingdom. From 1971 to 1976, twenty-six Caspian horses were exported to the United Kingdom to form the European foundation herd.

In 1974, political turmoil caused near extinction of the breed. Horses from the United Kingdom and Iran helped rebuild the breed. It was at this time that enthusiasts in the United States became interested in the breed and horses were exported to America from the United Kingdom, Australia, and New Zealand.

Caspian

Physical Description: Because of his diminutive appearance, the Caspian is often incorrectly labeled as a pony. But the Caspian is all horse—with limbs, body and head in proportion to each other. The overall impression should be of a well-bred, elegant horse in miniature. The Caspian has large almond-shaped eyes and small but graceful tipped-in ears set on top of a wide head with a vaulted forehead. The muzzle is small and well shaped, and the neck is long and arched with a slender throatlatch. Caspians' limbs are slender but strong with extremely hearty hooves that seldom need shoeing. The breed's mane and tail are full and thick; the tail is carried high in movement. All solid colors are found. The breed stands 10 to 12 hh. The Caspian moves with a floating, ground-covering trot and a rocking, easy-to-sit canter.

Uses: Because of their trainability, incredible strength, and gentle nature, Caspians excel in carriage driving, particularly scurry driving where speed and handiness is a bonus. They make good children's mounts in gymkhana, dressage, and jumping.

BREED IN BRIEF

DANISH WARMBLOOD: DENMARK

The Danish Warmblood, originally called the Danish Sports Horse, became a breed in the 1960s. Breeders crossed Frederiksborg stock with Thoroughbred. The mares from this crossing were bred with Trakehner, Wielkopolkski, Selle Français and more Thoroughbred stallions. The Danish Warmblood's distinguishing brand is a crown over a wave, a nod to the country's monarchy and coastline. Today, this strong and athletic breed excels in dressage and eventing. The Danish Warmblood stands 16.2 to 17 hh and is found in all solid colors.

Of Note: Caspian genes are so strong that they are found to be a recessive factor. A normal-size mare with even a small amount of Caspian blood can produce a throwback foal with Caspian features.

CHINCOTEAGUE PONY

Alternative Name(s): None
Region of Origin: United States

History: On the ocean side of the Chesapeake Bay lie the Assateague and Chincoteague Islands. The horses are believed to have roamed here since the 1700s, which is astounding considering that the only forage available is salt marsh cordgrass and seaweed, food that any domesticated horse would refuse to eat. But the Chincoteague Ponies are highly adapted to these island rations, being creatures of natural selection. Any pony unable to adapt to this harsh environment would not survive.

How the ponies came to the islands has always been a subject of romantic myth. Many say that a Spanish ship sunk, and the ponies on board swam to shore. Some French enthusiasts of the Pottok Pony even go a step further, surmising that the often colored ponies of Assateague could be related to the breed, possibly imported by Spanish navigators during the sixteenth and seventeenth centuries as pack-horses for the New World.

It is more likely that they were ponies belonging to Virginian settlers who let the horses roam free on the islands. The farmers eventually gathered the ponies to work their fields, and they all helped each other by scheduling specific days for herding the ponies, branding them, and breaking them for work. This became a regular occasion, and by the late eighteenth century, pony and sheep penning events were held annually.

In the 1800s, the roundup became more popular among island residents. The whole Chincoteague community anticipated the day, and eventually the penning events became a festival. The celebration became something of a tourist attraction, which was a boon for local businesses. In 1909, the last Wednesday and Thursday of July became official Pony Penning Days.

After two tragic fires struck the islands in the 1920s, citizens founded the Chincoteague Volunteer Fire Company. The department needed a way to keep financially afloat. The answer was to take the money raised at the Pony Penning Days carnival and use it to buy equipment and supplies for the fire department.

Chincoteague Pony

The department, in turn, would be responsible for caring for the herd and overseeing the penning.

When most of Assateague was sold to a private citizen in 1920, the citizens' pony penning activities moved to Chincoteague. In 1925, they decided to swim the ponies across the channel instead of ferrying them by boat. During this time, the fire department took over the ponies' care. In 1939, the department released twenty Mustangs into the herd to help create genetic diversity; a year later, they added some Arabians. In 1943, the US government purchased Assateague Island and transformed it into a national park.

Physical Description: Because of the influx of Mustang and Arabian blood, Chincoteague Ponies can have many different characteristics, including a dished profile, a straight profile, or a broad forehead. In general, they are stocky, compact ponies with sturdy features. The breed stands from 12 to 14.2 hh in every coat color, including pinto.

Uses: Chincoteagues are said to be kind and willing riding ponies.

Of Note: Any horse fanatic has probably read Marguerite Henry's 1947 novel *Misty of Chincoteague* (and may have also seen the movie), which accurately describes the island's ponies and the annual penning. Because of the resulting adoptions, the ponies are now found throughout the United States.

CLEVELAND BAY

Alternative Name(s): None
Region of Origin: United Kingdom, England

History: The Cleveland Bay is the oldest native horse breed in the United Kingdom, even predating the development of the Thoroughbred. The breed developed in the Yorkshire and Cleveland areas in northeast England. Most of the foundation bloodlines of the Cleveland

Cleveland Bay

Bay aren't known. However, it is documented that, aside from the Darley Arabian, the Godolphin Barb was used to refine the breed. Further Barb blood may have also had a hand in the origination of the breed, as Barbs were said to have come into the Port of Whitby in Yorkshire.

In medieval times, the breed (then called the Chapman Horse) was a packhorse for monasteries, convents, and peddlers. As roads developed, the Bays were used as coach horses. The early rough roads required steady horses, but as the roads improved and could accommodate faster horses, people later added Thoroughbred blood. Consequently, the breed split in two: the Yorkshire Coach Horse and the extremely similar Cleveland Bay. By 1938, the Yorkshire Coach Horse was extinct. Cleveland Bays were imported to the United States in the mid 1800s as coach horses. In 1930, Matthew Mackay-Smith revived interest in the breed by importing stock as foundation for hunters.

The Cleveland Bay population plummeted in the United Kingdom during World War II, in which they

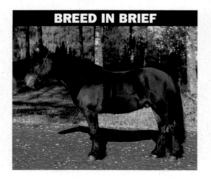

BREED IN BRIEF

DØLE GUDBRANDSDAL: NORWAY

The Døle Gudbrandsdal developed in the mountain valleys of Norway. The breed closely resembles the British Dale and the Fell Pony; all three breeds probably came from the same ancient ancestors. Døle Gudbrandsdal was traditionally used as a packhorse to carry goods over hill and dale. Today, a lighter type of Døle Gudbrandsdal is bred—one that is a good all-around riding horse. More than half of the horses in Norway are of this breed. Døles are very fast at the trot and extremely agile. They are powerful-looking horses, with some feathering on their legs. They stand from 14.2 to 15 hh and come in all solid colors, but the most common are brown, black, and bay.

DØLE TROTTER: NORWAY

The Døle Trotter's origins are believed to date back between 400 and 800 BC. This breed was further developed in the mid-nineteenth century, when people crossed it with the lighter Døle Gudbrandsdal and with Swedish trotting breeds. The Døle Trotters are small and fast horses, greatly prized as harness racers. In Europe and the former USSR, the diagonal trotting gait is preferred in harness racing; so the Døle Trotter races in that gait as well. The looks of the Døle Trotter are similar to those of the British Fell. The Døle Trotter stands 15 hh and is found mostly in black and brown and sometimes in bay.

were used as artillery and supply horses resulting in many losses. In 1950, there were only four or five stallions left in England, and ardent breeders, including the Queen of England, have since worked hard to ensure the breeds' survival.

Physical Description: The Cleveland Bay is an elegant, full-bodied horse, fancy enough to pull a carriage but athletic enough to be used for riding. The movement varies: some horses have long sweeping movements, while others are more upright in their stride. They are also very hardy horses and easy keepers. The Cleveland Bay is a uniform breed, very similar to the Morgan, where each one looks much like the other. The breed stands from 16 to 16.2 hh. They are almost always dark bay with black points (main, tail, and legs).

Uses: Although the Cleveland Bay is best known as a coach horse, he is also a great riding horse. The breed is used for general riding activities such as dressage, and their steady temperaments make them ideal trail horses. People often drive the horses in combined driving events, a kind of equestrian sport similar to eventing.

Of Note: Why almost all Cleveland Bays are bay in color is a mystery. Perhaps a herd of horses happened to be of the same color, and the people of North Yorkshire continually bred to them until a type was fixed. There are a few rare genetic throwbacks, where foals will be born chestnut.

CLYDESDALE

Alternative Name(s): None
Region of Origin: United Kingdom, Scotland

History: The Clydesdale is a 300-year-old breed that originated in Scotland's Clyde Valley, now known as Lanarkshire, and is the youngest of all the United Kingdom's heavy breeds, finding its full development only

Clydesdale

DON: RUSSIA

The Don was the prized mount of the fearless Cossacks of Russia and was developed from Mongolian steppe horses, Thoroughbreds, Arabians, and Russian Orlovs. As cavalry horses, they had great stamina and strength, and so were chosen as the foundation breed for the Budenny. The Don has a straight profile and strong hindquarters. Today,

Dons are used for all English pursuits and for endurance riding. They are also driven four abreast in a traditional Don quadriga: in this formation, the two horses in the center trot forward while the outer two are bent to the outside with fixed reins, resembling a bird's wings. They often have a burnished look, usually in chestnut or brown coloring. They stand 15.3 to 16.2 hh.

---

in the last 150 years. Docile native mares of Lanarkshire were put to large English horses (of the type that carried armored knights into battle), as well as Flemish horses, imported by the Duke of Hamilton, in the late 1700s. This combination gave the breed height and a strong build. Two horses are recognized today as the breed's foundation stallions: Prince of Wales and Lord Darney.

The Clydesdale was put to work in the streets of Glasgow and on farms in the Scottish countryside. They were used in the city streets up until the 1960s, drawing milk and vegetable delivery wagons. Once a plentiful breed, by the 1970s, the Clydesdale's numbers dwindled into the hundreds in the United Kingdom. Ardent breeders throughout the world brought the horses back from the brink of extinction, although their numbers still remain low in the United Kingdom.

Imported throughout the world, Clydesdales helped build cities and towns in Canada, Australia, and America. They first came to America in 1840; today, the United States has more of them than any other country.

Physical Description: The Clydesdale is a flashy, high-stepping horse with a long stride. His strong and rather large feet give him a strong underpinning. Historically, the horses' feet were so big that farmers found them to be too large to fit in plough furrows; therefore Clydesdales were often worked in towns rather than on farms. Although considered a working horse, the breed is very elegant with a high-set neck, flashy white stockings, and a bold white face. Clydesdales stand between 16 and 19 hh and are usually bay with white stockings, feathers, and wide blazes. But brown, roan, black, and gray are also acceptable. Flecks of white throughout the coat are found in all the colors. The sabino color pattern also occurs in Clydesdales, with white extending well up the leg and even onto the underbelly.

Uses: Clydesdales are primarily used for driving and riding, but some crossbred horses are prized eventers.

Of Note: Today, the Clydesdale is an extremely familiar breed in the United States due to the breed's use in advertisements by Anheuser Busch, the makers of Budweiser beer. The company has proudly displayed its six-horse teams hitched to the classic Studebaker beer wagons since the Great Depression era. Today, teams of Budweiser horses are kept and displayed at Anheuser Busch theme parks throughout the United States and are often the subject of the beer company's popular commercials.

COLORADO RANGER HORSE

Alternative Name(s): Rangerbred
Region of Origin: United States

History: The Colorado Ranger Horse might have all the characteristics of his Appaloosa cousin, but he's really "a horse of a different color." This breed's origin goes beyond the American shores, tracing its beginnings to the blood of Middle Eastern horses.

It all started with American General and President Ulysses S. Grant. An excellent horseman, Grant appreciated a good mount. At one point during his presidency, he sailed to Europe and the Middle East. He spent some time in Turkey, where he met the Sultan Hamid of the Ottoman Empire. They developed an amicable relationship, and when Grant was ready to return home, Hamid sent with him a farewell gift of two young stallions, an Arabian named Leopard and a Barb named Linden Tree. The gray horses returned to America and became part of Grant's stable, where he used their Oriental bloodlines to help diversify his stock.

In 1879, the two stallions were loaned to Rudolf Huntington, a renowned breeder of trotting horses on the East Coast. For more than a decade, Leopard and

Colorado Ranger Horse

Linden Tree were used in hopes of establishing a new line of trotters. When Huntington lost his funding for his breeding program, just before the turn of the century, the horses were sent to the Colby ranch in Nebraska and the A.C. Whipple family ranch in Colorado. The stallions covered range stock mares of mixed breeding, resulting in foals born with spots. They grew up to have amazing cow sense. Thus, a new working cow horse was born, dubbed the Colorado Ranger Horse.

Because the gene pool breeders were working with was small, linebreeding was common, and Rangerbred horses seemed to come out with spots of all colors and patterns, including leopard, blanket, and snowflake. Most of the first Rangerbred breeders didn't place much emphasis on whether or not the horses had color.

Colorado State University took interest in the new Rangerbred and, with breeder Mike Ruby, helped fund a program to produce working horses. Ruby used two stallions, Patches #1 and Max #2, offspring from the Leopard and Linden Tree breedings.

Ruby's resulting crop of foals had the stamina and refinement of their Arabian and Barb ancestors. They

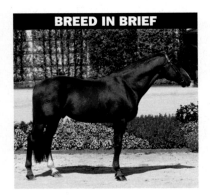

BREED IN BRIEF

EINSIEDLER: SWITZERLAND

The Einsiedler is the pride of Switzerland and a very old breed. The breed's origins date back to native Schwyer stock from the tenth century, and the first studbook began in the 1600s. The breed was later crossed to Spanish, Friesian, and Turkish horses, as well as to Anglo-Norman, Thoroughbred, Swedish Warmblood, and Holsteiner horses later on. Traditionally, the horses were used in the cavalry, but they are now used for sporting, such as dressage and jumping. The Einsiedler is bred at the Federal Stud at Avenches. Einsiedlers there are also called the Swiss Warmblood, and they are branded with a Swiss Cross. The powerful Einsiedler has a fine head and long elegant legs. The horses stand from 16 to 16.2 hh and are found in all solid colors.

FALABELLA: ARGENTINA

The Falabella is a rare miniature horse named for the family who started his breeding at Recreo de Roca Ranch near Buenos Aires, Argentina, in the mid-nineteenth century. The result of selective breeding, the Falabella was created by crossing Shetland Ponies with a small Thoroughbred stallion. The breeders then downsized further by breeding the smallest to the smallest. Falabellas are used for driving and are shown in hand. They stand 30 inches high, when measured at the withers.

---

also had the level-headedness of their dams, the range mares. In 1938, the breed officially received its name as the Colorado Ranger Horse, and Mike Ruby founded the Colorado Ranger Horse Association that same year.

Physical Description: The Rangerbred generally has stock horse conformation, with a level topline, powerful haunches, clean legs, solid hooves, and a smallish profile. While some horses have spots, many bear solid coats. The Rangerbred stands from 14.2 to 16 hh.

Uses: The Rangerbred is a prized western horse, used in ranching and other work, as well as competitions.

Of Note: The sole requirement for inclusion in the Colorado Ranger Horse Association is a pedigree that traces to either of the foundation stallions, Patches #1 or Max #2.

CONNEMARA PONY

Alternative Name(s): None
Region of Origin: United Kingdom, Ireland

History: This pony springs from the rocky coastline of Western Ireland in County Galway, from a region called Connemara, which means descendents of the sea. The ancestor of the Connemara Pony is believed to have looked a lot like a Shetland Pony. That pony roamed the area for centuries before mixing with the Celtic ponies that arrived with raiders in the fifth and sixth centuries AD. In the centuries that followed, other breeds came to Ireland's West Coast and mixed with the ponies there. The foundation blood for today's Connemara pony is thought to come from the now-extinct Spanish Jennet and Irish Hobby breeds. The Connemara is believed to have inherited his smooth gaits from these two horses.

Impoverished farmers valued this breed, and almost everyone owned only one—a mare. In the household, she had to do it all (farming, driving, riding, and breeding) because each farmer couldn't afford to feed more than one horse. Farmers who needed horses caught them from the nearby area. The smart ponies found

Connemara Pony

they could do a little work for a little better food, so they were fairly easy for the farmers to train.

The nineteenth century saw the influx of Arabian blood to further refine the breed. But by the end of the century, the Connemara's numbers had declined drastically because of the Great Potato Famine. Farmers had more to worry about than the continuation of the breed, and many stopped breeding or bred without discretion. The government intervened, by introducing Welsh blood to help revive the Connemara's bloodlines. Later, Hackney, Thoroughbred, and Norfolk and Yorkshire Roadster breeds were also added, and the breed standard was set.

Ireland has kept close tabs on the breed since 1924. During annual inspections, the best ponies are selected for Class 1 certification. These ponies must fulfill all the required characteristics to be chosen. Many Connemaras are crossbred, especially with the Thoroughbred, to create a taller animal.

Physical Description: The environment is largely responsible for shaping the Connemara's conformation. The breed's jumping ability developed from moving over rough and uneven ground and from jumping stone walls in search of food. The rocky ground made for nimble ponies with tough hooves that rarely need shoeing. The Connemara's athletic features include round haunches and long hind limbs. Colors are brown, dun, black, gray, chestnut, and sometimes roan. Heights range from 12.2 to 15 hh. The Connemara has a long, ground-covering stride and slightly higher knee action. The pony's neck has what the Irish call a length of rein, meaning the neck has a good shape and length, similar to a horse's.

Uses: The breed is valued as an English sport horse.

Of Note: The Connemara first came to America in 1950, imported by wealthy Americans who rode the ponies in Ireland.

CRIOLLO

Alternative Name(s): Argentinean, Argentine Criollo, Crioulo (in Brazil), Costeño and Morochuco (in Peru), Corralero (in Chile), Llanero (in Venezuela)

Region of Origin: Argentina

History: The word criollo originally referred to both humans and animals of purebred Spanish ancestry born in the Americas. The Criollo breed is descended from the horses of the explorers, from the time of Christopher Columbus to the Spanish conquistadors of the 1500s to Don Pedro Mendoza, the founder of Buenos Aires, Argentina, in the 1530s. It is said that many of these horses were of Barb decent, although there are Sorraia and Garrano origins in their blood as well. Like the Mustangs of North America, these cavalry horses escaped, were turned out, or were stolen and ran free on the South American grassland area stretching north, south, and west from the delta of the Rio de la Plata near Buenos Aires, known as the Pampas. The ones that survived the harsh climate eventually adapted to life on the Pampas.

Over the next 400 years, the baguales, or feral horses, became much stronger. Their numbers grew

Criollo

into the thousands, and many natives and settlers reported seeing giant herds surging over the grasslands. The native peoples of South America captured some of these horses, although they allowed most of the animals to run free on the plains until they were needed for work.

With more settlers moving in and more cattle operations setting up, the face of the wild Pampas changed. In 1806 and again in 1825, an influx of European invaders into the region had an impact on the horse population. Imported French Percherons and English Thoroughbreds were introduced into the feral horse herds and increased the size of the Criollo.

The pure Criollo was beginning to disappear. In 1917, the Sociedad Rural de Argentina organized to protect the Criollo. The society searched and discovered that a native South American tribe had a small herd of 200 purebred horses, which became the cornerstone of the breed's recovery.

Physical Description: On average, the Criollo stands at 14.1 hh and is a robust horse with a powerful

BREED IN BRIEF

FINNISH: FINLAND

Until the nineteenth century, the Finnish horse was developed for forestry and agricultural work by crossing native ponies with warmblood breeds. There used to be two distinct types of Finnish: light for saddle work and heavy for draft work; today, the lighter type is the most common one. The breed possesses a short square head on a close-coupled body. Because of the horse's strong hindquarters, the breed is useful for both riding and harness work. The Finnish stands 15.2 hh. All solid colors are found, although the most common coloring is chestnut with a flaxen mane and tail.

conformation, which reflects the breed's tough, vigorous nature. He has sloping, strong shoulders with a heavy crest and upright neck. The Criollo has short, strong legs with ample bone, flat knees, and good hocks and pasterns. The back is short, the croup is sloping, and the hindquarters are brawny. Many coat colors are found in the Criollo breed, but dun is the most common. Many horses bear primitive markings, including dorsal stripes and zebra stripes on the legs.

Uses: The Criollo is a well-sought-after ranch horse and endurance horse.

Of Note: The breed was originally referred to as the Argentinean, and the name eventually changed to the Argentine Criollo. Today the horse is just known as the Criollo, since the horses of Brazil and Uruguay have been determined to be of the same type and ancestry. The Criollo is considered the national horse of Argentina.

CURLY HORSE

Alternative Name(s): Bashkir Curly
Region of Origin: United States

History: In 1898, a father and son found the American Curly living free in the arid central foothills of Nevada. They later captured the unusual curly coated horses and brought them home. Most American Curlies today descend from this herd of three horses.

How the breed came to exist in America and how it developed the nickname "Bashkir" remains a mystery. Theories suggest the horses were brought to the continent by early settlers, perhaps Russians, Vikings, Spanish conquistadors, or an American settler in the 1800s. The name Bashkir possibly arises from misinformation in the 1930s suggesting that all horses with curly coats were named after an ancient Russian breed from the Bashkortostan region. However, this

Curly Horse

was later disputed because the Russian breed with a curly coat is the Lokai, not the Bashkir. There is some evidence that curly-coated horses have been in North America for at least 200 years. Modern blood typing shows that the Curly is not a distinct breed; his genes comprise many breeds, such as the Morgan.

Physical Description: The Curly is born with a coat of tight curls everywhere, even in the ears; the coat settles down as the horse matures. The characteristic curly coat is passed along about 50 percent of the time when a Curly horse is crossed with a non-curly-coated one. The conformation of the Curly is different from other horses; the overall look of the animal is stout and round. The Curly is a hardy breed, able to withstand harsh climates. Another discerning feature is this breed's gait. Most Curlies have a running walk. The horses' average height is 14.2 to 15.1 hh, appearing in all solid colors.

Uses: The sure-footed and nimble Curly is a top gymkhana and western sport horse. The breed's stability and trail manners make them ideal trail horses, but they are also found in English sports, such as dressage.

BREED IN BRIEF

FREDERIKSBORG: DENMARK

In 1562, the Frederiksborg was developed in Denmark at the Royal Frederiksborg Stud to create cavalry horses that would also be good riding horses. The stud crossed Spanish horses to English horses with Norfolk Roadster blood. The resultant elegant, high-stepping horses were in great demand among the privileged classes. So many Frederiksborg horses, including breeding stock, were exported worldwide that their numbers at the stud dwindled, and it closed in 1839. The breed revived again in the 1920s, but Frederiksborgs are still rare throughout the world. The Frederiksborg is often used as a foundation breed for the Danish Warmblood. The Frederiksborg stands 15.2 to 16 hh and is usually chestnut colored.

FREIBERGER: SWITZERLAND

The Freiberger developed in the western part of Switzerland, a very hilly region that required a nimble and powerful mountain horse. The Freiberger stems from a foundation stallion named Vaillant, foaled in the late 1800s, which had Norfolk Roadster, Thoroughbred, and Anglo-Norman blood. Today, Freibergers are prized as harness horses; they are also used by the Swiss Army as pack horses. They are bred at the Federal Stud at Avenches and are branded with the Swiss Cross. The Freiberger is a short-coupled breed with a thick, crested neck and muscled hindquarters. The breed stands 14.3 to 15.3 hh and is only found in bay or chestnut coloring.

Of Note: The Curly's coat is also considered to be hypoallergenic; anyone allergic to the dander of regular horses is less likely to be affected around a Curly.

DALES PONY

Alternative Name(s): None
Region of Origin: United Kingdom, England

History: The Dales Pony is a native of the upper dales (valleys) of the eastern slopes of northern England's Pennines Range, from the High Peak in Derbyshire to the Cheviot Hills near the Scottish Border. Small Dales herds can still be found on the upper dales of the rivers Tyne, Wear, Allen, Tees, and Swale.

The pony's origins are with the ancient Pennine Pony, as well as Scotland's extinct Scottish Galloway. During the Roman occupation of the area (AD 43–410) until the mid-nineteenth century, a lead-mining industry flourished. Although small, the ponies were strong, and soon native ponies were selected to work the mines, carrying heavy loads of lead through the countryside from Northumberland and Durham to smelt mills.

The Dales soon were sought after for their ability to carry heavy loads over unforgiving countryside without flagging—traveling up to 200 miles in one week. They also were put to work on small farms throughout the region. Breeders added the Norfolk Trotter, the Yorkshire Roadster, and possibly also the Friesian into the Dales herds to give the breed a little more substance and speed.

The infusion of trotting horse blood had a unique side effect: the Dales developed into a pony with pretty knee action and the ability to race along at the trot. The endurance and agility of these ponies made them a favored mount on local hunts in the eighteenth century. In the 1850s, the Welsh Cob stallion Comet was bred to Dales mares to improve their gait. Later that century, when roads in northern England were

Dales Pony

FRENCH TROTTER: FRANCE

During the nineteenth century in the region of Normandy, breeders developed the French Trotter from Norman, Thoroughbred, and Norfolk Roadster stock; Standardbred was added at a later period. French Trotters do not pace, but they do move in diagonal pairs. The French Trotter, also known as the Norman Trotter, has a straight profile, flat withers, and a long, well-muscled back. The horses are used for trotting races, under saddle, and in harness. The studbook for the breed was established in 1922. French Trotters stand 16 to 16.2 hh, and they are found in all solid equine colors.

---

improved, the demand increased for faster animals on mail and stage coaches. The quick little Dales Pony met that requirement.

Physical Description: Modern Dales Ponies are not too different from the native horses that roamed the Pennines centuries prior. Standing about 14 hh, they are stout, muscular, and beautiful. They are usually black or dark brown, although occasionally bay, gray, or even roan horses are found. The breed features an expressive, well-shaped head with a straight profile, small muzzle, and wide-set eyes. They are refined through the throatlatch and have short, arched, upright necks, set on a good sloping shoulder. The legs are clean, with good substance and feathering below the knee. Strong loins and hindquarters give the Dales excellent power, and a well-sprung rib cage allows ample room for lung capacity, which aids endurance. The horse's hooves (which are large, round, and open at the heels) are usually of blue horn and very strong. A signature of the Dales is his full, lush mane and tail, which grow very long and thick.

Uses: The Dales is a versatile pony and is used for riding and driving and all English sports.

Of Note: The Dales Pony found a home as a military horse in the eighteenth and nineteenth centuries. In the mid-twentieth century, World War II took its toll on the breed. Countless horses were used as artillery horses and many died. Thanks to the Dales Pony Society, which formed in 1963, numbers have increased.

DARTMOOR PONY

Alternative Name(s): None
Region of Origin: United Kingdom, England

History: Dartmoor is an area of moorland in the county of Devon in southwest England. This beautiful yet desolate area is home to the wild-roaming Dartmoor Pony. The Dartmoor has lived in southern

Dartmoor Pony

England for centuries and is one of Britain's native breeds. Because the area was rich in tin during the Middle Ages, hardy Dartmoors were used to carry heavy loads across the moor to neighboring villages.

In 1535, Dartmoors and other native British ponies suffered a blow at the hands of King Henry VIII. He passed a law to eliminate "nags of small stature." Due to the weight of the armor worn by soldiers, it was necessary to breed large horses of war—infused with draft blood—as only a large horse could carry the weight of both soldier and armor. Anyone using a stallion under 14 hh was fined severely. Six years later, another law was passed that prohibited the use of any horse under 15 hh. Smaller ponies were slaughtered. Queen Elizabeth I later annulled these laws.

The Dartmoor Pony's sure-footedness and strength, despite his small size, made him popular for many jobs. However, breeders during mining of the eighteenth century wanted to make the Dartmoor a better pit pony and introduced Shetland Pony blood into their population. Area mines eventually closed, and while some Dartmoors were kept to work on farms, most were turned loose, to find their way on the moors.

In 1988, the Dartmoor Pony Moorland Scheme was founded in order to help the purebred pony of the region rebound.

Physical Description: The compact and powerful Dartmoor Pony stands between 11.1 and 12.2 hh. Most ponies are brown, bay, black, gray, chestnut, or even roan, but any pinto-colored ponies represent a part-bred pony, and therefore are ineligible for registration. The Dartmoor has a small head and wide-set, intelligent eyes and alert ears. Dartmoors have short legs with good, flat bone, and hard, flinty hooves. They have good angular shoulders, which produce nice movement at all gaits. The ponies have lush manes and tails.

Uses: They are prized as children's mounts and as driving ponies.

Of Note: In 1951, the Dartmoor area was officially made a national park. There are many ponies found in Dartmoor National Park; some ponies are owned and protected by farmers and are usually identified with unique brands.

DÜLMEN

Alternative Name(s): Dülmener
Region of Origin: Germany

History: The Dülmen pony is the last of Germany's native breeds. As far back as 1316, herds of the wild mouse-colored ponies were documented as living in

Dülmen

an area called Merfelder Bruch. This hilly meadow and woodland area, situated in Westphalia, has been home for the ponies for some 700 years, with small herds thriving throughout pockets of the land. The Lord of Dülmen was given dominion over the land at that time, and he and the nearby farmers who shared the common land watched over the local ponies.

In the 1800s, when the common land was divided, the areas available to the ponies shrank. Around 1840,

FURIOSO: HUNGARY

The Furioso takes his name from his foundation stallion, a Thoroughbred by the same name who was brought to Hungary in 1841 to breed to Nonius mares. Because the Nonius and the Furioso are related, they resemble each other to an extent, but the Furioso displays Thoroughbred characteristics even more strongly.

A sure-footed and honest jumper, the Furioso is a prized steeplechase mount and riding horse. In fact, the Furioso has been used in many disciplines. In the past, the horses were also used for farm work. The strong and compact Furioso horses stand at approximately 16 hh, and they are most frequently found in the colors of black, chestnut, and bay.

the Dukes of Croy became the landowners of a portion of Merfeld, and they inherited the last remaining ponies. At this time, only about twenty Dülmen remained. The dukes set aside a sanctuary area for the ponies, where they could live in relative isolation and safety. The dukes released other breeds of horse and ponies into the herd, including British native ponies and some Polish horses, in hopes of bolstering the breed's numbers. The feral ponies still relied solely on their instincts to survive, yet their numbers grew in the next decades.

Today, the nature reserve, which is owned by the Duke of Croy and known as Wildpferdebahn Merfelder Bruch, allows the Dülmen ponies to carry on as they have for centuries—foraging for food and shelter on their own. One of the reserve's missions is to preserve the area that the ponies live in, while working to keep the breed's bloodline pure from further crossbreeding. Due to the Dülmen ponies' isolation, geneticists are particularly interested in learning how this feral breed's behavior differs from that of domestic horses.

Physical Description: Small, close-coupled, and sturdy, the Dülmen has a primitive look about him. The breed is generally a mousy dun color with black points, although some of the ponies with mixed breeding can be dark bay, brown, chestnut, or even black. They have small pony-type heads, short necks, lightly boned legs, and short backs. They can be round in the barrel and have little wither. The ponies usually stand 12 to 13 hh.

Uses: Dülmens are good riding ponies and, are smart and willing to work, despite having wild origins. They are even good for pulling carts or working on farms.

Of Note: Each year, on the last Saturday of May, the ponies are gathered for inspection; the colts are sold at a public auction, and the mares are returned with only one or two stallions.

DUTCH WARMBLOOD

Alternative Name(s): KWPN (referring to the Royal Warmblood Studbook of the Netherlands)
Region of Origin: The Netherlands

History: For many years, the Netherlands was well known for producing top-notch work and carriage horses. But after World War II, when farms became mechanized and horses were no longer needed to work the land, the Netherlands changed its equine breeding goals to suit the new market—that of the avid sportsman. Equestrian sports were on the rise, and like many

Dutch Warmblood

GALICENO: MEXICO

The Galiceno comes from ancient Spanish bloodlines, which Hernán Cortés brought with him from Spain in 1519. The breed developed in the Spanish province of Galicia in Mexico. This horse has the smooth, running walk found in many old Spanish breeds. In Mexico, the nimble, strong Galiceno is a prized ranch horse. Although the Galiceno is a small horse, 12 to 14 hh, and is often used as a child's mount, his strength and power make him capable of carrying an adult all day. Galicenos were not imported into the United States until 1958. An attractive horse with a refined head, a short, compact back, and tough, well-formed legs, the Galiceno usually can be found in bay, black, chestnut, or pinto.

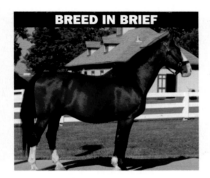
BREED IN BRIEF

GELDERLANDER: THE NETHERLANDS

The Gelderlander is the ultimate carriage horse—elegant, fast, and strong. The breed originated in the Gelder Province of the Netherlands during the nineteenth century. The breed was developing from the Cleveland Bay, the Thoroughbred, the Hackney, the Anglo-Norman, and the Arabian. The Gelderlander is one of the foundation breeds for the Dutch Warmblood. In fact, the number of Gelderlanders has decreased in recent years as they have been absorbed into the Dutch Warmblood. Still the Gelderlander remains popular in competitive carriage driving. The horses stand 15.2 to 16.2 hh and are found in all solid equine colors; the most popular color is chestnut.

---

other Europeans, the Dutch chose to keep up with this new way of breeding by developing a sport horse of their own called the Dutch Warmblood.

Breeders selected two lighter farm horses, the Gelderlander and the Groningen, to help establish their new breed. The two horses were perfect choices to use in building a sport horse: the Groningen had the impressively strong hindquarters needed for jumping and collection, and the Gelderlander had a beautiful action highly desired for the basic movement for all sports. The resulting offspring were further refined with Thoroughbred blood, which created a horse with more sloped shoulders and a longer stride. It also gave the breed a longer neck and shortened the back of the carriage breeds. To even out the temperament, the carriage/Thoroughbred crosses were bred with the German Oldenburg, the Hanoverian, and the Prussian Trakehner.

Physical Description: The Dutch Warmblood has attributes from a mix of many breeds. The Thoroughbred blood gave this sport horse a longer neck, flatter shoulders, lighter bone, longer movement, and speed and stamina. The body is deep, and the neck is muscular. The hindquarters are strongly muscled, and the hocks are stout, features of the breed's German warmblood and Trakehner forefathers. The straight profile is a direct result of the breed's farm and carriage ancestry; depending on breeding, however, this is not currently a hard and fast conformational rule of thumb, as some Dutch Warmbloods have quite elegant heads. The breed stands 15 to 16.2 hh and is found in all solid colors, primarily chestnut and gray.

Uses: The Dutch Warmblood shines in the sporting arena, raking in top honors in jumping, dressage, eventing, and competitive driving.

Of Note: Oddly enough, the Dutch managed to breed a very successful and athletic horse within a short period of time. Breeding in Holland is controlled by the Koninklijk Warmbloed Paardenstamboek Nederland (the Royal Warmblood Studbook of the Netherlands). This association puts horses through a rigorous selection process. Stallions are assessed on conformation and performance, including jumping, cross country, and harness (if the horse is to be part of the warmblood carriage section). Mares are also tested, with emphasis on conformation, temperament, and movement.

ERISKAY

Alternative Name(s): None
Region of Origin: Northern Scotland, Isle of Eriskay

History: In the Outer Hebrides of northern Scotland lies the Isle of Eriskay, ancestral home of the Eriskay pony. The origins of the Eriskays, considered the last surviving remnant of the ponies that once

Eriskay

GEORGIAN GRANDE: UNITED STATES, OHIO

The Georgian Grande was developed in the 1970s by breeder George Wagner Jr., in Ohio. His aim was to re-create the old-style, bigger-boned Saddlebreds. To do this, he crossed draft breeds with Saddlebreds. Today's bloodlines can include Saddlebred mixed with draft and Gypsy breeds and Friesian. The resulting horses are very versatile. They can be found in all English and western pursuits, as well as in harness.

The horse is built uphill and has a well-shaped head with a broad flat forehead. The long well-muscled neck is a breed hallmark. The horse stands 14.2 to 17 hh and is found in all colors.

---

lived in the Western Isles, goes back to the ancient Norse and Celtic equines of the region. In fact, likenesses of similar ponies can be found etched into ancient Pictish stones in Scotland.

Until the mid-nineteenth century, crofters used the ponies to carry peat and other goods throughout the island. Then, as the islands became more populated, residents crossbred the ponies with bigger breeds to produce larger draft ponies for field work. Although the number of purebred Eriskays decreased, some residents continued to raise them. Machinery's popularity led to a further drop in the number of Eriskay ponies on the island and, by 1970, fewer than twenty remained. Since then, dedicated residents on the island have fought to save the breed.

The Eriskay pony remains on the Rare Breeds Survival Trust's (UK) critical list. Some 500 ponies are believed to remain in the area. The Eriskay Pony Society in the United Kingdom encourages the increase of the breed's population in a number of different ways, including giving advice to mare owners who wish to breed, helping mare owners determine which stallion to use, and transporting the stallions around the country for as broad of use as possible. The society occasionally buys young stallions of useful bloodlines to make sure they are kept for future use; they also have frozen semen from a number of stallions.

Physical Description: The Eriskay is a long-legged, refined pony with a large head and a high-set neck. The horse has a dense, waterproof coat with slight feathering on the back of the legs.

Uses: The breed is versatile and easygoing—useful for any discipline, including trail riding, jumping, and driving. The breed stands 12 to 13 hh. Foals are born black or bay, and most turn gray as they mature. The darker horses have lighter hair around the muzzle and eyes.

Of Note: Although the Eriskay is technically one of the United Kingdom's native ponies, the breed is not recognized by the National Pony Society (UK), which maintains that there are only nine breeds specified in its Memorandum and Articles of Association.

EXMOOR PONY

Alternative Name(s): None
Region of Origin: United Kingdom, England
History: Uncultivated heath and moorland cover about a quarter of Exmoor, an area on the Bristol Channel coast of Devon and Somerset in southwestern England. Some moors are enveloped by fragrant heather, others by grasses and sedges. This is the home of the Exmoor Ponies, which roam freely on the moors.

The Exmoor Pony is the oldest and purist of the British native pony breeds. Evidence shows that the Exmoor traces directly back to the ancient Pony Type 1. As early as 1000 BC, three ancient ponies crossed with European native horses throughout the region. Due to the moorland's isolation and climate, along with time and natural selection, the pony developed into a tough, enduring equine.

In 1818, the Crown sold the Royal Forest to industrialist John Knight, and there was a dispersal sale for the ponies. The last warden of Exmoor, Sir Thomas Acland, took thirty of the ponies and established the Acland Herd (now known as the Anchor Herd), whose descendants still roam Winsford Hill. Local farmers who had worked with Acland also bought ponies at the dispersal sale and founded several new herds, which helped keep the ponies' bloodlines pure.

The Exmoor Pony Society was formed in 1921 to preserve and protect the purebred Exmoor. The breed's numbers increased until World War II, at which point the pony's fortunes changed dramatically.

Exmoor Pony

fighting chance. Two types of Exmoor still exist: the Acland type and the Withypool type (a slightly larger, darker pony with a straighter profile).

Physical Description: Exmoors are usually dark brown with black manes, tails, and legs, although dun and bay coloring is also seen. The muzzle and eyes are a light fawn color called mealy. The breed stands 11.3 to 12.3 hh. The Exmoor has wide-set, expressive eyes, a straight profile, small ears, and a deep jaw with a good throatlatch. The pony's compact body has an ample barrel, powerful haunches, short back, and round croup. The legs are short with hard hooves. The pony moves with little knee action, providing a smooth way of going.

Uses: The Exmoor is prized as a child's mount and as a harness pony.

Of Note: Over the centuries, the pony developed two unusual characteristics: a heavy upper brow, known as a hooded eye, and an ice tail (a fan of top tail hairs that allow rain to sheet off the body and keep the belly drier, similar to the Icelandic Pony's). The Exmoor grows a dense coat during the winter, with an underlayer of downy hair that insulates the pony from wind and rain. The top layer also has special oils, which act as extra waterproofing. Both the ice tail and the winter coat are shed in spring, leaving the Exmoor sleek and gleaming.

FELL PONY

Alternative Name(s): None
Region of Origin: United Kingdom, England

History: The Fell is one of Britain's native breeds. It first appeared on British soil as the ancient Pony Type 1. The ponies further evolved in the fells, which are the hills surrounding the Lake District in Cumbria in northern England. The weather, with its rainy

Troops practiced shooting on the moors, often using the ponies as targets. Many ponies were also killed for food, as Britain's rationing forced people to turn to horsemeat. A brutal winter followed, and by spring of 1948, there were only about fifty Exmoors left.

However, concerned breed enthusiasts helped ensure that the existing ponies would be given a

PURITY OF WHITE

In Western civilization, the color white most often represents purity—of spirit, of intention, of self. The British and American suffragists fighting for a woman's right to vote adopted white as one of their colors, to signify the purity of their cause. On March 3, 1913, lawyer Inez Milholland Boissevain attired completely in white rode at the head of a suffrage parade in Washington, D.C., astride a beautiful white horse *(left)*.

GOTLAND: SWEDEN

The Gotland pony (also known as the Skogsruss) originated on the Swedish island of Gotland, from which it takes its name. The Gotland pony is thought to be the most ancient breed of Scandinavian origin. Like the Huçul and the Konik, which it resembles, the Gotland is a descendent of the ancient Tarpan. Gotlands lived wild for centuries on the island. Later, Gotlands were crossed with an Arabian x Gotland stallion named Ollie, which greatly influenced the breed. The short-coupled, sturdy pony is found throughout Scandinavia. The breed stands 12 to 12.2 hh and is found in all equine colors. The Gotlands have been used on small farms in Sweden, but today, they are primarily children's mounts.

summers and long cold winters, created ponies that could thrive under such challenging conditions.

Fell ponies are often mistakenly referred to as miniature Friesians, but while the Fell may share some similar characteristics with the Friesian, the latter does not appear to be a major player in the Fell's history. The now-extinct Scottish Galloway, a pony ridden by Scottish raiders, is also considered very similar to the Fell pony. The Scottish Galloway may even have been absorbed into the Fell native pony stock. It can be confusing to separate the early history of the two, as galloway was the old name given to any draft-type pony.

Farmers in Cumbria needed a utility animal to work the hilly landscape. The larger draft horses may have been perfect for working large, flat swathes of land, but they weren't much use on the smaller,

sloping plots. Fells, however, found working on the slopes to be second nature, and their willingness made them perfect for the farmer's uses, pulling hay rakes and mowing machines and hauling carts. Many farmers used the ponies for shepherding, riding the ponies to the sheep and back again. During the Industrial Revolution, Fell ponies were used to pack coal out of the fells. Their calm and unflappable nature, in addition to their strength and agility, made them easy to handle, and it wasn't unusual to see twenty ponies with only one person in charge of the entire group.

In 1893, a studbook was established in the United Kingdom to record and register ponies. The UK Fell Pony Society was founded in 1916 by hill farmers, Fell Pony admirers, and the Earl of Lonsdale.

Physical Description: The levelheaded Fell retains many of his unique prehistoric Forest Pony characteristics. The breed's thick forelock and mane and feathers on the legs help shed water away from the skin, and the large hooves help prevent the pony from sinking into the soft ground. The sloping shoulders and rounded hindquarters that first made the Fell nimble in the boggy forest floor now make the pony move with comfortable, easy gaits. The Fell stands 12.2 to 13.3 hh. With his strong haunches and muscular thighs, the Fell is a heavy-boned sturdy pony, capable of carrying heavy weight. The most common color is black, though brown, bay, and pale gray are also seen. White markings are acceptable, although a small star or socks are preferred to blazes or stockings.

Uses: Although some Fell Ponies are still used as general work ponies, many are used for driving and riding. One particular sport in which the nimble breed shines is in combined driving.

Of Note: England's Queen Elizabeth II adores this English pony so much that she is a patron of the Fell

Fell Pony

Florida Cracker Horse

Pony Society. She donates to the association, consults with the society's council members, and invites them to her palaces and homes for the occasional chat and tour. Her ponies are used after the hunt to carry the stags off the moors in Scotland at Balmoral.

FLORIDA CRACKER HORSE

Alternative Name(s): Chicksaw Pony, Seminole Pony, Prairie Pony, Florida Horse, Florida Cow Pony, Grass Gut

Region of Origin: United States, Florida

History: The search for the fabled Fountain of Youth sent the Florida Cracker Horse into American history. In 1521, Ponce de León brought horses that were a mixture of African Barb and Spanish descent from Spain to Florida to search for the elusive fountain. He never found it, but he did set up breeding stations in the region. The herds grew, and horses were let loose or escaped, which created many herds of roaming feral horses.

The native Seminole tribe quickly made use of the feral horses, capturing and breaking them to ride. When settlers arrived in the 1700s, they followed suit, using the horses in their cattle operations in Florida's flatlands. The cattlemen snapped bullwhips over the horses' heads to collect the cattle. The whip made a loud cracking

BREED IN BRIEF

GRONINGEN: THE NETHERLANDS
The Dutch Groningen developed over centuries from the Friesian and the Oldenburg for farm and harness work. The Groningen takes its name from a northwestern province of Holland, which is where the breed originated. One of the foundation breeds for the Dutch Warmblood, the Groningen is still a prized harness horse. Groningens have long heads, well-muscled necks, and long and prominent withers. Their legs are short, strong, and heavily muscled. Groningens usually range from 15.2 to 16.2 hh and can be found in all solid equine colors.

HACKNEY PONY: UK, ENGLAND

The Hackney Horse and the Hackney Pony are officially considered one breed and are registered in the same stud book. (*See Hackney Horse, page 99.*) While it took many centuries to develop the Hackney Horse, it only took one man to develop the Hackney Pony. An Englishman by the name of Christopher Wilson bred a harness pony with the same type and movement as the horse, but with pony characteristics. Wilson first used Fell pony mares and later added Thoroughbred and Welsh bloodlines. He then put the mares to a small, beautiful Hackney Horse named Sir George, who stood under 14 hands. Sir George is the patriarch of the Hackney Pony. Like the Hackney Horse, the pony has an elegant look, with a small head, delicate muzzle, and well-shaped ears and comes in black, brown, bay, and chestnut, and some spotted varieties. The pony stands from 12 to 14 hh.

---

noise, and soon the name Cracker stuck, despite the fact that the horses were also known by other names.

The Cracker horses were nimble and fast, but some of them possessed an easy-to-sit gait called a coon rack, a kind of single foot or amble gait. They also had good cow sense and could easily control cattle. In the early twentieth century, tractors began to replace the Crackers, causing their numbers to diminish. Things were not helped during the Dust Bowl era of the 1930s, when ranchers brought cattle down from the Midwest and imported an internal parasite called the screwworm. The infected cattle needed to be roped and restrained for medication—a job for the bigger, stronger Quarter Horse, rather than the slight Cracker.

Several ranching families continued to breed the Crackers. A rancher presented a herd to the Department of Agriculture and Consumer Services in Florida, which placed some of the horses in Withlacoochee State Forest at the PK Ranch property and others at the Agricultural Museum in Tallahassee. Six mares were later released on the 21,000-acre Paynes Prairie Preserve State Park, an original home to the wild Crackers.

Physical Description: The Florida Cracker is a small riding mount, standing from 13.2 to 15 hh, with refined features in all colors. The breed has a pretty head with a fine muzzle, wide-set eyes, and an intelligent face. The horse has a well-formed neck and clean throatlatch, a short back, an ample rib cage, and a nicely sloping croup. The Cracker's slender legs and light build give the horse a delicate appearance, and he is known for his comfortable stride. Although most modern Crackers are gaited, some of them are not. The association does not advertise them as a gaited breed, instead describing them as "good travelers."

Horses with gaits might have the flatfoot walk, running walk, and amble.

Uses: The horses are prized endurance horses and are used for traditional western ranching pursuits. They are also used in pleasure riding.

Of Note: In 2008, Florida lawmakers voted to make the Cracker Horse the state's official heritage horse.

FRIESIAN

Alternative Name(s): None
Region of Origin: The Netherlands

History: The ancient Friesian, one of Europe's oldest breeds, gets his name from the Friesland Province in the north of the Netherlands. Records of this horse date as far back as AD 100 by the Roman historian Tacitus. It's hard to believe, but the Friesian was considered quite an ugly breed in the early days. Stock was bred to Arabs and Andalusians, a scheme that brightened up the breed's appearance considerably.

The Friesian was imported to North America in the seventeenth century, but the horse eventually disappeared as a result of crossbreeding. In 1974, the breed was reintroduced. Friesians are still considered rare in North America, but their numbers are growing due to the demand for these romantic, fairy tale-like horses.

Physical Description: With his dark coat, long flowing mane and tail, and lightly feathered legs, the Friesian's striking appearance leaves an impression. All Friesians are black; the only white coloring allowed is a small star marking. The Friesian stands 14.3 to 16 hh. The long arching neck is rather upright and ties in to rounded withers. The horse's head is rather expressive and well sculpted, with tiny elegant ears. Its powerful shoulders, sloping quarters, strong hindquarters, and

Friesian

compact body create a movement that is long reaching, powerful, and supple.

Uses: The Friesian is one of the best carriage horses in the world, and in the early days the horse was used to pull carriages and funeral hearses, as well as in circuses. Due to the horse's expressive movement, athleticism, and kind temperament, the Friesian excels in dressage. The breed is also used for riding and driving.

Of Note: The Friesian almost became extinct worldwide at the turn of the twentieth century, as many Friesians were crossed to other breeds to create a faster horse for trotting races. In fact, only three purebred stallions were left. While World War II very nearly destroyed some breeds, it actually brought the Friesian back from the brink of extinction. Due to fuel shortages, Dutch farmers turned back to horses for transportation and fieldwork.

GYPSY VANNER HORSE

Alternative Name(s): Vanner Horse, Tinker Horse, Gypsy Horse, Gypsy Cob, Irish Cob, Colored Cob, or Drum Horse

Region of Origin: United Kingdom

History: After World War II, Romany (Gypsy) families began selectively breeding horses with the

---

BREED IN BRIEF

HOKKAIDO: JAPAN

Horses are not native to Japan; horses were brought to the country in the third century AD from Central Asia. The rare Hokkaido is one of three Japanese breeds and mostly likely originates from Chinese pony breeds. The Hokkaido breed lives in its eponymous island in the north of Japan on wide grassland. Samurai used the horses in *yabusame*, (once a Samurai sport and now a Shinto ritual) in which the warriors rode their horses while shooting arrows at targets. Today, a number of people in the area continue to use the Hokkaido for farmwork and for pulling sleds.

The Hokkaido is a stout animal, with some feathering on the legs. The breed stands 13 hh and can be found in all colors.

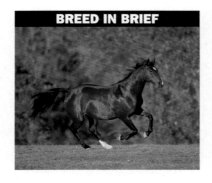

HOLSTEINER: GERMANY

The Holsteiner is Germany's oldest warmblood and was a foundation breed for the Hanoverian. The breed's origins began in the fourteenth century with the burly Marsh Horse of Schleswig-Holstein, a knight's mount later used as a coach horse. In the 1800s, Arabians, Barbs, and Spanish horses were crossed to the Holsteiner. After World War II, Thoroughbred blood was added to lighten the breed's frame and create a nimbler sport horse. Although these horses are now bred largely for jumping, they are also successful in dressage and as competitive carriage horses. The Holsteiner is a tall yet refined warmblood with strong hindquarters. The breed stands 16 to 17 hh and is found in all solid colors.

---

endurance and strength of a draft, but with elegance and an appearance reflective of the colorful Gypsy culture. The Romany people—who love color, strong patterns, and lots of hair—wanted a flashy, magical-looking horse to match their culture.

To create this breed, they chose horses with genes that included the Shire, the Clydesdale, the Dales Pony, and the Friesian. One of the foundation stallions, Sonny Mays, produced many foals with color. Colored horses in the United Kingdom were not common after World War II, and Gypsies say that Sonny Mays is responsible for most of the colored horses there today. Black stallions and mares played a key role in color too, as the Gypsies believed black horses would produce deeper black and white babies.

Although there are many Romany-bred horses, the Gypsy Vanner Horse, as the breed is called in America (where Gypsy Vanners were first imported in the 1990s), is selectively bred. In the United Kingdom and Ireland, Gypsies tend to differentiate between their selectively bred horses (called Proper Cobs) and their less fancy horses (called Common Cobs).

Physical Description: A Vanner Horse should have the look of a small draft horse: short coupled with rounded hindquarters and a short, yet well-formed, neck. It has an abundance of feathering starting from the knee and hock that flows down over the hooves. It's beauty with a purpose: the hair protects the legs from the elements. The mane and tail are long and flowing. Because the breed's original use was as a caravan horse, the Vanner moves with a snappy, bright trot, but he can extend the foreleg if needed. The breed possesses a beautiful, bounding canter. The Vanner stands 13.2 to 15.2 hh and is not a color breed, yet color is one of the attributes that set him apart from other breeds. Color patterns include:

Gypsy Vanner Horse

piebald (black and white), skewbald (a combination of brown, red, and white), blagdon (a solid color with white splashes underneath); and odd colored (all other colors).

Uses: Vanners are known for their versatility. Vanner Horses can be found in carriage driving, combined driving, English and western pleasure riding, hunter and jumper competitions, and trail riding. Although they will probably never reach the upper levels of the jumping or dressage worlds, they are willing to do anything for the humans in their lives.

Of Note: Because of their intelligence, unflappable nature, and affection for people, Gypsy Vanners are used in equine therapy and equine-assisted education programs.

Hackney Horse

HACKNEY HORSE

Alternative Name(s): None
Region of Origin: United Kingdom, England

History: The Hackney Horse was perhaps the most sought-after carriage horse in England. The breed was first developed in the late eighteenth century in Yorkshire and Norfolk from crossing Thoroughbred stallions to local mares known for their trotting talent. The first Hackneys were long-distance riding horses (averaging 18 miles in one hour) with a ground-covering stride; they were used by European breeders to help improve and refine breeds such as the Dutch Gelderlander and the Selle Français. Gentleman farmers found the breed to be the perfect multitasker: farm horse during the week and elegant coach horse on the weekend. When roads became smoother and easier to traverse, the horses really hit their stride. The faster Hackney soon passed right by the slow, sturdy coach horses. There were two Hackney types: a heavier one and a lighter. Breeders began to focus on the lighter, high-stepping Hackney, since that's what gentlemen preferred to drive (and even race) in the parks and around town.

It's fair to say that more people are more familiar with the Hackney Pony than with the Hackney Horse; after all, 95 percent of Hackneys in North America are ponies. Many people may have seen these dashing little

BREED IN BRIEF

HUÇUL: POLAND

The Huçul hails from Poland's Carpathian Mountains and is a descendant of the primitive Tarpan. This strong and hardy mountain breed was commonly employed in agricultural work. Like many horse breeds, the Huçul's numbers were greatly reduced during World War II. In 1972, the Huçul Club was founded to help reestablish the Huçuls and a studbook was created a decade later. Today, there are more than 1,000 Huçuls to be found throughout eastern Europe. The breed stands 12.1 to 13 hh and can be found in dun, bay, and pinto colorings.

IOMUD: TURKMENISTAN

The Iomud was bred by the Iomud tribe in southern Turkmenia from Turkmene horses. A cousin of the Akhal-Teke, this rare desert breed—which can subsist on small amounts of water and feed when necessary—has also been influenced by the Arab, the Kazakh, and the Mongolian. Although it has primarily been a saddle horse, the muscular and hardy Iomud has been used in harness and in racing as well. The horse has an attractive head with small ears, large eyes, and a slightly Roman nose. The Iomud also has a muscular neck, fairly prominent withers, and strong legs. The manes and tails of these horses are sparse. The Iomud ranges from 14.2 to 15.2 hh and can be found in gray, bay, and chestnut.

---

ponies exhibited in the road pony classes commonly found at county fairs, pulling road bikes behind them with their drivers dressed in jaunty jockey silks. The pony was developed after the horse, but they are classified as the same breed. (*See Hackney Pony, page 96.*)

Physical Description: The Hackney Horse ranges in height from 14 to 16 hh. Both pony and horse possess an extremely elegant appearance with a small head, delicate muzzle, and well-shaped ears. The Hackney breed has a beautifully shaped head, a high set-on neck, and well-sloped shoulders. The breed features a compact body with a short loin and a level croup. The tail is either long or cobbed (docked), or it appears to be cobbed. The natural movement of the both horse and pony is very high and distinctive. Shoulder action is fluid and easy; the hocks in the hind legs should come up under the body and rise up high. Common Hackney colors are black, brown, bay, and chestnut, and there are even some spotted varieties.

Uses: Hackneys are used in driving and in all English sports, including saddle seat, jumping, and dressage.

Of Note: Every breed group has a broad spectrum of opinions, and the Hackney group is no different. Many aficionados argue that the Hackney Horse and the Hackney Pony should be considered different breeds. But the only real difference between the two is that one is a horse and one is a pony. They are judged by the same criteria in the show ring.

HAFLINGER

Alternative Name(s): Avelignese
Region of Origin: Austria and Northern Italy

History: High in the southeastern Alps of northern Italy, just south of the Austrian border, lies the municipality of Hafling—a favorite haunt for skiers, hikers and mountain climbers, and the original home of the beautiful and charming Haflinger. Late Medieval writers were the first to document the horses' existence, writing about Oriental-type horses living in the southern Tyrolean Mountains. This area was later relinquished to Italy by Austria on September 10, 1919. These native mountain horses were family farm horses. They were ridden by children, they took the family to church, and they pulled the family wagon. And because the mountain trails were steep and narrow, they were used as packhorses.

In the late 1800s, the government bred horses that could be used as military packhorses. In 1874, they crossed a native mare with a half-Arabian stallion named 133 El' Bedavi XXII, which produced a chestnut colt named 249 Folie, who was born with the perfect conformation for a packhorse. Folie became the foundation stallion, and the lineage of all Haflingers to this day must trace back to this stallion. Mares were added to the stud in 1897, and in 1898 the government officially declared the breed, retaining the name Haflinger.

In 1958, American steel tycoon Tempel Smith sent Leo Lightner to Europe to find a unique breed of horse for his stable, Tempel Farms, in Wadsworth, Illinois. Lightner found the breed that the farm is now famous for—the Lipizzan. But he also suggested that Smith consider bringing in the Haflinger. Lightner bought thirteen Haflingers, the first to come to America.

Physical Description: With a flowing, thick mane, forelock, and tail, the Haflinger has a romantic look. His color ranges from flaxen on the liver chestnuts to cream on the lighter chestnuts. Ideally there is no white on the legs or on the body; however, very often the coat will lighten toward the feet, giving the appearance of white. Although the horse is broader and shorter than other riding breeds, the Haflinger is not short-coupled like a draft. He looks more like a riding horse—well

Haflinger

proportioned in three equal pieces, and with large powerful hindquarters. The eyes are large and expressive, and the head is elegant. The coat is soft and tends to be thick. Because of their pack animal roots, Haflingers have a steady, sturdy climbing ability. The slight knee action and bounding canter come from the mountainous habitat. Height ranges from 13.2 to 15 hh.

Uses: Although originally prized as driving and pack animals, Haflingers are considered all-rounders, taking to whatever discipline their owners choose. The horses can be found competing in dressage, jumping, western sports, and combined driving. The charming Haflingers are patient with new riders and handlers, yet they still have the athletic talent to help competitors meet their goals.

Of Note: There are seven Haflinger stallion lines, all derived from Folie in some way. The Haflinger is very hardy and an easy keeper. Owners have to be aware of keeping their horses on lush grass because they can quickly become overweight.

BREED IN BRIEF

ITALIAN HEAVY DRAFT: ITALY

The Italian Heavy Draft originated in northern and central Italy and is one of the most popular draft horses throughout the country. The breed began with the Belgian in the nineteenth century, but Percheron and Boulonnais genes, as well as Breton blood, further developed the Italian Heavy Draft. Although the horses have been used as draft animals, their heads are very fine, with a broad forehead, small ears, and large eyes. They have heavy manes and short, straight backs and are very muscular. The Italian Heavy Drafts are also used as harness horses. These horses range in height from 15 to 16 hh and can be found most commonly in liver chestnut coloring, with a flaxen main and tail, or in roan.

BREED IN BRIEF

JAVA: INDONESIA

The Java, which originated on the island of Java, is one of the eight native ponies of Indonesia. It is one of the largest and strongest due to an influx of Barb and Arabian blood. The Java pony has been used to transport goods and pull the heavy two-wheeled cart that is called a *sados*, which the Javanese use as a taxi for people as well as for goods. This muscular pony has a heavy head, a short neck, a long and straight back, and long legs. The Java stands 12.2 hh and is found in all equine colors.

The other native ponies of Indonesia include the Bali, the Batak (*page 67*), the Deli, the Padang (*page 132*), the Sandalwood (*page 141*), the Sumba/ Sumbawa (*page 144*), and the Timor.

HANOVERIAN

Alternative Name(s): None
Region of Origin: Northern Germany

History: In 1714, King George I of England—originally the elector of Hanover—sent several English Thoroughbreds to Germany to refine the native stock. His son, George II, carried this idea further, formally establishing a stud at Celle in 1735, with the purpose of starting a breeding program for superb working horses and top cavalry mounts. Fourteen black Holsteiners, as well as some Thoroughbred blood later on, were used to create this new breed.

As World War II drew to an end, the Russian army invaded Germany, sending the inhabitants fleeing before them. Among the refugees were Trakehner horses and handlers. They found sanctuary in Lower Saxony, and some of the breeding stock ended up in Celle. With the infusion of Trakehner blood, as well as Thoroughbred blood, the Hanoverian changed from a bulky horse, perfect for pulling a plow or a caisson, to the mount of choice for competitive riders. The influx of genes gave the Hanoverian his outstanding movement and further refinement.

Cavalry mounts and plow horses were no longer needed, and so the horse community had to change. Luckily, sport riding was coming into fashion, and savvy Hanoverian breeders answered the call by changing their breeding philosophy toward this new phenomenon.

Mares and stallions of any breed can enter an inspection, and those who meet the guidelines are included in the Hanoverian breeding program. The original Celle stallion station that began in 1735 is still the center of Hanoverian breeding.

Celle has always been a huge operation, standing 100 stallions alone in the 1800s and more than 200

Hanoverian

today, with about 8,000 breedings taking place every year. The American Hanoverian Society was founded in 1978. Germany's Hanoverian breed association is called the Verband Hannoverscher Warmblutzuchter, and this association is the mother organization of the American Hanoverian Society.

Physical Description: The Hanoverian has a fine head with expressive features and a long and elegant neck set into big, well-angled shoulders. The body of the horse is rectangular and equally divided into shoulders, rib cage, and haunches. Because the horse is bred for sport, the conformation of the haunches is particularly important, as they are the power source of the horse. The croup should be slightly sloping, and the back should be strong and well muscled, with the hocks well angled and broad. The legs and cannons are short, with large bones and joints. Bay, black, chestnut, brown, and gray are the standard colors, and white markings are common. The Hanoverian stands from 15.3 to 17 hh.

Uses: The Hanoverian is one of the most popular warmblood breeds. While the breed was originally prized for how deeply the horse could plow a furrow, today's Hanoverian is assessed on movement, trainability, and performance potential. The Hanoverian is a renowned mover, covering the ground with plenty of spring and impulsion. This gorgeous movement and presence make these horses specialists in jumping and dressage.

Of Note: The brand on the breed's lower-left hip is a modified *H* shape with two horse heads on the top. This design was inspired from the crossed horse heads on the roofs of the breeding farms in Lower Saxony. Hanoverians are named using the first letter of their sire's name.

HIGHLAND PONY

Alternative Name(s): None
Region of Origin: United Kingdom, Scotland

History: Roaming over the Scottish Highlands is a sturdy pony that has thrived among the ancient rock and coastline, the lochs, and the windswept valleys for centuries. The Highland's ancestor is the ancient Pony Type 2 (*see page 48*), which lived in the Highlands perhaps as long as 10,000 years ago. Over the centuries, the prehistoric pony received an influx of different blood, depending upon which army was conquering the area. The Galloway pony featured quite prominently in the Highland Pony's ancestry in the eighteenth and nineteenth centuries. Although the basic blood within the ponies remained similar, their regional isolation resulted in different strains of ponies, including the Islay, Rhum, Mull, and Barra.

The Highland Pony has a long, illustrious relationship with the Crown. From Queen Victoria to Queen Elizabeth II, there has been a loving association between the ponies and the royal family. Queen Victoria was a patron of the breed society and had personal interest in both showing and breeding Highlands.

Highland Pony

JUTLAND: DENMARK

An ancient breed, the Jutland originates from the Forest Horse and has been known in Denmark's Jutland peninsula for centuries. Historically, the Jutland was used in many ways, including as a knight's mount, as a farm horse, and as a coach horse. British coach and draft horses were introduced to the Jutland's bloodline in the 1800s. The resulting horse has a boxy head on an arched neck and slight feathering on the legs. The Jutlands have pudgy but compact bodies and are considered enduring, sturdy, and energetic. These draft horses range from 15 to 16 hh, and they are usually chestnut colored, with a flaxen mane and tail. Brown and black coats have been seen but they are rare among the Jutlands.

KABARDIN: CAUCASIAN MOUNTAINS

The Karbardin was bred in the northern Caucasus and is related to the Karabakh. Because of geographic positioning, the breed was created from Mongolian horses and eastern breeds from Turkey, Iraq, Iran, and Kurdistan. The breed developed in the mountains, which is why these horses are very hardy and sure-footed. The Kabardin has a pacing gait and possesses enormous stamina. The horses stand 15 to 15.2 hh, and they are found in black and bay. Kabardins have primarily been used as saddle horses.

Queen Elizabeth II currently has one of the largest (if not *the* largest) working studs of Highland Ponies, and during the season, you can usually see more than twenty of her ponies out on the hills at Balmoral Castle. With two stallions (Balmoral Dee and Balmoral Moss) and a broad-based line of mares, the Balmoral Stud consistently produces quality foals for the future of the breed.

Physical Description: There are no stringent breeding rules, so Highlands also differ from other breeds by having a larger gene pool to draw from. This breed stands 13 to 14.2 hh. Highlands have pretty heads that feature alert eyes, small ears, broad muzzles, and deep jowls. The horse's neck is slightly arched and well developed and tied into a long sloping shoulder. The powerful pony has well-muscled hindquarters, a deep chest, and a broad rib cage. His legs are blessed with solid bone and muscle and sturdy dark hooves. His fetlocks bear silky feathering.

The ponies are found in a variety of coat colors, including dun, gray, brown, black, chestnut, and cream. Some of the ponies will have dorsal (back), zebra (leg), or shoulder stripes. Any extra white coloring usually signifies outside blood, so that coloring is not favored.

Uses: These native ponies were originally bred to help farmers to perform various chores, such as packing game out from the mountains, plowing the fields, and hauling wood from the forestland. The Highland Pony was used by warring clans and later in the army, even during the twentieth century, in World War II. Highland Ponies have been nearly synonymous with the sports of stalking and shooting. The breed has become a popular trekking and packing pony, as well as a beloved mount for children and adults.

Of Note: The ponies' great strength and natural sure-footedness assures that they'll always have a job carrying hunters up the steep, rocky Scottish hillsides and traveling back down in special pack saddles carrying stags weighing up to 225 pounds.

ICELANDIC HORSE

Alternative Name(s): None
Region of Origin: Iceland

History: In the Land of Fire and Ice roams Iceland's greatest equine treasure: the Icelandic Horse. The Icelandic was most likely brought to Iceland in the ninth century by Viking settlers from Norway and the British Isles. Although the breed shares characteristics with the Mongolian horse, little is actually known about its ancestry. Because of the remoteness of the area where the horses originated, the Icelandic Horse has remained a pure breed, unchanged for more than a thousand years. Inhabitants have valued the horses for their all-around ability, working for their owners and providing transportation. In fact, they were the only form of transportation until roads were built in the 1800s. The Icelandic horse is still greatly treasured in his homeland.

Physical Description: An elegant animal with a flowing double-sided mane and long lush tail, the sure-footed Icelandic Horse is best known for his unusual gaits. The horse's five comfortable gaits include the walk, trot, canter, and gallop, as well as the tölt (a beautiful four-beat gait, similar to the rack, conducted in a high step). The tölt is a very smooth gait and can be as slow as the walk or as fast as the gallop. The breed comes in more than a hundred colors and combinations, except spotted, and there are Icelandic names for all of them. Most horses are brown, black, bay, or chestnut. The breed stands from 12 to 14 hh.

Uses: Although small, an Icelandic horse can carry a full-size adult across long distances. The breed is also used in dressage and jumping. In Iceland, many

Icelandic Horse

people still use the horses for sheepherding, local transportation, and competition.

Of Note: Some Icelandic Horses also have a pace, which is a two-beat gait in which the legs on the same side move together.

IRISH DRAUGHT HORSE

Alternative Name(s): None
Region of Origin: Ireland

History: It's unknown how the breed came to be in Ireland or what its ancestry is. The Irish Draught Horse Society in Ireland maintains the horse's ancestry is most likely made up of whatever horses were available to farmers and able to adapt to the work.

In 1905, the Department of Agriculture for Ireland decided this equine type needed to be finalized and so provided quality stallions for farmers and even offered money for stallions of original type to create a foundation stock. In 1971, the breed's registration was handed to the Irish Horse Board, which made a rule that each foal had to have one foundation stallion in his pedigree to be considered a Registered Irish Draught. In 1978, the board closed its books to new stallion bloodlines, and the type turned into a breed.

It soon came into vogue to cross the Irish Draught with the Thoroughbred to create the sportier Irish Hunter and other crosses. These horses were so successful that Irish Draught mares were no longer producing purebred foals. With few purebred mares being born, the breed's survival was in jeopardy. This, coupled with the fact that there were only five foundations stallions (Comet, Prince Henry, Young J.P, Kildare, and Brehon) meant breeders were in danger of creating a shallow gene pool.

In recent years, there has been an increase in the number of purebred Irish Draught foals produced. For several years, the Irish Horse Board operated an

BREED IN BRIEF

KAGOSHIMA: JAPAN

The Kagoshima (which was formerly known as the Kyushu) is one of Japan's three horse breeds (the Hokkaido and the Kiso are the other two breeds). The Kagoshima breed developed on the southernmost island of Japan, Kyushu, in Kagoshima Prefecture. The origins of the Kagoshimas can be traced to China, to Korea, and to Mongolia. The Kagoshima, considered wild until recent times, is a thick and sturdy pony, with hard, rounded hooves, which are often of bluish horn, and short legs. Kagoshimas stand 13 hh and can be found in all colors. Until the early decades of the twentieth century, the horses were used for farmwork and as pack animals. Today, the number to be found in Japan is few.

KARABAIR: UZBEKISTAN

The Karabair developed in Uzbekistan and is the oldest of the Central Asian breeds. This horse stems from eastern breeds and primitive steppe horses. The tough Karabair is used in the national game of kokpar. Karabairs are also ridden and raced. Although the breed shares many of the attributes of the Arabian, including great stamina and speed, it has a straighter profile and topline. The Karabair stands 15 hh and is usually found in bay, chestnut, and gray. Sometimes they are can be found in palomino, dun, black, and pinto.

Irish Draught Foal Grant scheme to encourage pure breeding of Irish Draught mares. At present, the organization Horse Sport Ireland operates the Irish Draught Colt Retention Scheme to encourage the retention of quality colts as prospective stallions. Horse Sport Ireland also operates the Irish Draught Rare Bloodline scheme to encourage the maintenance of genetic diversity within the Irish Draught breed.

Physical Description: Strong and sound are key characteristics of the Irish Draught, particularly in the hooves and legs, which are two of the five top breed inspection requirements along with temperament, movement, and type. Unlike other stocky draft horses, the Irish Draught is more like a riding horse with a high-set neck, clean withers, and long legs. The breed's long ears and broad profile give the horses a noble look. Heights range from 15.1 to 16.3 hh, and any solid color with white markings is common, although socks above the knees or hocks are not ideal.

Uses: The Irish Draught is a powerful horse. He has bold, strong, ground-covering movements. He is also a natural jumper. Because of his kind nature, he is a good mount for the amateur rider.

Of Note: The Irish Draught is one of two native Irish breeds. The other is the Connemara Pony.

IRISH SPORT HORSE

Alternative Name(s): Irish Hunter, Irish Draught Sport Horse

Region of Origin: Ireland

History: Although seemingly a crossbred animal, the Irish Sport Horse (ISH), the preferred mount for sports in Ireland, is considered a breed. The main composite breeds of the ISH are the Irish Draught and the Thoroughbred. (In the past, the Irish Draught cross with Thoroughbred was referred to as the Irish Hunter.) Varying proportions of these breeds result in the production of ISHs. The ISH draws speed and endurance from his Thoroughbred ancestry and soundness and unflappability from his Irish Draught ancestry. The sloping croup from the Irish Draught breeding adds to the horse's jump ability. Recently, there has been an infusion of Continental warmblood breeds into the bloodline.

Physical Description: All solid colors are found; pintos are rare. The ISH stands from 15 to 17 hh.

Uses: Temperament, versatility, and substance combine to create a horse that appeals to both amateur and professional riders. The Irish Sport Horse can be found competing throughout the world at the highest levels. The ISH is a talented hunter and jumper and is highly prized in eventing.

Irish Draught Horse

Irish Sport Horse

Of Note: Because of their calm and easygoing temperaments and their ease of training, the ISHs are popular police horses in Ireland and Great Britain.

KIGER MUSTANG

Alternative Name(s): Kiger, Kiger Horses, Steens Mountain Kigers
Region of Origin: United States
History: Oregon is a state of unparalleled beauty. Its western border, the Pacific Ocean, provides a gorgeous rocky coastline, while its eastern Idaho border identifies the mountainous side of the state. Further defining Oregon's unspoiled identity are two rugged ranges, the Pacific Coast Range and the Cascades. It is in these stark,

unpopulated foothills that a small band of wild horses called the Kigers thrived, isolated from civilization.

Spanish horses were brought to America by settlers and explorers from the 1500s through the 1700s. Throughout North America, these horses were allowed to run free, and from time to time, other horses would enter the gene pool. Every breed from Thoroughbreds to draft horses ran with the feral horses, and their blood showed up soon after in new feral generations.

The pure Spanish feral horses became less common, and by the 1900s, they had been all but replaced by a tough, hardy little horse—the Mustang—that roamed the plains and prairies, the scrub and the desert. In some areas of the country, particular herds

BREED IN BRIEF

KARABAKH: AZERBAIDZHAN

The rare Karabakh is related to the Kabardin and developed in the Karabakh Mountains of Azerbaidzhan from Mongolian ponies and eastern breeds, including the Arabian and the Akhal-Teke. The Karabakh has more Akhal-Teke and Arabian blood than does the Kabardin. This small, hardy, and sure-footed mountain horse has been

used primarily as a riding horse. The Karabakh has a refined head set on an arched neck of good length. The Karabakh stands 14 hh and can be found in coat colors of dun, bay, and chestnut, with a darker tail and mane and a dorsal stripe. The influence of the Akhal-Teke bloodline is apparent in the coat of the Karabakh, which usually has a metallic sheen.

KATHIAWARI: INDIA

The Kathiawari, whose roots are believed to date back to the fourteenth century, was the foundation breed for the Marwari. The Kathiawari originated in the Asian plains of India, in the Kathiawar peninsula. Traditionally, wealthy families bred the Kathiawari horses; today breeding of the Kathiawari is controlled by the government at the stud at Junagadh. The breed is sturdier and squarer than the Marwari, and the Kathiawari's ears are shorter and curved a little tighter than those of the Marwari. The nose is also less Roman than the nose of the Marwari. The Kathiawari is used for riding and in harness. This hardy desert horse stands 13.3 to 14.2 hh and is found in all colors except black.

Kiger Mustang

flourished undisturbed, and in relative isolation. These were called Colonial Spanish Horses.

In 1971, Mustangs came under the jurisdiction of the government, with the Wild Free-Roaming Horse and Burro Act, which put the U.S. Bureau of Land Management (BLM) in charge as caretakers of the feral equines. Later in that decade, one of the BLM's wild horse specialists, E. Ron Harding, was present at an Oregon gather of horses. Harding came upon a rather large band of wild horses, uniform in color, from Kiger Gorge, near Steens Mountain. These horses bore beautiful coats of dun, accented by jet-black manes and tails. Harding knew he had discovered something special and surmised that these feral creatures had to be of Spanish descent. It was decided to divide the herd and re-release them into southeastern Oregon. Part of the herd went into the Kiger HMA (herd management area), while the other went into the Riddle HMA.

Physical Description: The Kiger Mustang has a stunning, burnished dun-colored coat, often highlighted by a thick, black mane and tail. The Kiger is also found in other variations, including red duns, grullas, and a unique "claybank" coloring (which is a pale sand color, highlighted by a dorsal stripe). Bearing more evidence of its Spanish heritage, the Kiger Mustang is a compact, close-coupled horse, with upright carriage and a beautifully high-set neck on its sloping shoulders. The Kiger's face is noble and expressive, with hook-tipped ears and a straight or slightly convex profile. The Kiger has strong bone and dense hoof walls, which makes the breed sound and sure-footed for the trail. Kiger Mustangs, like their other American Mustang brethren, are not tall horses. Most range from 13.2 to 15.2 hh. But what they lack in stature, they make up for in magnificent presence.

Uses: The breed is a prized trail and endurance horse.

Of Note: Kigers are now bred in captivity. Horses born domestically are simply called Kiger Horses or Steens Mountain Kigers.

KNABSTRUPPER

Alternative Name(s): Knabstrup
Region of Origin: Denmark

History: Denmark's Knabstrupper is among the magnificent breeds that have been categorized as Baroque. The Baroque breeds descended from horses of the Middle Ages, such as the Neapolitan Horse, the Barb, and the Iberian horse, represented today as the Sorraia.

The Knabstrupper was developed in Nordsealand, Denmark, by Major Villars Lunn, who decided to put a blanketed chestnut mare of Spanish breeding to a Frederiksborg stallion in 1812. This first breeding resulted in a colorfully spotted foal and became the basis for a new breed. A small gene pool and a fire at the breeding farm diminished the Knabstrupper's numbers greatly, and the breed was nearly lost for good. In 1947, however, the Danish stud farm Egemosegaard attempted to reestablish the breed. In 1971, breeder Frede Nielsen brought three Appaloosa stallions to Denmark to infuse new blood into the breed.

Physical Description: Although Knabstruppers are often spotted as Appaloosas are, they are only connected to that breed through the same ancient color gene. The most popular spotted pattern is the full leopard, which is solid white with black or brown spots. Other markings include the blanket, the snowflake, and the few spot pattern (which is an almost-white horse with a couple of spots). There are three types of Knabstruppers: the Baroque, the Sport Horse, and the Pony.

Uses: Due to his crossing with European warmbloods, in particular the Danish and the Trakehner, the Sport Horse excels in English sports, such as dressage, jumping, and eventing. The Baroque type, which is more traditional in build and resembles the old-style circus horse, is a prized harness horse. The Pony type, a smaller version of the Baroque, is a popular child's mount.

Of Note: Historically, the Knabstrupper was cherished by royalty and noblemen and used for leisure pursuits and in festivals. Knabstruppers were also used as officers' cavalry horses. After World War II, the Knabstruppers were commonly used in circuses throughout the world.

LIPIZZAN

Alternative Name(s): Lipizzaner
Region of Origin: Spain, Austria, and Slovenia

History: The beautiful and athletic Lipizzan is another of the Baroque breeds that developed during a time of great appreciation of the arts, including the riding arts. The Lipizzan was bred in the sixteenth century by the Habsburg Empire that ruled Spain and Austria. The Habsburg rulers wanted to create a beautiful and trainable cavalry mount, so in 1562 Emperor Maximillian II brought several Iberian horses to Kladrub (now in the Czech Republic). In 1580, his brother Archduke Charles II set up another stud at Lipizza near the Adriatic Sea (now in Slovenia), which gave the breed its name.

The Iberian horses from both studs were crossed with native Karst horses and other breeds from Europe, such as the Neapolitan horse. The stud at Kladrub was

Knabstrupper

HANSOM CABS

The horse-drawn hansom cab became the most popular London cab in the 1830s. Not until the late 1800s did it make a debut in New York City and Boston. This unique enclosed carriage could carry two passengers and had an elevated seat for the driver at the back, from which the driver could open and close the doors. At left, in 1896, a NYC driver poses before his hansom cab while his hardworking horse eats from a nosebag.

KENTUCKY MOUNTAIN SADDLE HORSE: UNITED STATES, KENTUCKY

Bred by the mountain folks of eastern Kentucky for 200 years, the gaited Kentucky Mountain Saddle Horse sprang from the Narragansett Pacer, Spanish Jennet, and gaited Galloway (all now extinct breeds). The resulting breed was perfect for farming, and the comfortable gait meant a rider could stay aboard for hours. The breed's natural four-beat gait is called an amble or a rack. The horses are found in the show ring and on the trail. Kentucky Mountain Saddle Horses are classed by height: 11 to 14.1 hh is class B; 14.2 hh and up is class A. The breed is found in all colors. Spotted horses are called Spotted Mountain Saddle Horses.

responsible for creating heavy carriage horses while Lipizza created fine riding and carriage horses. The stallion lines that continue today are Pluto, Conversano, Maestoso, Favory, Neapolitano, and Siglavy. Two other lines are recognized in eastern European countries and in North America: Tulipan and Incitato.

Dressage grew from the cavalry, and in 1735, Holy Roman emperor Charles IV established the Spanish Riding School of Vienna (which derived its name from the breed's origins). Today, the school's performances carry on the tradition of showcasing the best-trained horses.

Between 1958 and 1959, Tempel and Esther Smith imported three stallions and thirteen mares from Austria, Hungary, and Yugoslavia, the largest herd to come to the United States. Tempel Farms, located in Old Mill Creek, Illinois, continues to breed Lipizzans and hold classical dressage exhibitions in the manner of the Spanish Riding School.

Lipizzan

Physical Description: The Lipizzan is a classic Baroque horse type, with a convex profile, broad, deep quarters, and relatively low-set hocks that make it easier to raise the forehand. The Lipizzan is characterized by his sturdy body, brilliant action, and proud carriage. The horse's rectangular compact body—set off by a powerful, crested neck—presents a picture of strength. The breed features a broad back and loins, well-rounded powerful quarters, muscular shoulders, and short, strong legs with well-defined tendons and joints. The head is usually straight or slightly convex; big prominent eyes and small ears are set wide apart. A small muzzle balances a prominent jaw. The breed stands from 14.3 to 15.3 hh. Most Lipizzans are white due to the color preferences of the Austrian nobility, but bay horses sometimes occur. Lipizzans are born dark brown or dark gray and turn white between the ages of six and ten.

Uses: The Lipizzan possess the strength, athleticism, beauty, and temperament for classical dressage, as well as driving, cross-country, and jumping.

Of Note: Lipizzan colts are named for their father and mother: for instance, Conversano Dagmar (sire's name first, dam's name second). Numbers are added if there is a colt with the same name. Fillies are typically named after their mothers.

LUSITANO

Alternative Name(s): Puro Sangue Lusitano
Region of Origin: Portugal

History: The daring Lusitano is a national treasure in Portugal; his presence is synonymous with bullfighting. Portugal was once under Spanish rule but received its independence in 1640 after the Portuguese Restoration War. The Portuguese horses were also affected by Spain's influence at this time. The

Lusitano

Lusitano and the Andalusian/P.R.E. are historically the same breed, with Barb and Iberian horse (also called Jennet) breeding. But while the Andalusian/P.R.E has only two bloodlines, today's Lusitano has four bloodlines (two of them Spanish): Andrade, Veiga, Coudelaria Nacional, and Altèr-Real (bred only from the royal stud established in Alter do Chao).

In 1789, the French Revolution saw the decline in all things royal, including the riding of Baroque horses. In England, Thoroughbreds were making their mark, and dressage soon was eclipsed by horse racing and foxhunting. Napoleon's invasions of Spain in 1807 and 1809 nearly destroyed the Iberian peninsula and its horses. The situation in Portugal worsened when the Portuguese royal family was ousted in the early twentieth century. The government took control of the royal stud in Alter do Chao and began to indiscriminately breed the horses. In the 1940s, breeders

KERRY BOG PONY: IRELAND

Although a native pony, Ireland's Kerry Bog Pony, like Scotland's Eriskay, is not officially considered one of the nine native ponies of the British Isles. However, today the breed is recognized as Ireland's National Heritage Pony. The Kerry Bog Pony originated in the marshes of southwestern Ireland and contributed to the region's turf industry, which used rich bog land to burn as fuel. However, once mules and machinery gained popularity in Ireland, the ponies were no longer needed to haul the turf from the bogs. They nearly faced extinction as a result, with only 40 remaining when a local enthusiast named John Mulvihill stepped in and worked to rebuild the pony's numbers and regenerate interest in the breed. The Kerry Bog's numbers are now growing, with more than 200 ponies living in Ireland and in the United States. Kerry Bog Ponies are extremely elegant yet also strong. They're used for driving and as a child's mount. Standing from 10 to 12 hh, the Kerry Bog Pony is found in pinto, as well as all solid colors, including palomino.

KISBER FELVER: HUNGARY

In 1853, King Ferenc József I established a Hungarian royal stud in Kisbér to breed horses for the cavalry. The breeders chose horses for their military suitability to breed with Thoroughbreds. In addition, just before World War II, Trakehner blood was added. The numbers of Kisber Felvers (or Kisber Halfbreds) fell dramatically during the First and Second World Wars, and today they are seldom to be found outside their native Hungary. The horses, which are very athletic, are used for competition sport and agricultural work. The breed possesses great stamina and strength and stands 15.2 to 17 hh. Although they can be found in any solid color (such as gray, as seen here), Kisber Felvers are usually chestnut or bay.

---

sought to find horses with original Altèr-Real blood from the Royal Stud in order to rebuild the Portuguese bloodlines. These horses were to be used for traditional pursuits, such as ranching and bullfighting, and so they kept much of the original Iberian features, such as the dun coloring and the straighter profile.

The Portuguese Lusitano was officially created in the late 1960s, after Portuguese breeders opened a studbook that would set their Andalusians apart from Spanish Andalusians. The official name of this breed is Puro Sangue Lusitano (translated as pure-blooded Lusitano). The studbook was opened with Spanish and Portuguese horses.

Physical Description: The spirited Lusitano has a more convex profile and possesses a stronger and more muscular hindquarter than does the Andalusian/P.R.E. Due to the breed's traditional use in bullfighting, the Lusitano is athletic and energetic and tends to have an excellent canter. The use of these horses in bullfighting emphasized strong hindquarters, needed for the quick turns and bursts of speed. Those quarters help in the dressage world, creating a horse that is easily able to use the haunches and collect his gaits for upper-level work. Heights range from 15.2 to 16.2 hh. All solid colors exist, but the most common are gray or bay. Black, dun, chestnut, palomino, cremello, and perlino also appear.

Uses: In addition to bullfighting, Lusitanos are also used for traditional ranch work and are beginning to compete successfully in cutting and western pleasure as well as dressage.

Of Note: In Portugal, Lusitanos still rule the bullring, where their athleticism is literally a matter of life or death. In Portugal's equestrian bullfight, a mounted cavalier faces the bull in a small arena, with the goal of placing six darts in the bull's neck. The darts irritate the bull but will not kill it. Great dexterity on the part of rider and mount are crucial to avoid a goring. The nimble and hearty Lusitano, with the ability to move quickly around a charging bull, is a prized mount.

MARWARI

Alternative Name(s): Malani
Region of Origin: India

History: Spiritual reverence for the horse has long been a part of India's culture. Tombstones honoring fallen cavalry horses remain erected in the desert, and countless paintings and engravings celebrate the beauty of the horse. There are many festivals devoted to the horse, and welcome prayers are said to newborn foals. Among the celebrated horses is the Marwari, favored mount of Indian warriors called the Rathore Rajputs, who used the breed in their cavalry. Historically, the Marwari has been known by many names, but this has been the one most frequently used in recent times. This is largely because the bulk of the breeding has taken place in the former state of Marwar, in an arid region now called Rajasthan.

Physical Description: The Marwari comes in an array of colors. Gray, pinto, palomino, and bright sorrel are all commonly found, but never chestnut. Gray is considered the most holy color and is in the greatest demand; black is considered an evil color and thus undesirable. A popular but rare color is called nukra or cremello, which is a cream horse with blue eyes. The breed has uniquely shaped curved ears. The Marwari stands from 14.2 to 16 hh and is more full-bodied than the other desert breeds—slightly similar to the Spanish breeds. The Marwari's gracefully arched neck is often described as a scimitar. Like many desert breeds that have evolved the ability to glide over the sands, the Marwari moves with a floating, easy gait. Many of the horses are born with a pacing gait known as the apchal or revall.

Uses: The Marwari is used in rural India for farming and transportation. In the cities, the horses pull tourist carts and are ridden by police. The breed is also the centerpiece for festivals, particularly weddings.

Of Note: The most amazing feature of the Marwari is his curved ears. They often touch or cross in the middle giving an appearance of a spectacular head-dress. No one knows why the animals have this feature, but cave paintings show that horses with curved ears were in existence at least as far back as 2000 BC.

MINIATURE HORSE

Alternative Name(s): Mini
Region of Origin: Europe

History: The Miniature Horse is a truly man-made breed, one that traces its history back to the seventeenth century in Europe, when oddities and unusual animals were popular among the nobility. Less-refined Minis were employed as pit ponies working and living inside mines up into the twentieth century. Miniature Horses were imported to America in the 1930s to work in the coal mines.

Physical Description: The Mini's foundation breed is the Shetland Pony. Over the years, other breeds were included, such as the Hackney Pony, for refinement and movement, and the pinto for color. This selective breeding created the modern Miniature Horse. The Mini comes in various types. The horses are basically scaled-down versions of their full-size counterparts, such as Arabians, draft horses, Quarter Horses, and Paints.

The Miniature Horse is not a pony; he is proportioned just like a horse, lacking the short cannons, the

Marwari

BREED IN BRIEF

KISO: JAPAN

The Kiso is one of Japan's three horse breeds, the other two being the Hokkaido *(page 97)* and the Kagoshima *(page 105)*. The Kiso breed is most probably from Korean stock, horse stock that came to the country during the third century AD. The Kiso breed developed in the central part of Japan, in a mountainous region that is called Kiso-Sanmyaku, which is located near the Kiso River.

Although primarily employed as farm horses, Kiso horses have also been used in the Japanese military for centuries (including as mounts for the famous samurai warriors) and for transportation. The Kiso horses stand about 13 hh, and they can be found in all solid colors.

KLADRUBER: CZECH REPUBLIC

In the sixteenth and seventeenth centuries, in what is now the Czech Republic, the Kladruber breed was developed from Spanish horses as an elegant carriage horse suitable for royalty. Today, the Kladruber is still bred at the Kladrub national stud farm (formerly the imperial court stud).

This breed has an arched neck and powerful hindquarters. Very early on, Kladrubers came in many different colors of coat, including palomino and appaloosa, but because they were ultimately bred for royalty, they are now usually gray or sometimes black. Horses of this breed stand 16.2 to 17 hh. Kladrubers are used under saddle and in harness.

---

larger head, and other pony traits. The Mini stands about 34 inches at the withers. The horse can be any color and any type of conformation.

Uses: Miniature Horses are athletic, sturdy, and clever. They are excellent first-time horses for those who do not want to ride. They cannot be ridden, yet they can still introduce kids to the wonders and responsibilities of owning equines without the complications of keeping big horses.

Minis are usually shown in hand or in harness, in pairs or in single hitch. They can even be shown in jumping classes, where handlers will jog the horses over a set of fences.

Of Note: Minis make great companions for people who may not be able to handle full-size horses, such as senior citizens or disabled individuals. They have proven to be excellent horses in therapeutic programs. Miniature Horses can also be affordable alternatives for people who have always wanted to own horses. They need much less room to roam and much less feed to eat than larger horses.

Miniature Horse

MISSOURI FOX TROTTER

Alternative Name(s): Horse from the Ozark Mountains
Region of Origin: United States, Missouri

History: The Missouri Fox Trotter is a reflection of the breeders' desire to make work a pleasure. In the early 1800s, horse racing was a popular pastime with Americans, and pleasure horses often doubled as weekend racers. In the foothills of the Ozarks, this breed of horse was just in its infancy. After the Louisiana Purchase was made, hundreds of new residents flooded south to Missouri, with its beautiful rolling green hillsides, its rich forests, and its plentiful water.

The settlers used the native stock in the area to breed with the horses they brought with them. Their own horses—Arabians, Thoroughbreds, and eastern-bred Morgans—were combined in hope of creating the quintessential race/work horse. The resulting horses were willing workers, able to toil in the field and pull the family buggy during the week, as well as to act as weekend racers and sure-footed trail horses. They also had an interesting broken gait that made them easy to ride. Soon the settlers' horses were reflecting particular bloodlines that were in high demand, including the Brimmers, descended from Thoroughbred stock, and Jolly Roger and Old Skip, of Morgan/Thoroughbred origin. Later, with an infusion of American Saddlebred, Standardbred, and Tennessee Walking Horse blood, the stature of the Ozark horse grew, his conformation improved, and his distinctive gait refined itself into a fox-trot.

Physical Description: The Missouri Fox Trotter is a compact, medium-size horse, standing from 14 to 16 hh. The horse is close-coupled, with a short but well-proportioned body. His well-sprung ribs create a solid barrel shape. His sloped shoulder sits underneath slightly rounded withers, allowing the horse to

Missouri Fox Trotter

naturally execute the fox-trot gait. The powerfully built neck carries an honest head with a straight profile, small ears, and expressive eyes. With thick, full manes and tails, the breed is shown with a distinctive pair of braids woven right at the top of the bridlepath. The horse can come in a variety of coat colors, including chestnut, black, bay, gray, or pinto patterned. The fox-trot is a diagonal ground-covering gait in which the horse walks in the front and trots in the back. The horse also has an active, free-flowing walk and gentle canter to complement his trademark gait.

Uses: This is a popular mount for long-distance riding, ranch work, and pleasure riding.

Of Note: The Horse from the Ozark Mountains became the breed of choice for local doctors, for sheriffs, for postmen, and for other professionals to help them make their rounds comfortably and in style. Breeders looked to pass along that smooth-moving gait, certainly, but they also worked carefully to create a tractable, willing temperament in their pleasure horse, making him even more desirable as a leisure and workhorse.

BREED IN BRIEF

KONIK: POLAND

The Konik may be a cousin to the Huçul, but the Konik pony more closely resembles the Tarpan, the breed from which both of these pony breeds most likely descended. The dun coloring, dark dorsal stripe, and wither stripes of the Konik are excellent genetic markers of the relationship between the Konik and the Tarpan.

The Konik originated in Poland and is used in agriculture as a good light draft animal and as a riding horse for children. The Konik (which is a Polish term meaning "little horse") is a very sturdy and hardy pony. The Koniks stand from 12 to 13 hh. These horses are primarily bred at the state studs, though many small farmers also breed them.

Troika

The Russian troika is a sled harnessed with three horses abreast to resemble a flying bird. The two outer horses are check reined to the outside (one to the left and one to the right) to form the shape of the wings, while the middle horse forms the bird's head. The two outer horses canter and gallop, while the inside horse stays at a very fast trot.

MORAB

Alternative Name(s): None
Region of Origin: United States

History: The Morab, as the name suggests, is a combination of Morgan and Arabian blood. Gold-dust, a palomino stallion born in 1855 as the offspring of a Morgan stallion and an Arabian mare, turned out to be a magnificent show horse and quite fast, winning handily at the racetrack. He sealed his place in history, however, as quite a valuable stud. He, like his ancestor the Justin Morgan horse, was able to impart his characteristics on a new generation of colts and fillies, and soon many Americans sought out the horse that could produce quick, graceful carriage horses.

In the 1920s, the millionaire publishing mogul William Randolph Hearst began breeding horses for his expansive cattle operation in central California. He decided that a Morgan/Arabian cross would be ideal for his purposes—handy around cattle and attractive. Hearst even gave the breed the name Morab. Other ranchers were soon adding Morabs to their herds.

Later, the breed took a different turn. Martha Doyle Fuller, a breeder, decided the true calling for the horse was in the show arena. She began to selectively breed Morabs for type and flashiness. Her daughter, Ilene, took her work one step further and founded the first Morab breed registry in 1973.

Physical Description: From the Arabian side, the Morab gets an extra boost of endurance, stamina, and grace; from the Morgan side, the horse gets strength, power, and agility. The Morab is well muscled, yet refined. The horses may have a straight or slightly concave profile to their smallish heads. Morabs have wide-set, intelligent eyes; small, alert ears; and beautifully curved necks. The breed stands from 14.1 to 15.2 hh and is found in all solid colors and with or without white markings.

Uses: Today, the Morab is a treasured show horse, endurance horse, and dressage mount.

Of Note: The Morab is not just a cross between two different horses. Farms that mate Morab to Morab in second and third generation breedings know that theirs is an actual type, so much so that they can easily predict what type of Morab will result from what breeding. This goes a long way to dispel some of the criticism that the Morab is not a true breed.

Morab

THE COSSACKS

The Cossacks (from the Turkic word *kazak*, or "free man"), came from many regions, including the hinterlands of the Black and Caspian Seas and the Dnieper and Don regions. Famed for their fighting prowess as mounted warriors, they depended greatly on their horses and took care in their breeding *(see Don, page 82)*. In this undated photo, a group of traditionally dressed Cossasks, rifles slung over their backs, look ready for battle.

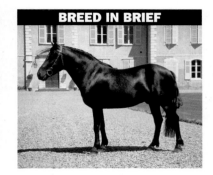

LANDAIS: FRANCE

The Landais, one of the three ponies that are native to France, developed in the Landes region in the southwest. Like most native ponies, this breed developed from ancient stock. Arabian blood was added for refinement in the eighth and early twentieth centuries; Welsh Section B blood was added after World War II.

As with many European and Asian breeds, the number of Landais fell drastically during WWII. Today, the Landais is bred at the national stud at Pau, and its numbers have increased. These ponies are used under harness and are prized as children's mounts. The breed stands 11.3 to 13.1 hh and can be found in the colors of bay, brown, chestnut, and black.

MORGAN

Alternative Name(s): None
Region of Origin: United States

History: The Morgan is considered to be the first-documented native breed in America. The breed can be traced back to one stocky little stallion who had the strength of a draft horse and the speed of a Thoroughbred. A logging horse named Figure was born in West Springfield, Massachusetts, in 1789, with unknown parentage. While some historians name True Briton, an English Thoroughbred, as the colt's sire, others point to Canadian Horse, Welsh Cob,

Morgan

Friesian, or even Norfolk Trotter parentage within this genetic mix.

When Figure was a yearling, he was given to a schoolteacher named Justin Morgan, who took the horse to Vermont. Deciding to put Figure to the most use, Morgan loaned him out to neighboring farms, where he worked as a plow and logging horse. Figure was also a racehorse. People soon discovered that the little stallion could outrun the fastest Thoroughbreds in match races. Like his townspeople, Morgan was always game to pit his little horse (about 14 hh) against other horses in competition—in speed races, log-pulling contests, and even trotting races on country roads. Despite his small stature, Figure continually beat the massive draft horses in pulling heavy loads—so much so that villagers made the pulling contests even tougher by sitting astride the large load, but Figure still easily hauled the weight.

As Figure's legend grew in Vermont, farmers began asking for the little stallion's stud services. Local mares of various breeding, including the English Thoroughbred and Norfolk Trotter, made up a good portion of his roster. The stallion put his stamp on future generations, creating his own likeness with amazing success. People began referring to Figure as "Morgan's horse," and eventually the name shortened and stuck.

Physical Description: The Morgan is a compact horse, with an expressive face; large, intelligent eyes; and a straight or slightly dished profile. The slightly crested neck blends into well-defined withers. The Morgan is known as being a sound horse with flat bone and good substance to his limbs, yet he is refined and almost delicate in appearance. The breed is powerful, with a short back, sturdy legs, and well-muscled haunches. Morgans stand from 14.1 to 15.2 hh. They have developed into two basic types: the old-style Morgan, stout and powerful, and the modern Morgan, elegant and refined. Both types are known for having

excellent stamina and vitality, as well as a good nature. The Morgan is usually bay, black, or dark chestnut, but palomino and buckskin can occur.

Uses: The Morgan excels in a number of equestrian disciplines including dressage, English, driving, and western.

Of Note: In 1945, Marguerite Henry published a fictionalized version of Figure's story, titled *Justin Morgan Had a Horse*. Walt Disney released a movie version in 1972.

MUSTANG

Alternative Name(s): None
Region of Origin: United States

History: Horses roamed America 10,000 years ago but vanished from the landscape until the Spanish conquistadors arrived in the sixteenth century with their horses of Barb decent. Many Indian tribes "liberated" horses and brought them further into North

Mustang

BREED IN BRIEF

LATVIAN: LATVIA

The Latvian is descended from the Latvian Forest Horse, which is native to the country of Latvia. There are three types of Latvians: the sturdy and strong draft type; the lighter-bodied harness type; and the Latvian Riding Horse, which was created by crossing the heavier types with warmbloods.

The Latvian Riding Horse is an all-round sporting horse that is highly prized in the English activities of dressage and show jumping. Latvian horses stand 14 to 15 hh, and they can be found most frequently in bay, black, and brown and sometimes in chestnut. They are powerful horses with good endurance.

LOKAI: UZBEKISTAN

The Lokai originated in the mountains of Tajikistan in the sixteenth century, bred by the Uzbek Lokai tribe. This horse is a mixture of Central Asian breeds, such as the Iomud and the Karabair. Later the blood of the Arabian was introduced.

The breed possesses enormous stamina and spirit. The Lokai has a straight profile and a well-muscled body set on slender legs and has been used as a riding horse. Lokais stand 15 hh and can be found in bay, chestnut, and gray coloring. The coat of the Lokai sometimes has a metallic sheen because of the the the horse's Central Asian breeding, which includes the blood of the shimmering Akhal-Teke.

America. As America evolved, horses from Europe were imported, and offspring accompanied the settlers moving west. Feral horse bands formed from escaped or abandoned horses. These feral horses came to be known as Mustangs, a derivative of the Spanish word *mestena*, which means "wild."

In the early 1900s, cattle ranching operations vied with Mustangs for grazing space on public lands. The feral horses were liabilities, nuisances for ranchers who leased public lands from the government. And so began the era of slaughtering Mustangs. Hundreds of thousands of horses were captured and shot. At the beginning of the twentieth century, more than 2 million feral horses roamed the West, by 1926 the number declined to half that. In 1971, the government passed the Wild Free-Roaming Horse and Burro Act, protecting Mustangs from slaughter. Congress established Herd Management Areas, and the Bureau of Land Management (BLM) began conducting gathers and offering excess animals for adoption.

Physical Description: There are no overall characteristics of the Mustang, as many different breeds of horses have contributed to the development of feral horses in various areas. Draft horses were popular in certain areas among settlers, and hot-blooded horses were more popular in others. So some Mustangs are large and full-bodied, while others are smaller and daintier in appearance. The abundance of or lack of forage also helps determine the horses' size. The Mustang ranges from 13 to 16 hh and comes in all colors, including black, bay, dun, palomino, gray, and spotted.

Uses: This is a superb western trail horse, also used for western and some English sports, such as dressage.

Of Note: In 1976, the BLM introduced its formal Adopt-A-Horse program, which allowed the public to buy a real Mustang, fresh off the range, for a small fee.

NAMIB

Alternative Name(s): None
Region of Origin: Africa, Namibia

History: Nobody knows exactly how these horses came to Africa's Namib Desert. Originally, there were no horses in southern Africa. They only started to

Namib

appear in the region in the seventeenth century, imported by Europeans.

The Namibs reside in Naukluft Park, the area around Garub, just 14 miles west of Aus. In the last twenty years, however, the herds suffered tremendously when drought struck the area, and many of the horses starved to death. In 1999 numbers were down to 89. Equine and humanitarian organizations brought in food and water to the horses, ensuring that some would survive. The population of the Namib horses is currently about 150.

Physical Description: Small and hardy, most Namibs stand about 14 to 14.2 hh, although a few can be found reaching 15 hands. While their conformation varies, they generally have large heads with convex profiles and small muzzles, large ears, wide-set eyes, and short necks. They have narrow chests, prominent withers, and average-length legs. Most of the horses are chestnut colored, although various bays and browns also exist.

Uses: At the current number, the horses will be kept in their feral state. However, if the number of horses reaches 200 (the number the Namibia Nature Foundation has set for overstocking), excess horses, carefully chosen by age, sex, and group, will be rounded up and relocated. Visitors to the desert can watch the horses from a hide near the drinking hole.

Of Note: Namib feral horses are unique because of their isolation, mysterious origins, unique gene pool, and ability to survive in a climate that would kill most ordinary horses. Their ability to survive by changing their behavior and the way they conserve and expend energy is a marvelous demonstration of the enduring adaptability of the horse.

NATIONAL SHOW HORSE
Alternative Name(s): NSH
Region of Origin: United States

History: The National Show Horse breed began as a result of saddle seat riders wanting to be competitive in half-Arabian show classes. When Gene LaCroix, an Arabian breeder, began to truly take note of the winners of these classes, he determined that the Saddlebred-Arabian cross came out on top fairly consistently. The blend of the two breeds became very popular in the half-Arabian classes, and even on the AHSA show circuit.

Around 1980, breeders chose two horse breeds with extraordinary show qualities, the Arabian and the American Saddlebred, and joined them to create a unique show horse. The Arabian contributed his class, beauty, stamina, and expression; and the Saddlebred offered brilliance, animation, flash, and grace.

Judges responded enthusiastically to the size, gaits, athleticism, and brilliance of the National Show Horse. In the its short existence, the breed has gained a very steady following. Today, more than 15,000 horses are registered with the breed association.

Physical Description: While the horse started out as a crossbred, after multiple generations of breeding for type, he now has distinctive characteristics all his own. The NSH is a picture of grace and nobility. He stands 15 to 16 hh and has an attractive head and arched neck, set high on deep, sloping shoulders. The breed is seen in all colors, including palomino and pinto.

Uses: The National Show Horse moves with a lively activity and high action, making the breed a perfect mount for English pleasure pursuits such as saddle seat, English pleasure, and equitation. These

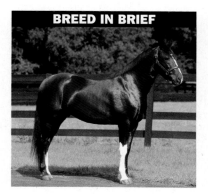
BREED IN BRIEF

MANGALARGA MARCHADOR: BRAZIL

In 1740, Portugal's Joa Francisco developed the Mangalarga Marchador in Brazil when he crossed an Altèr-Real stallion named Sublime with Spanish Jennets, Criollos, and Andalusians. The resultant breed has a smooth gait called the marcha. Depending upon how the horse moves his legs during the gait, the movement is called either the Marcha batida (in which the legs move diagonally) or the Marcha picada (in which the legs move laterally). The Mangalarga is believed to be the closest relative to the extinct Spanish Jennet. Mangalargas possess the elegance and strength typical of the Spanish breeds; they are prized endurance and ranch horses. These horses stand 14 to 16 hh and are found in chestnut, bay, gray, and roan coloring.

MAREMMANA: ITALY

The Maremmana is named for the Tuscan province of Maremma, Italy. This horse was used primarily to herd cattle, although the breed was also used as a draft horse. The Maremmana most probably began with Spanish, Arabian, and Barb blood.

Versatile and agile, Maremmanas are still ridden by Italian cowboys, who are known as butteri. The Maremmana also does well in jumping competitions. This hardy horse has good stamina and good endurance.

The Maremmana has a plain but noble head set on a short neck; the withers are high and well muscled and the legs are solid. These horses stand 15 to 15.3 hh, and they can be found in all solid colors.

National Show Horse

classes demonstrate the horse moving in all gaits, including the slow gait and the rack, which are four-beat gaits with extreme action.

Of Note: The slow gait and rack of the NSH are so smooth that the rider is able to sit in the saddle rather than posting.

NEW FOREST PONY

Alternative Name(s): New Forester, Forester
Region of Origin: United Kingdom, England

History: In England, near the coast in southwest Hampshire, lies the beautiful New Forest. The New Forest Pony, one of the native breeds of the United Kingdom, existed here as early as AD 1016, which means the pony was recorded roaming the area six decades before King William I (William the Conqueror, AD 1066–1087) founded the forest.

Native ponies mixed with transient stock (Welsh ponies, Arabians, and Thoroughbreds); because of this influx, the Forester's genes are more diverse than those of any other British breeds. The distinctive type of the New Forest Pony was shaped by his environment as well. The ponies became hardier because they survived on what little nutrition could be found in the native plants. Yet in a true sense, the ponies have shaped their home as much as it has shaped them. Living to graze freely in the forest, the horses (along with the cattle and swine) will eat what is palatable to them and leave what they can't eat. As a result, the forest vegetation is defined by what the animals leave behind.

Physical Description: The New Forest Pony can range from 11 to 14.2 hh. Foresters come in all colors except palomino and pinto. The New Forest Pony is more horselike than the other native breeds. Viewed from the front, the shoulders slope in and form withers,

New Forest Pony

which is more characteristic of a horse than a pony. The Forester must lift his shoulders and his hocks as he moves; this is functional because a low mover would have trouble in a land filled with heather and gorse.

Uses: Today, New Forest ponies are popular throughout the world as leisure and driving horses, and they excel in dressage and jumping.

Of Note: While the New Forest Pony lives from day to day, roaming free throughout a portion of southern England, the breed is not exactly feral. Still, the horses are untamed and allowed to roam loose. The ponies of the forest share their existence with the local villagers.

NEZ PERCE HORSE

Alternative Name(s): None
Region of Origin: United States

History: When the Lewis and Clark Expedition of the American West crossed the Bitterroot Mountains into eastern Idaho in 1805, Meriwether Lewis noted in his journal that the Nez Perce tribes had very grand horses. "Their horses appear to be of an excellent race: they are lofty, elegantly formed, active and durable: in short many of them look like fine English horses and would make a figure in any country."

The Nez Perce tribe is attempting to resurrect these horses today by crossing the Akhal-Teke with the Appaloosa. The tribe didn't feel the modern stock type Appaloosa was a true representation of the breed

Nez Perce Horse

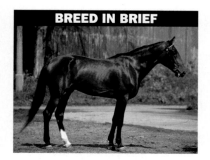

METIS TROTTER: RUSSIA

The Metis Trotter (also called the Russian Trotter) is a harness racehorse that was created before World War I. The breed is a cross between imported American Standardbreds and Orlov Trotters. Although the Metis Trotter is faster than the Orlov, American and European trotters are faster still.

The characteristics of the Metis Trotter are very similar to those of the American Standardbred, with a straight profile, a long and muscular neck, and long legs. The Metis Trotters stand 15.3 to 16 hh. Although they can be found in the colors of gray, black, and chestnut, bay is the color most frequently seen.

---

they had in the 1800s, which was longer and leaner. They felt the desert Akhal-Teke had those qualities and would mix well with the Appaloosa.

Physical Description: The Nez Perce Horse, which has a long neck, is more slender and has thinner withers than the modern Appaloosa. The Nez Perce is very strong and sure-footed and is a quality endurance horse. Both solid and Appaloosa coloring are found along with the burnished coat, a characteristic of the Akhal-Teke. Heights range from 14.2 to 15 hh.

Uses: This horse's uses include endurance riding, competitive trail riding, dressage, jumping, western pleasure, cutting, reining and even driving

Of Note: The breeding program began in 1994 with four donated Akhal-Teke stallions and Appaloosa mares that the tribe already possessed. The tribe now has seventy horses. They sell some to keep numbers manageable, but the rest of them are used in the tribe's Young Horsemen Project, a program to help reintroduce horses to tribal youths, ages fourteen to twenty-one.

NOKOTA

Alternative Name(s): None
Region of Origin: United States, North Dakota

History: The Nokota horse is descended from generations of Mustangs that once lived in the rugged Little Missouri badlands of southwestern North Dakota. When the Lakota tribe's medicine man and leader Sitting Bull was forced to surrender to the US Army in 1881 at Fort Buford, North Dakota, the Lakotas' horses were confiscated. The tribe's horses were sold to three of the fort's traders. The Marquis de Mores, who was a flamboyant French aristocrat and pioneer rancher of western North Dakota, admired the stamina of the Lakota tribe's horses and purchased 250 of them from the Fort Buford traders.

The marquis founded the town of Medora, the gateway to what is now the Theodore Roosevelt National Park. He allowed his newly acquired horses to run free on his land, intending to start a large breeding operation with them. However, when the marquis died in 1896, his ranch foreman rounded up as many horses as he could and sold them off. Not every horse left the area, though, and many wandered into the badlands.

During the 1980s, horsemen Leo and Frank Kuntz of Linton, North Dakota, began buying the horses to save them from slaughter or crossbreeding. They recog-

Nokota

nized that the park horses looked different from modern breeds and seemed to form a common physical type. By 1990, the men had begun calling the horses Nokotas, signaling their North Dakota origins. Support for Nokotas has built slowly, through the efforts of devoted enthusiasts. Unlike many feral horses, Nokotas are curious, rather than wary.

Physical Description: The Nokotas have large eyes; broad foreheads; and thick, long manes and tails. They have straight or slightly concave profiles on medium-size heads, with large eyes and hooked ears. They are large-boned with a good angle to the shoulder, well-defined withers, a sloping croup, strong legs, and tough hooves. The Nokota stands 14.2 to 15 hh. The breed's signature coat is often roan (blue or strawberry-red, but some black roans are also seen), as well as bay or overo pinto.

Uses: The Nokota is used in a range of activities, including English riding and trail riding. Nokotas are talented western horses, often used on working ranches.

Of Note: A research report submitted in 1989 concluded that Nokota horses are descended from early twentieth-century ranch and Indian stock, a type of horse that had been considered obsolete since the 1950s.

NORWEGIAN FJORD

Alternative Name(s): Fjord
Region of Origin: Norway

History: Norway is a country with high mountains. Fjords are narrow inlets of sea surrounded by steep hills and mountains, and the country's fjords come very far inland. During the Viking conquests, there were no roads and the only transportation was by boat, so the Vikings carried their horses in boats as a matter of necessity. When the boats had to be drawn over land, the horses did the work. Norwegian hill farmers used the horses as little draft animals. There wasn't much forage on these hills, and, over thousands of years, this created an animal that could be described as an easy keeper.

Named for the region's landscape, the Norwegian Fjord is one of the world's oldest horse breeds. He is thought to have existed in western Norway for more than 4,000 years and to have been domesticated as early as 2000 BC. Evidence shows the Fjord was developed as a breed by the Vikings some 2,000 years ago.

Physical Description: A few horse breeds have features that make them stand out. There is one breed in particular whose silhouette rendered in a folk art picture or carving

is enough to make it identifiable. With tiny pricked ears, arched neck, crescent-shaped two-toned mane and dun coloring, the Norwegian Fjord is unmistakable.

The Fjord's well-muscled hindquarters are a product of the breed's mountain upbringing. Sure-footed and strong, the horse can carry can carry a 200-pound adult with no problem or pull a cart up a winding, steep path. The Fjord has smooth gaits with high knee action. The breed's large eyes, short ears, and slightly dished face create a very pretty head. But the Fjord's most striking feature is his mane—black in the center, white on the outside. Breed enthusiasts clip the mane in a crescent shape, so that it stands up. If left uncut, the Fjord's coarse mane would flop over on both sides of the neck. To accentuate the black center, the white hairs are trimmed half an inch lower. Fjords appear in all dun colors; however, yellow dun is very rare. Buckskin, is most plentiful. Gray dun, or grulla, is also common. Many horses have primitive zebra striping on their legs and withers; most have a dorsal stripe. The breed stands 13.2 to 15 hh.

Uses: One of the biggest misconceptions about Fjords is that they are ponies and therefore children's mounts. The versatile Fjord is top-notch under harness and excels in English pursuits and on the trail.

Of Note: There are many fjords along the west coast of Norway, where the horses have lived for thousands of years. There are straight drop-offs from the roads into the

Norwegian Fjord

Oldenburg

fjords—thousands of feet down and scary. The Norwegian Fjord is a very sure-footed and quiet family horse because his native habitat created those characteristics.

OLDENBURG

Alternative Name(s): Oldenburger
Region of Origin: Germany

History: For the past forty years, European warmblood breeders have been in the vanguard of selective equine breeding. With the decline of horse-drawn transportation and agriculture, these breeders have risen to the challenge of producing top-quality saddle and sport horses for riders around the world. And one of the best known of all the warmbloods is the Oldenburg.

The Oldenburg horse differs slightly from his warmblood counterparts in that all the stallions and mares used in the breeding program are privately owned. For other warmblood associations, such as the Hanoverian and Westphalian, state studs own or control a great deal of the breeding stock. Thanks to the Oldenburg's historic development and the generosity of one man (the breed's namesake), a state stud has never been necessary.

The Oldenburg horse became popular in the seventeenth century through the endeavors of Count Johann XVI von Oldenburg (1540–1603) and Count Anton Günther von Oldenburg (1583–1667). Count von Oldenburg needed good cavalry horses. He started up small

HARNESS RACING
This May 1, 1908, photograph shows prize-winning trotter Strange Leaf with his driver Louis Frank. The caption for the image reads "Speedway Parade." This was probably the Harlem River Speedway, a three-mile dirt track built in 1898 along the bank of the river to accommodate New York City's buggy races. Harness racing of one kind or another has been popular for millennia, as ancient images of Roman chariot racing clearly illustrate.

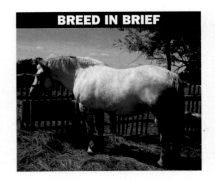

BREED IN BRIEF

MURAKÖZI: HUNGARY

The hardworking Hungarian Muraközi was developed in Muraköz in southern Hungary in the nineteenth century from the Noriker, the Ardennais, the Percheron, and the Arabian. Muraközis were so popular for farming that they became the most plentiful breed in Hungary. After World War II, however, the breeding stock was severely depleted. Hungary imported more Ardennais horses from France and Belgium, and the Muraközi rebounded. The Muraközi is a strong-bodied, clean-legged draft horse with no feathers. Because its Arabian blood, the breed features an alert expression, small ears, and large nostrils. Muraközis stand 16 hh and are found in bay, black, gray, brown, and chestnut coloring, with flaxen manes and tails.

---

breeding farms throughout the provinces of Oldenburg and East Friesland. These original Oldenburgs were based on native Friesian stock, with Turkish, Neapolitan (Italian), Danish, and Andalusian influences.

Count Gunther wanted a horse to fill his penchant for riding classical dressage and to promote the breed as top-carriage horses. He brought back horses from Naples, Spain, Barbary (which now encompassed the African states of Morocco, Algeria, Tripolitania, and Tunis), and England. The count freely lent these stallions to local farmers. In return, the farmers took great pride in helping further the quality of the Oldenburg breed. In 1950, the objectives of agriculture, cavalry, and carriage gave way to those of sport, and more Thoroughbred blood was mixed in.

Physical Description: The Oldenburg is a compact horse, quite refined and elegant, with a high rounded movement. His long front legs and powerful hind end help create the essential balance for dressage and jumping. The Oldenburg's neck is set high out of his shoulders, with a good length, and finishes into a slender throatlatch. Oldenburgs tend to have lovely heads. Heights range from 16.1 to 16.2 hh. Colors are bay, black, brown, gray and chestnut.

Uses: The Oldenburg is used for eventing, jumping, dressage, and competitive carriage driving.

Of Note: In June 1820, a studbook opened and a law passed stating that Oldenburg stallions had to be government tested and approved. Horses that passed were identified by a symbol—a large *O* topped with a crown, branded on to their left hips. In 1897, two Oldenburg breeding associations began in Germany. In 1923, the two societies joined to form the Verband der Züchter des Oldenburger Pferdes (Oldenburg Horse Breeders' Society), which today is based at the Oldenburg Horse Center in Vechta, Germany.

ORLOV TROTTER

Alternative Name(s): None
Region of Origin: Russia

History: The Russian Orlov Trotter was developed in the late eighteenth century by professional breeder Count Aleksey Orlov, a favorite of Russian empress Catherine the Great. The count wanted to create a fast horse for the Russian sled pulled by three horses, known as a troika (*see page 116*). The count crossed a gray Arabian stallion bred in Turkey (named Smetanka) to a Danish mare (named Isabelline). They produced Polkan, who was bred to a Dutch mare, which produced Bars I, who was the first Orlov Trotter and the breed's foundation stallion.

For a time, Orlovs were widely considered to be the best harness-racing horses in Europe. However, American Standardbreds were faster, and the Russians answered by crossing Standardbreds with Orlovs; the resulting horses were named Russian Trotters.

Orlov Trotter

Although Russian Trotters proved to be faster than Orlovs, they had less beauty and stamina.

In 1959, Soviet leader Nikita Khrushchev gave several horses to the American secretary of agriculture, Ezra Taft Benson, after grain from America was sent to the Soviet Union for famine relief. In 1997, the International Committee for the Protection of the Orlov Trotter was created. There are currently only a handful of Orlovs in the United States.

Physical Description: Despite their massive appearance, Orlov Trotters are fast and very forward-moving horses. They are muscular and strong and able to work for long periods of time. The horses have large yet beautiful heads, with expressive eyes. Their arched, high-set necks give Orlovs a noble presence. They stand 15.2 to 17 hh and are usually gray; foals are born dark, but as they mature, they lighten until they are nearly white. Black, bay, and chestnut varieties are also found.

Uses: A popular driving and riding horse in Europe, the Orlov also excels in dressage.

Of Note: Balagur, an Orlov ridden by Alexandra Korelova, is a top-placing Grand Prix dressage horse that competed in the 2008 Olympics at age eighteen.

PASO FINO

Alternative Name(s): Paso

Region of Origin: South America and the Caribbean, Colombia and Puerto Rico

History: The breed's earliest ancestry includes Barbs, Andalusians, and gaited Spanish Jennets, which came to Santo Domingo (Dominican Republic) with Christopher Columbus to be used as conquistadors' mounts throughout the 1500s. The blood of the Spanish Jennet (now extinct) dominated the future of the Paso Fino, and the Jennet's unusual gaits were passed down to become the Paso Fino's hallmark. *Paso fino* means "fine step" in Spanish.

Paso Fino

BREED IN BRIEF

MURGESE: ITALY

The modern Murgese, which is also known as the Murgesi, was developed in the 1920s from horses that had lived in the Murge district of the Puglia region of Italy since the fifteenth century. During the fifteenth and sixteenth centuries, the horses in the region most likely served as cavalry mounts. Today, the Murgese—a strong, heavily muscled horse with short movement—is used primarily as a light draft horse that can be ridden.

Murgeses stand 15 to 16 hh. They are most frequently chestnut in color, although they can be found in all solid colors. Because there are no strict breed regulations, the Murgeses are not as uniform in appearance as are other breeds.

NONIUS: HUNGARY

The Nonius breeds takes its name from its foundation sire, a French stallion known as Nonius Senior. Born in Normandy in 1810, Nonius Senior was was a mixed breed, with Norfolk Roadster and Norman blood. Nonius Senior ended up in Hungary in 1813, taken from France by Hungarian cavalrymen after Napolean's defeat in Leipzig. Over the following twenty years, Nonius Senior was bred to mares of several different breeds, and his descendents were interbred with Spanish Neapolitan horses and later, for further refinement, with English Thoroughbreds.

Today, the steady and stoutly built Nonius is used under saddle and in harness. These horses stand 15 to 15.2 hh and are usually bay or brown.

The horse developed further in Puerto Rico and Colombia. It came to Puerto Rico in 1509, when the island's first governor, Juan Ponce de León, brought the horses in from nearby La Española (now Hispaniola). The Paso Fino was highly prized as a plantation horse and racehorse. Two of the most famous Paso sires are Manchado and Dulce Sueño.

Physical Description: The Paso Fino is a slight but elegant and highly refined animal, standing from 13.2 to 15.2 hh. All colors and markings are found, including pinto and palomino.

Uses: Today, the Paso Fino is shown in the breed's traditional tack and is renowned as a competitive trail horse, possessing both speed and stamina.

Of Note: Although the Paso Fino walks and canters, he does not trot. His natural yet highly stylized gaits include the *paso fino* (the slowest), the *paso corto* (the preferred gait, which is as fast as the trot), and the *paso largo* (the fastest). Each foot in the comfortable gait strikes the ground independently and in an even rhythm.

PERCHERON

Alternative Name(s): None
Region of Origin: France

History: The Percheron's development began in the Perche region in Normandy in AD 732, when marauding Moors left behind Barb horses after their defeat in the Battle of Tours. Massive Flemish horses were crossed with the Barbs to give the Percherons their substance. In the 1800s, the French government began breeding Percherons for cavalry purposes at the famed Le Pin National Stud in Le Haras du Pins in Normandy, which still exists.

Percherons are now being crossed with other breeds (the most popular choices are Thoroughbreds, various warmbloods, and Spanish breeds) to create dressage horses, eventers, hunters, and jumpers. People want a larger riding horse, and by crossing these traditional riding breeds with the Percheron, they can get a bigger, heavier horse. The Percheron is an attractive horse that has some of the best of both aspects of the light and heavy breeds and is sought after by riders.

Physical Description: French Percherons are born black; they turn gray by the age of three. British and American Percherons are gray or black. All have legs without feathers. The horses stand 16.2 to 17.3 hh.

Percheron

Peruvian Horse

Uses: The Percheron breed is very popular in the discipline of dressage. The horses are also used for riding and driving activities.

Of Note: Because of their influx of Barb blood, Percheron horses are more energetic than other draft breeds usually are. The French prized Percherons as coach horses, calling them Diligence Horses, as *diligence* was the French word for "stagecoach." These horses had to pull a load quickly and with elegance, which is why Percherons also have the reputation for being lovely movers.

PERUVIAN HORSE

Alternative Name(s): Peruvian Paso, Peruvian Stepping Horse

Region of Origin: Peru

History: Conquistadors brought the Peruvian horse's ancestors from the Iberian Peninsula to Peru, where they developed further without the influence of foreign breeds. Although the Peruvian Horse (also known as the Peruvian Paso) and the Paso Fino share previous parentage (the Andalusian, Barb, and gaited Spanish Jennet) and both breeds are gaited,

BREED IN BRIEF

NORIKER: AUSTRIA

The stately and beautiful Noriker is the most popular and plentiful horse in Austria. The breed derived from horses brought to the area by the Romans in the last century BC; in fact, the name stems from the Roman vassal province of Noricum, which is now Austria. In the 1500s, monks oversaw the breeding. Later on, the Noriker was crossbred with Spanish horses. The Noriker is a popular draft horse; but, despite his large head and broad conformation, this horse is also used for riding. Norikers stand 16 to 17 hh and are found in black, brown, and chestnut varieties. The breed's most common color is a liver chestnut, with a flaxen mane and tail. These horses appear in several coat patterns, including brindle, dappled, and spotted.

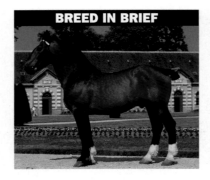
NORMAN COB: FRANCE

The Norman Cob was originally created by the Romans, who crossed bidets (small horses native to Brittany and Normandy) with large-bodied pack mares that they brought to Normandy. Later breeders crossed the Norman Cob with the Arabian and the Barb, as well as with the English Thoroughbred and the Norfolk Roadster.

Today, the Norman Cob is bred at the French state studs in Le Pin and St. Lô in Normandy, yet there is no studbook presently kept for this breed. Although heavy bodied, the Norman Cob has a lively movement and a dignified presence. Norman Cobs stand 15.3 to 16.3 hh, and they can be found in the colors of chestnut and bay.

---

they are not the same breed. The Peruvian was selectively bred for four centuries. Breeders chose the smoothest, the strongest, and the most beautiful horses to carry on the ancient genes. By selecting for gait, conformation, and character, breeders created a horse that people valued for transportation. Plantation owners also rode the Peruvian horses over their sugar and cotton fields.

In the 1900s, breeding declined due to the development of roads. Many breeders gave their horses away to farmers. Political unrest made matters worse, and few Peruvians were being bred in their homeland of Peru. However, in the past few decades, the breed has had a resurgence, and now the Peruvian is the National Horse of Peru.

Physical Description: The Peruvian is a medium-size breed, more muscular and bolder in appearance than is the Paso Fino. The Peruvian has a highly arched neck and a head with a straight profile. His mane and tail are thick and luxurious. The Peruvian stands from 14.1 to 15.2 hh. These horses can be found in all solid colors, including buckskin, roan, and palomino. The Peruvian is shown in traditional tack: the montura de cajón, or box saddle, with accessories (cinch, stirrups, crupper, and breeching) and braided head gear.

Uses: The horses are prized pleasure-riding mounts and are also found in traditional ranch work.

Of Note: This breed has the same type of four-beat gait as the Paso Fino, however, the slower gait is called the *paso llano* and the faster, more lateral gait is the *sobreandando*. The Peruvian also has a unique way of moving called the *termino*, in which the horse lifts his knee up high and swings the hoof outward. The Peruvian is said to possess a natural brilliance and energy, which enthusiasts call "brio."

PONY OF THE AMERICAS

Alternative Name(s): POA
Region of Origin: United States

History: Considered the ultimate family pony, the Pony of the Americas has captured the hearts of American children since his inception in the early 1950s. The POA's forefather, the accidental offspring of a Shetland stallion and an Arabian/Appaloosa mare, had a colorful spotted coat and refined conformation. When the foal was born, he bore wild Appaloosa markings—complete with what looked like a dark handprint on his hip. A local lawyer paid the farmer for both the mare and the foal and took them to his own farm, naming the pony Black Hand for his unusual markings. Other breeders thought he would be an ideal kids' horse, able to compete in a variety of disciplines and be a solid equine citizen. A group of breeders finally decided on calling the breed Pony of the Americas, named for the first pony type from the United States. They established Black Hand as the foundation stallion.

Physical Description: The POA is proportioned more like a horse than like a pony, having an ample clean leg under a robust body. The breed's coat patterns are as unique as two fingerprints, and often a pony's coat will change with age. One of the most common patterns is a blanket, in which the pony has white over his croup and hips that is sometimes dotted by spots. Others are leopard patterned, speckled with spots over the horse's entire body. Roan-type Appaloosa markings also occur. The Pony of the Americas stands 11.2 to 14 hands at the withers. The POA's head is small and dished, reminiscent of that of his Arabian forebears.

Uses: The POA is a top children's mount in both English and western pursuits.

Of Note: From the beginning, an essential characteristic that every POA must possess is coloring.

Pony of the Americas

While solid Appaloosas compete in the Appaloosa Horse Club's breed shows, POA enthusiasts staunchly require that their horses bear spots to show in breed shows. A pint-sized pony with flashy markings is every young horse-lover's dream. The pony's breed registry allows for several different breed crossings, including Appaloosas, Welsh, Connemara and Shetland Ponies, American Quarter Horses, Thoroughbreds, and Morgans. Breeders have done well in keeping the gene pool diverse, while retaining those lovely spotted patterns. Solid-colored horses and ponies of unknown breeding can be registered as breeding stock, as long as they pass an evaluation and are determined to have no pinto coloring and not be of gaited stock.

POTTOK PONY

Alternative Name(s): Basque
Region of Origin: France and Spain

History: Bridging two countries, the small horse known as the Pottok has been around for thousands of years. This ancient breed of pony is found in the Basque region of southwest France and Spain, and some people even believe the Pottok's descendents can be found across the Atlantic Ocean in America.

BREED IN BRIEF

NORTH SWEDISH HORSE: SWEDEN

The North Swedish Horse stems from ancient Forest Horse ancestors that roamed throughout Scandinavia. Crossbred greatly with Norway's Døle Gudbrandsdal, the two are now closely related to each other. The Swedish Horse's breeding was fairly indiscriminate until the nineteenth century, when standards were set down. North Swedish Horses are popular harness horses for farming and forestry work. This breed is a solidly built draft type, with feathering on the legs. The horses stand 15 to 15.3 hh and are found in all solid colors.

PADANG: INDONESIA

The Padang pony, one of the eight ponies of Indonesia, was developed by the Dutch, during the seventeenth century, at Padang Mengabes in Sumatra. The breed was developed primarily from the Batak by crossing that breed with Arabians.

Although small in stature, the Padang is a strong pony and a nimble one. These horses are used under saddle as well as in harness. Padangs stand 12.2 hh, and the ponies can be found in all colors.

The other seven ponies from Indonesia are the Bali, the Batak (*page 67*), the Deli, the Java (*page 102*), the Sandalwood (*page 141*), the Sumba/Sumbawa (*page 144*), and the Timor.

Pottok (pronounced pot-tee-ok) comes from the Basque word *pottoka*, meaning "little horse." The Basque countryside straddles the western Pyrenees Mountains that define the border between France and Spain, all the way down to the coast of the Bay of Biscay. Many historians date the breed's origin to prehistoric times, as paintings found on the cave walls throughout the area depict similar-looking horses. The Pottok is believed to have descended from the Magdalenian horses (14000–7000 BC), possibly having been a relative of the extinct Tarpan. Over the centuries, the horse has adapted to his mountain home. Although small in stature, these nimble and hardy creatures thrived in the rugged countryside: The altitude gave them excellent endurance, and their feet and legs became extremely tough from traversing the difficult terrain.

Throughout the seventeenth century, some Pottok Ponies were used for smuggling trade between

Pottok Pony

France and Spain; their sure-footedness and dark coats, which camouflaged them at night, made them the ideal smugglers' horses. Later, Pottoks hauled wagons in coal mines in Italy and northern France in the nineteenth and twentieth centuries.

Most Pottoks roamed freely in the Pyrenees Mountains until recent history, but crossbreeding and the loss of habitat have greatly reduced their numbers during the twentieth century—with less than a couple hundred purebred mares in existence by the late 1900s. In 1970, when an official studbook was created, the French administration recognized the Pottok.

Although the Pottoks currently live in a semi-feral condition, they all have owners. A horse reserve in Bidarray, a small village at the base of the Pyrenees in the former Basque Province of Basse-Navarre, was developed to protect the breed and the pony's native environment.

Physical Description: The Pottok is quite a horse-like pony. His head bears a straight profile, small ears, large and intelligent eyes, and wide nostrils. The Pottok has an upright, short neck, straight back, well-defined withers, and a sloping croup. There are three types of Pottok: the solid-colored Standard and the Piebald (or pinto) types stand 11 to 13 hh; and the Double stands 12.2 to 14.2 hh. The Standard and the Double are found in chestnut, brown, and bay coloring. The Piebald is black and white; chestnut, white and black; or white and chestnut. They have medium-set tails and thick, bushy manes.

Uses: Harness and as an all-round child's riding pony.

Of Note: Traditionally, Pottoks are gathered from the Pyrenees on the last Wednesday of January, then branded for identification and either sold or returned to the mountains.

PRZEWALSKI HORSE

Alternative Name(s): Asiatic Wild Horse, Takhi, Kertag

Region of Origin: Russia, Mongolia

History: There are millions of horses all over the world, with some 200 breeds in existence. All but one are the result of man's influence. The Przewalski (pronounced sha-val-ski) horse is the only equine left in the world that hasn't been domesticated.

Nineteenth-century interest in the horses was mainly spurred on by the desire for collection. Zoological gardens had become fashionable, and wealthy men and women wanted to have rare animals in their possession. They were willing to go to great lengths to get them. In 1889, one of them, Ukrainian Friedrich von Falz-Fein, sent collectors to capture a stallion and two mares. The Duke of Bedford, who was

Prezewalski Horse

PINDOS: GREECE, MAINLAND

The Pindos (which is also known as the Thessalonian) is descended from an ancient breed called the Thessalonian. The Pindos horses developed in the mountains of Thessaly and Epirus on the mainland of Greece.

In Greece, the tough little Pindos are ridden and utilized as packhorses in the mountains, in the forest, and on the farm. Outside their native country, Pindos are very uncommon.

A plain horse, the Pindos has a long profile and a thick neck. The Pindos horses stand 12 to 13 hh and are to be found in the colors of bay, black, and brown. These hardy horses are survivalists, able to ride out the difficult times and get by on a minimal amount of food.

PLATEAU PERSIAN: IRAN

In 1978, the Royal Horse Society of Iran gave the name Plateau Persian to a group of ancient breeds that had developed in the central plateau of Iran, a harsh mountainous area. The list of breeds covered by the name Plateau Persian includes the Persian Arab, the Darashouri, the Jaf, the Basseri, the Bakhtiari, and the Shirazi.

These breeds share a common desert ancestry, and so all of them have dished faces with large nostrils, compact and muscular bodies, rounded quarters and high tails, great strength and great endurance, and Arabian-like movement. Plateau Persians stand 14 to 15 hh, and they can be found in the colors of chestnut, bay, brown, and gray.

then chairman of the Zoological Society of London, wasn't to be outdone, and he sent Carl Hagenbeck, a German animal collector, to bring him seventeen colts and fifteen fillies. These collections later saved the species from extinction.

The Przewalski population was small but growing until World War II, when overhunting threatened the breed's survival. By 1968, the breed was considered extinct in the wild. In 1977, the Foundation for the Preservation and Protection of the Przewalski Horse (FPPPH) was established to help zoos with breeding and reintroduce the horses to their native environment. FPPPH also opened a studbook.

Physical Description: The Przewalski stands 13 to 14 hh. He has a large head that bears a Roman nose, high-set eyes, and a protruding profile. His mane stands up and ends between the ears with no forelock. The breed's conformation is very asinine, with chunky features such as short quarters, a straight back, and no withers. The tail hairs start below the dock, much as a donkey's does. Przewalskis are mostly dun with black points, a cream stomach, and a dorsal stripe. Sometimes they're red dun with white markings. The breed is tough, going a long time without water and existing on very meager rations.

Uses: Apart from a picture of an 1880s Russian Cossack riding a Przewalski shown in Erna Mohr's book *The Asiatic Wild Horse* (2nd ed., 1971), no one has ever trained one.

Of Note: The Przewalski differs from modern-day horses by possessing sixty-six chromosomes rather than sixty-four. When mated with other horses, Przewalskis will produce fertile offspring with sixty-five chromosomes. By contrast, a horse that is mated with a donkey (sixty-two chromosomes) will produce sterile offspring with sixty-three chromosomes.

ROCKY MOUNTAIN HORSE

Alternative Name(s): Rocky
Region of Origin: United States, Kentucky

History: In the early 1900s, a type of gaited horse emerged in the eastern part of the state of Kentucky, near the foothills of the rugged Appalachian Mountains. A colt with a chocolate-colored coat and a flaxen mane and tail was born in the early 1900s in Kentucky's Appalachians. He was brought to the eastern portion of the state as a breeding stallion to cover Kentucky-bred mares. His foals bore his unique coloring, and they had a wonderful nature and unique four-beat gait. It became clear that these horses, isolated by the Kentucky Rocky Mountains' geography, were an actual type. During the late 1980s, a genetic researcher observed 100 foundation horses to establish the breed's genetic identity. He found five unique markers indicating pure bloodlines that were a result of the breed's isolation in the Appalachian foothills. Because of this seclusion, the horse developed a unique appearance and an unusual way of walking.

Rocky Mountain Horse

Physical Description: Rocky Mountain Horses usually stand from 14.2 to 16 hh. The Rocky has a wide chest; a graceful, sloping shoulder; and a compact frame. He possesses a straight profile, with kind, expressive eyes and well-proportioned ears. Rockys also bear a graceful arched neck, proportionate to the body and set at an angle for naturally upright carriage. Many people say it is apparent that the Rocky has Spanish forebears, since his appearance is often similar to that of Iberian horses. All dark colors accompany a flaxen mane and tail—although most are a rich dark chestnut, almost chocolate, color.

Uses: Today, the Rocky Mountain Horse is being used for pleasure, trail, competitive, and endurance riding. The breed's sure-footedness and sensibility on the trail have help make it a popular choice for riders participating in annual events such as competitive trail riding and endurance. Rocky Mountain Horses also perform in the show ring in a variety of disciplines.

Of Note: With extraordinary natural endurance, the Rocky is sure-footed over rough terrain. A Rocky must have a naturally ambling four-beat gait, with no lateral pacing. This means that, when the horse single-foots or racks, you can count four distinct hoof beats that are similar to the cadence of the walk.

SABLE ISLAND HORSE

Alternative Name(s): None
Region of Origin: Nova Scotia, Sable Island

History: The feral Sable Island Horses live off the coast of Nova Scotia, Canada, on Sable Island, a remote and windy sandbar of land. There are many theories about how the horses came to live on the island, including tales of shipwrecks, but history points to early French colonists called Acadians,

Sable Island Horse

who purchased the imported horses from Boston merchant Thomas Hancock (John Hancock's uncle) in the mid 1700s.

The Acadians planned to start a farming settlement on Sable Island, but the British deported them during the Great Acadian Expulsion between 1755 and 1763, during which 10,000 French settlers were forced off the Maritimes and sent as prisoners to other British colonies. Their horses remained on the island and continued to thrive with no help or hindrance from civilization, aside from the occasional

BREED IN BRIEF

PLEVEN: BULGARIA

The Pleven horse is a Bulgarian breed of Anglo-Arab origin that was initially developed at the end of the nineteenth century, at the Klementina state stud (now known as the Georgi Dimitrov Agricultural Center), near the city of Pleven, in northern Bulgaria. The Pleven gained official recognition a half century later, in 1951.

The Pleven horses are strong and sporty-looking, with muscular necks, fairly long backs, high withers, and strong legs. Their gaits are long and flowing. Plevens are prized competition jumpers and dressage horses; they are used for agricultural work, as well. These horses stand from 15 to 16 hh, and they are always chestnut colored.

POITEVIN: FRANCE

The Poitevin (also known as the Mulassier Poitevin) originates from the ancient primitive Forest Horse of northern Europe, as well as from Dutch, Danish, and Norwegian draft breeds. Originally, the Poitevins were used for draining the marshes of Poitou on the Atlantic coast of France during the seventeenth century.

Today, the Poitevin horses are primarily used to breed with the 16-hand Baudet de Poitou (also called the Poitevin jackass or donkey) to create huge mules. The heavy and plain Poitevin has very large feet and heavy feathering on the legs. Poitevins stand 16 hh; they are usually dun colored, although bay and brown varieties also can be found.

---

roundup to provide horses for the coal mines on nearby Cape Breton Island.

In the early 1800s, saddle and draft horses were brought to Sable Island for use on the island's life-saving stations, which helped victims of nearby shipwrecks. Later, in the 1900s, more domesticated stock horses were introduced to the feral herds to create greater genetic diversity. Fewer than 300 Sable Island Horses remain on the island, which is now a restricted wildlife preserve.

Physical Description: Sable Island Horses, like Mustangs, are diverse in size and conformation. Some are compact horses with straight heads, while others are more refined with dished faces. Rather than any Spanish influences, their genetic diversity appears to include the blood of draft and riding animals found in eastern Canada. Sable Island Horses stand under 14 hh and generally can be seen in dark colors, with some white markings.

Uses: Although most Sable Island Horses are feral, some have been ridden and are said to be very tough and sure-footed.

Of Note: Although Sable Island Horses have been protected by the Sable Island Regulations since 1961, the breed is not under the care of any association or organization. Today, the station is manned by a handful of people who collect weather data and conduct research. The Canadian government is considering closing the station, which might also remove the breed's protection.

Selle Français

SELLE FRANÇAIS

Alternative Name(s): Cheval de Selle Français
Region of Origin: France

History: Le Cheval de Selle Français (or French Saddle Horse) is a warmblood horse type that was developed in the state studs in Le Pin in Normandy in the 1800s. Unlike with most warmbloods, which were crossed with draft types and Thoroughbreds, breeders crossed the steady Anglo-Norman saddle type with the Norfolk Roadster and the Thoroughbred to create the Cheval de Selle Français. Other French blood was later mixed in. Norman breeders produced types for a variety of needs.

The Anglo-Norman was used as a draft and riding horse. The second half-breed was a fast-trotting horse; later this horse became the French Trotter.

After World War II, the emphasis on breeding was with saddle horses rather than farm and cavalry horses, and many regional breeds began to resemble one another. In 1958, the French government brought these breeds under one name: le Cheval de Selle Français.

Physical Description: The Selle Français is an athletic horse, with an elegant profile. The muscular breed is often described as a substantial Thoroughbred. Until the 1980s, the breed was split into five classifications: three medium-weight horses (small, 15.3 hh; medium, up to 16.1 hh; large, over 16.1 hh) and two heavyweights (small, under 16 hh; large, over 16 hh). Horses are found in all colors, but chestnut and bay are the most common. The breed stands from 15.2 to 17 hh.

Uses: The breed is now divided into the lighter racehorse and the show jumper. It excels in show jumping and is also successful in eventing and dressage.

Of Note: The Selle Français studbooks are open, but the breed's genealogy is usually Thoroughbred, French Trotter, and Anglo-Arab. A number of Selle Français horses are run in what the French call "other than Thoroughbred" races.

SHAGYA-ARABIAN

Alternative Name(s): Shagya
Region of Origin: Austria and Hungary

History: More than 200 years ago, Austro-Hungarian breeders were searching for the perfect cavalry mount: one with great stamina that was also biddable, strong, and beautiful. The Bedouin Arabian horses were legendary for their endurance and their desire to please people. So the monarchy imported a desert-bred stallion from Syria named Shagya and bred him to Asian-type mares, Thoroughbreds, Lipizzans (most notably the Siglavy line), and Spanish horses. Select offspring were bred back to Arabians, and soon the ideal horse was produced: one that had all the attributes of a desert Arabian but with more bone and a steadier demeanor.

Physical Description: The Shagya-Arabian stately and substantial looking, standing 15 to 16 hh. All solid colors exist. The breed is 99 percent Arabian, but that extra 1 percent that stems from various breeds gives

Shagya-Arabian

RACKING HORSE: UNITED STATES, SOUTH

The Racking Horse originated on southern plantations throughout the United States, where the breed's single-foot gait (meaning that only one foot strikes the ground at a time) enabled riders to move from field to field quickly and easily. Breeders incorporated Walking Horse blood but did not like the result.

The Racking Horse's one-footed gait is uniquely its own. Most Racking Horses are shown in English classes.

The U.S. Department of Agriculture declared the Racking Horse to be its own breed in 1971. Four years later, the Racking Horse became Alabama's official state horse. The breed is found in many colors, including spotted, and stands on average at 15.2 hh.

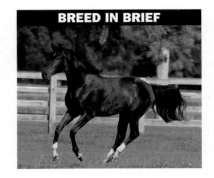

RHINELANDER: GERMANY

The Rhinelander is a German warmblood that developed from the old Rhineland heavy farm horse, which is no longer recognized today. After the Second World War, breeders decided that they would lighten this heavy draft horse with the Thoroughbred, the Trakehner, and the Hanoverian to create a sportier riding horse.

The Rhinelander is an athletic and strong horse with a plain, yet noble, head. Rhinelanders stand at 16.2 hh and are found in all solid colors. The Rhinelander is considered to be a good horse for an active amateur rider. Breeders are continuing to improve the conformation of the Rhinelander and establish better action. This breed is also valued for its even temperament.

the horse a strong hindquarter and a well-laid back shoulder. He has a more rectangular frame, which provides better flexibility and lateral movement. His head is without a dished face, more like the faces of desert Arabians. The straighter profile and prominent nose bone keep air passages open. The Shagya was bred to be tractable and trainable, a must for his original military use, so the breed is very steady.

Uses: Today, the Shagya appeals to people who are looking for a sportier Arabian for dressage and eventing. The breed excels in dressage, jumping, eventing, endurance racing, and driving. In Europe, the horse is also a popular western mount.

Of Note: This is not a plentiful breed; there are only about 3,000 in the world. There were many more before World War II, but the eastern European countries didn't prioritize horse breeding after the war. The Shagya is still very popular in Hungary, where there is a current resurgence in breeding.

SHETLAND PONY

Alternative Name(s): None
Region of Origin: United Kingdom, Scotland

History: Off the coast of Scotland lie the Shetland Islands, the native habitat of the smallest pony in Britain: the Shetland Pony. Little is known about the

Shetland Pony

pony's foundation blood. It's thought that Shetland Ponies evolved from the Scandinavian tundra and were possibly brought over by Viking raiders. They were a favorite among royalty.

The Shetlands made their mark in the nineteenth century as coal miners. Men, women, and children all worked together in deep underground mines until 1847, when laws were passed to prohibit women and children from working in such conditions—so Shetland Ponies took over their jobs. The small size and tough nature of the ponies made them the perfect choice for coal mines, and many spent their whole lives underground. Ponies were used until the 1990s, with the final pit pony retiring in 1994.

Physical Description: These tiny turbo ponies are built to last. Their thick manes and tails protect them from harsh island weather, and they possess sturdy legs and hooves, wide backs, and broad hindquarters. The breed stands 10 hh at the withers and is found in black, brown, chestnut, bay, and pinto. Although the stocky, small Shetland is often considered the "classic" Shetland, two significant types are established within the breed: a heavier-boned pony with a longer head, and a lighter pony with high tail carriage and a smaller, pretty head.

Uses: Throughout the world, children and small adults in English and western classes ride Shetland Ponies. People also jump and ride them in gymkhanas. Although the Shetland is strong enough to carry a third of his weight, larger adults don't often ride the ponies because of the unbalanced picture it creates. The breed is, however, a popular harness pony for adults. They are wonderful foundation stock, as well. A Shetland crossed with an Appaloosa created the Pony of America; a cross with the Hackney produced the American Shetland.

Of Note: Pound for pound, the Shetland is probably the strongest of all equines, with the ability to carry a large man. The breed is pressed into service for just about every aspect of equine work: harness, packing, riding, mining, novelty, and even circus tricks.

SHIRE

Alternative Name(s): None
Region of Origin: United Kingdom, England

History: The Shire is one of the two British draft breeds. The Shire made his first appearance on British soil in his original form of the Great Horse, which William the Conqueror brought from France in 1066. In the early seventeenth century, Dutch contractors, who were helping drain the fens in the east of England, brought with them their native horses, the Friesian and the Flemish Horse. These horses remained in the area and were bred to the descendants of the Great Horse. This resulting breed was called the English Black, and later the English Cart Horse. The breed's name changed again in the late 1800s to the Shire,

Shire

BREED IN BRIEF

RUSSIAN HEAVY DRAFT: RUSSIA

The Russian Heavy Draft is the smallest of all the draft breeds. Originally called the Russian Ardennes, this breed was developed in the Russian state studs in the Ukraine during the twentieth century.

The Russian Heavy Draft originated from three draft breeds: the Swedish Ardennes, the Belgian, and the Percheron. The Russian Heavy Draft is a compact, clean-legged draft horse with a heavy crested neck and a short head that features a slightly dished face. The horse is very muscular and powerful. Russian Heavy Drafts stand 14.2 to 14.3 hh, and they can be found in several color varieties, including roan, bay, and chestnut.

SALERNO: ITALY

The Salerno is an Italian warmblood formerly known as the Persano, in honor of the stud where the breed originated. The breed began in the eighteenth century in the Campania region of Italy. The lineage includes Neapolitan, an Italian horse with Spanish and Barb blood, and Thoroughbred blood, which was introduced in the 1900s.

At one point in time a prized cavalry horse, the Salerno is now a prized competitive jumper. The breed possesses the elegant and noble features that are common to Spanish-bred horses, such as a refined head, a thick mane and a thick tail, and powerful hindquarters. The Salernos stand 16 hh, and they can be found in all solid colors.

---

possibly in honor of the horse's development in the Fen county of Lincolnshire and use in the counties of Leicestershire, Staffordshire and Derbyshire.

The strong and unflappable Shire was the perfect choice for hauling cargo throughout London. Breweries, in particular, used the breed to make delivers to the many pubs throughout the city. This continued for many years, up until modern times.

Physical Description: The Shire is the largest of the draft breeds, standing up to 18 hh and weighing roughly a ton. The horses' colors include black, brown, gray, and bay, with white feathers on the legs.

Uses: Today, Shires can be found doing a variation of their traditional role: working smaller hobby farms, pulling brewer's drays in parades, and giving wagon rides. Shire crosses are used in the jumping arena and the dressage arena.

Of Note: Although the majority of breweries have ceased making deliveries using the heavy horses, Young's Brewery in Wandsworth, London, retains this tradition. Ten black Shires are housed in the old Victorian stables in the back of the brewery. As the brewery has done every morning since 1581, it sends out two drays pulled by teams of two horses. They make their way through the busy streets of Wandsworth to deliver beer to the local pubs.

SKYRIAN

Alternative Name(s): Skyros, Alogaki
Region of Origin: Greece, Skyros Island

History: Skyrian ponies are found on Skyros, the largest and most southerly island of the Sporades, located to the east of mainland Greece in the Aegean Sea. The north of the island provides green and fertile land, ideal for agriculture, while its southern side is mountainous and often dry in summer. Skyrian horses

have remained true to their original type. Their isolation has kept their gene pool pure.

In the 1950s, threshing equipment arrived on the island, followed by Jeeps and combine harvesters. Unfortunately, that also meant that the pony that had spent years as a valued partner on the farm quickly became obsolete. Even worse for the ponies, the government supplied grants to farmers to promote keeping sheep and goat herds, which had a drastic impact on the Skyrian pony's grazing land. As a result, the ponies' numbers diminished.

The area's first census, taken in 1993, showed a total population of 121 equines. A law on the island prohibited the removal of any purebred Skyrian from the island. Concerned individuals started to take notice. In 1996, Silvia Dimitriadis Steen was approached and asked to provide lodgings for two stallions and two mares on her estate on the Ionian island of Corfu. She became interested in the horses and visited the island of Skyros. And so began the Silva Project, a refuge on the Silva estate, which is located on the Kanoni peninsula near Corfu Town.

Another supporter was Alec Copland, a veterinary faculty member at the Edinburgh University of Scotland. Copland visited the island to investigate the ponies himself. He found that Skyrian ponies were quite similar to Exmoor ponies in their conformation, way of going, and temperament. Skyrians were just a bit smaller. Exmoor Ponies are the closest descendents to Pony Type 1.

In 2004, another census taken on Skyros revealed only ninety horses left on the island. Of those, only fourteen were purebred ponies. The Silva Project is working hard to lift the government's exportation ban, in hopes that breeding operations and pony refuges can assemble on the mainland and purebreds can become a viable commodity outside of Greece.

Skyrian

Physical Description: The Greek name for the Skyrian pony is Helliniko Alogaki, meaning "little Greek horse." There is no Greek translation for the word *pony*, meaning that the breed is called a horse in the English translation despite the Skyrian's short, average stature of 10 hh. The breed is dark brown or bay, with a fawn-colored muzzle and eye area (often called "mealy" coloring). Like the Exmoor pony, the Skyrian has a deep-hooded eyes (called toad eyes), a wide set, and an intelligent small head with a slightly convex profile. The horses have short but slender necks, narrow chests, slender bodies, and lightly muscled hindquarters. They have low-set tails and hard, black hooves.

Uses: These ponies were used on farms for harvesting grain, as well as riding ponies, pack horses,

---

SANDALWOOD: INDONESIA

One of the eight ponies native to Indonesia, the Sandalwood developed from Arabian blood on the islands of Sumba and Sumbawa. The breed is named after the aromatic sandalwood that is native to these islands.

Historically, Sandalwood ponies were used for racing in Thailand and as children's mounts in Australia. Today, Sandalwoods are used under saddle and in harness. On average, the Sandalwoods stand 13.1 hh, and the ponies come in all colors.

The other seven ponies that are from Indonesia are the Bali, the Batak (*page 67*), the Deli, the Java (*page 102*), the Padang (*page 132*), the Sumba/Sumbawa (*page 144*), and the Timor.

and cart ponies. They also have been used as breeding partners for mules.

Of Note: Genetically, the Skyrian's lineage traces back to prehistoric horses. Alexander the Great may have taken Skyrian ponies with him when he left Macedonia to conquer the world. Some people have also said that the Skyrian is very similar to the equine etchings depicted on the Parthenon frieze.

SORRAIA

Alternative Name(s): Sorraiana
Region of Origin: Portugal

History: The Sorraia is the last remnant of the ancient native horses of Portugal. Considered extinct for many years, the Sorraia was discovered near the river Sorraia in 1920 by Portuguese scientist Ruy d'Andrade. Cave paintings and bones have proven that the Sorraia horses have existed for thousands of years. The breed is the primary ancestor of the Lusitano and is thought to share the same origins as the North African Barb, due to the prehistoric land bridge that existed at Gibraltar and linked Africa to Spain.

There are fewer than 200 horses worldwide, with some herds in Germany and Portugal, including a herd that still runs freely in Portugal's Sorraia Horse Natural Reserve.

Physical Description: The Sorraia is a small, compact horse with strong sloping quarters and a convex profile. The conformation of the Sorraia is very different from those of modern breeds; he has a great resemblance to the ancient Tarpan, especially in his head and the color and texture of his coat. The breed is always dun or grulla (a gray version of buckskin) in coloring, with barred markings on the legs and a stripe down the back. Foals are born with zebra striping all over their bodies. Fully grown, the horse stands 13 to 15 hh.

Uses: The Sorraia is a prized dressage horse and working ranch horse.

Sorraia

SARDINIAN: ITALY

The Sardinian originated on the Italian island of Sardinia. The breed stems from Barb and Arabian horses that were sent to the island in the sixteenth century by King Ferdinand of Spain (1452–1516) in order to establish a stud there. These horses were crossed with local horses. In the 1900s, Arabian blood was added to improve the island's horses, which had declined in quality after Spain turned Sardinia over to the House of Savoy in the early 1700s.

The Sardinian is a hardy, sure-footed breed that is prized nowadays as a mounted police horse. Sardinians stand 15 to 15.2 hh, and most of them have coats that are either brown or bay.

Of Note: There are only two purebred Sorraias in the United States.

SPANISH MUSTANG

Alternative Name(s): None
Region of Origin: United States

History: The Spanish Mustang of the United States is the original Native American breed. The breed descended from the horses of the conquistadors and the Native Americans and developed on the plains of the American West, growing stronger and thriftier through natural selection. The breed differs from the BLM (Bureau of Land Management) Mustang in that it shows little ranch or draft horse influence. Bob Brislawn from Oshoto, Wyoming, is credited with bringing the Spanish Mustang back from near extinction. He gathered horses on Native American reservations and worked with the Bureau of Land Management to find the best Spanish Mustangs and took them to his 3,000-acre Cayuse Ranch. Brislawn's son, Emmet, now runs the ranch. Spanish Mustangs are still raised and sold there.

Physical Description: The breed possesses features that are very Spanish; it has a compact and muscular build, rounded hindquarters, and a low-set tail. The neck is arched and set high out of the withers, and the head is straight or concave. Many of these horses are gaited, which is a common Spanish attribute. There is a wide range of colors, including pinto, buckskin, roan, and black. They stand 13.2 to 15 hh.

Uses: Spanish Mustangs are sought after as ranch horses and as endurance and trail horses.

Of Note: The breed possesses tremendous stamina and hardiness.

STANDARDBRED

Alternative Name(s): None
Region of Origin: United States

History: The earliest harness racers in the Americas were the Narragansett Pacer and the Canadian Pacer, which raced in New England in the eighteenth century. When English Thoroughbreds were crossed with several other breeds, including the Norfolk Trotter, the Hackney, the Morgan, and the Canadian Pacer,

Spanish Mustang

Standardbred

SUMBA AND SUMBAWA: INDONESIA

These two ponies are native to Indonesia and are very similar. They developed, however, on different islands: Sumba and Sumbawa, respectively. Owing to their Mongolian ancestry, they are very primitive looking, with overlarge heads and small bodies. Their coloring is usually dun, with zebra striping on the legs. They are very small, only 12 to 12.2 hh, but they are strong enough to be ridden by adults. The Sumba is ridden in native games of throwing the lance.

The other seven ponies of Indonesia are the Bali, the Batak (*page 67*), the Deli, the Java (*page 102*), the Padang (*page 132*), the Sandalwood (*page 141*), and the Timor.

---

it resulted in the Standardbred—a horse that became known for its two distinctive racing gaits.

The original trotting races in the eighteenth century were simply held in fields, and the horses were actually raced under saddle. However, farmers and breeders began to take their match races seriously, and trotting races were held on official courses by the mid-1700s (this time with the horses in harnesses). Breeders quickly moved to improve their trotters, selecting bloodlines that could produce even more speed from their horses.

In 1788, a gray Thoroughbred stallion named Messenger was brought to America from England. The Thoroughbreds during this era actually had similar bloodlines to the Norfolk Roadster, and Messenger was a prime example of this breeding. He became one of the most influential sires of racers, producing both runners and trotters, with the trotters possessing great action, speed, and heart.

In 1849, Hambletonian, a descendent of Messenger, was born out of a crippled bay mare that was bred to an ungainly, belligerent stallion named Abdullah. Sold as a castoff, Hambletonian astounded everyone when he proved to be a prolific sire. His offspring, too, were natural trotting athletes.

Physical Description: The Standardbred is both like and unlike his cousin, the Thoroughbred. Although typically a little shorter (standing from 14.2 to 16 hh) and slightly stockier than Thoroughbreds, Standardbreds have similar refined legs, powerful shoulders and hindquarters, and medium to long backs. The breed's unique facial profile is straight or sometimes squarish in appearance. Most Standardbreds are bay or brown, but other coat colors occur, including chestnut, strawberry roan, and gray. Racing Standardbreds have two gaits: the trot and the pace.

The pace is a lateral gait, in which both legs on one side of the body move together.

In the trot, the horse's legs move in diagonal pairs. Because pacers usually trot when they're not racing, they are trained and raced with hopples, which are straps that link the front and back legs on each side to keep them moving in unison. Ex-racers are used in traditional pursuits, such as jumping, dressage and western pleasure.

Uses: The Standardbred is the fastest-trotting horse in the world. With lightning speed performed at a two-beat gait, this horse (known mostly as a harness racer) has also been used to improve other breeds of racing trotters and pacers around the globe. Today, nearly all trotters and pacers can trace their lineage back to Hambletonian, a remarkable stud of humble origins.

Of Note: The name Standardbred was first used in 1879; and it simply came about because harness racers had to prove that they could trot a mile within the standard time of two minutes and thirty seconds to be registered. Modern Standardbreds often race a mile in as little as one minute, fifty seconds.

SUFFOLK PUNCH

Alternative Name(s): Suffolk Punch Heavy Horse, Suffolk, the Suffolk Horse, the Suffolk Draft
Region of Origin: United Kingdom, England

History: Prior to the 1600s, the people who lived in East Anglia, on England's southeast coast, eked out a living from the swampy land called the fens. When these swamps were finally drained in the seventeenth century, the landowners faced a prosperous future because the peaty soil found underneath the water was the perfect environment to raise crops. But a special sort of horse was needed to work such

Suffolk Punch

heavy soil. The farmers needed a breed that was strong, well mannered, and easy to keep. The Suffolk Punch developed over the years to fill the farmers' needs and consequently helped give rise to Britain's agricultural revolution.

The word *punch* was a slang term for a jolly, solid, hearty character—a term that describes the Suffolk breed of heavy horse perfectly. Today, the breed is formally known as the Suffolk Punch Heavy Horse.

The breed may have been influenced by the Norfolk Roadster, the Norfolk Trotter, or the Norfolk Cob, and the horses' size may have come from Belgian draft blood. Most likely, they came from a scrubby group of horses found in the area, and farmers just bred from the best ones available.

Eventually, one stallion would make his mark on the breed forever—Thomas Crisp's Horse of Ufford, born in 1768. The 15.2 hh stallion was striking in appearance, with a very pretty head and a bright chestnut coat. Today, the breed's male lineage can be traced back to this one stallion, which gives the Suffolk a uniform look.

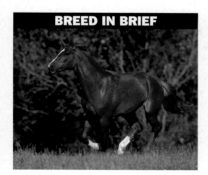

BREED IN BRIEF

SWEDISH WARMBLOOD: SWEDEN

The Swedish Warmblood originated from Swedish cavalry horses in the seventeenth century, in the royal stud of Flyinge and the stud of Stromsholm. Later breeders developed the Swedish Warmblood further, refining it with the Thoroughbred, the Hanoverian, the Trakehner, and the Arabian.

The Swedish Warmblood is a well-proportioned horse with strong hindquarters and a small, elegant head. The horses have muscular shoulders and legs. Swedish Warmbloods are top-class dressage horses, show jumpers, and eventers. They stand 15.2 to 16.3 hh, and they can be found in all solid colors, though they are most often bay, brown, chestnut, or gray in color.

TARPAN (MODERN): CENTRAL ASIA

One of three ancient breeds, the Tarpan was hunted to near extinction in the eighteenth century; the last one died in the late 1800s. Twentieth-century zoologists, Heintz and Lutz Heck, brought the Tarpan back by crossing feral stock with the breed's relatives, the Konik and the Huçul and later the Przewalski Horse. This genetically re-created horse—which is referred to as the modern Tarpan or the Heck Horse—looks very primitive, with a large head, thick neck, and strong jaws capable of chewing woody forage.

Today, there are around 100 worldwide, several in North America. The breed is extremely hardy and very resistant to disease. They are small horses, standing 12.2 to 13.2 hh. All Tarpans are grulla (a mouse dun) in coloring, with primitive markings, such as a dorsal stripe. The semi-erect mane is flaxen, with a dark stripe down the middle. Although many modern Tarpan are feral, quite a few are ridden and driven.

Physical Description: Suffolk Punch horses are always colored chestnut (this archaic spelling of chestnut is used in the horse's registry), but they come in seven different shades: dark brown, liver or mahogany, dull dark, light mealy, red, golden, lemon, and bright chestnut. They have very few white markings, generally restricted to a small snip or star and white fetlocks. The Suffolk is a small draft horse, standing between 16.1 and 17.1 hh. The breed is clean legged, meaning the horse's legs have no feathers. Often described as a round-bodied horse, the Suffolk has short legs, despite a huge girth and wide front end

Tennessee Walking Horse

and hindquarters. This close-to-the-ground muscular stance made the horse perfect for tilling heavy soil. The Suffolk Punch's low-set shoulders helped him lean into his load.

Uses: Young horses mature early, and they can start training at two years of age. They learn easily and are ready for work by the age of three. The breed is used on hobby farms and for pulling wagons in rural museums.

Of Note: The Suffolk Punch was hailed as the perfect draft horse. Horses could work all day with only one break for a quick feeding and were easy to handle. The Punches' featherless legs were best for working in the sticky mud, so very little grooming was needed at the end of the day. The farmers simply turned all their horses out into one big field, threw down a few piles of hay, and left it to the horses to sort themselves out.

TENNESSEE WALKING HORSE

Alternative Name(s): Walker

Region of Origin: United States, Tennessee

History: The Tennessee Walking Horse is an American original, developed in central Tennessee in the late 1800s. The horse's genealogy includes a mixture of breeds that settlers brought with them, such as Morgans, Narragansett Pacers, and Canadian Horses. Soon the settlers were buying each other's horses or their neighbors' stud services, and they began to combine these bloodlines.

The resulting horses took many fine characteristics from this large gene pool, resulting in saddle horses that were robust yet elegant, with upright conformation and unusual gaiting ability. One characteristic of this new breed was the animal's smooth gait, which was later called the running walk. This gait replaced the ordinary bouncy trot and made it easy for farmers, deliverymen, and doctors to sit for hours in the saddle.

It was this trademark, the running walk, which became the Tennessee Walker's claim to fame. The running walk is something that the Walker is born with—and it can't be trained into another breed of horse. In this four-beat gait, the horse simply feels as though he is sliding across the ground effortlessly. His head bobs gently, and he will overstep, meaning that his back hoof prints will make marks ahead of where the front ones hit. And the running walk is just that: a speedy gait that can reach up to 20 miles an hour, which a Walker can perform for miles without tiring.

Physical Description: Tennessee Walkers have a distinct look. They're usually dark bay or even black, although they are seen in a variety of colors including chestnut, roan, palomino, gray, and even sabino. The horses stand fairly tall, between 15 and 16 hh. The Walker sports a large, intelligent head with a straight profile, swiveling ears, ample nostrils and bright eyes. The breed features a proportioned, slightly arched neck that comes up from a sloping shoulder. The horse also has a short, proportioned back; powerful haunches; and long, clean legs with strong hooves.

Uses: Tennessee Walkers are shown in English and western competitions, and they are prized mounts for leisure and trail riding.

Of Note: Besides the running walk, the Tennessee Walking Horse also performs two other gaits. The breed has a four-beat "flat walk" that is more of a marching, medium walk, as well as a canter that is similar to a gallop in hand with a rolling, rocking-horse motion.

BREED IN BRIEF

TERSKY: RUSSIA

Russia's quick and athletic Tersky was created in the first half of the twentieth century at the Tersk and Stavropol studs in the northern Caucasus region of the former Union of Soviet Socialist Republic. The breed originated from Arabian, Thoroughbred, and Don horse breeds. Terskies were developed for racing and steeplechasing.

Nowadays, the Tersky primarily is used for competition jumping and for eventing. Terskies are also noted endurance horses. The breed is well muscled with a fine, straight profile, a muscular neck, and long, slender legs. Terskies stand approximately 15 hh, and they are usually gray or bay, though there are some that are chestnut in color.

THOROUGHBRED

Alternative Name(s): None

Region of Origin: United Kingdom, England

History: Throughout equine history, few breeds have affected the horse world quite like the Thoroughbred. The Thoroughbred first made his mark as a racehorse, shaping a sport so favored by the gentry that racing was dubbed "the sport of kings." With his famous speed and stamina, it wasn't long before the Thoroughbred found his way into other sports including hunting, jumping, dressage, and eventing. The Thoroughbred has contributed his bloodlines for many breeds, including the European Warmbloods and the Quarter Horse.

Out of the 103 Oriental stallions used to develop the Thoroughbred horse, only three bloodlines flow through today's modern-day horses: the Byerly Turk, the Godolphin Barb, and the Darley Arabian. These foundation stallions were bred in the seventeenth century to sturdy English ponies and Scottish Galloways.

Ninety percent of all Thoroughbreds have the blood in their veins of Eclipse, the most important offspring from the Darley Arabian (foaled in 1764). Although Flying Childers, son of the Darley Arabian, was given the title of the first English Thoroughbred, Eclipse was given the title of the best representative of the breed. Eclipse alone makes the Darley Arabian the most important sire of all three foundation stallions.

Thoroughbred

Physical Description: Thoroughbreds are quite tall, standing above 16 hh on average. Their heads are long and narrow, with wide-spaced eyes. Their necks are longer and thinner than those of most breeds and are tied into high withers and a curved back. Their sloped shoulders are fully muscled. Much like the legs of human runners, the Thoroughbred's legs are very lengthy, making for a big stride. The bones of the upper hind legs are particularly long, and the hips are very wide. This type of bone structure makes way for well-defined muscles.

Throughout more than 300 years of careful breeding, this conformation has made the Thoroughbred the perfect runner. The horses can cover more than 20 feet in one stride with power up to 40 miles per hour. The long, slender hind legs are bred to provide springing strength, bending and straightening as the horse gallops. The breed's longer-than-average neck helps the horse propel himself forward and creates a smooth and fluid running stride. Thoroughbred colors vary from bay, chestnut, black, gray, and the occasional roan, with varying white markings on the legs and faces.

Uses: The Thoroughbred's wonderful conformation and elegant movement were so admired that the breed was used as foundation stock for American Saddlebred and Quarter Horse breeding, and to enhance all the warmblood breeds. The Thoroughbred is found outside the racecourse in all equine English sports, including show jumping, dressage, and eventing.

Of Note: It wasn't until the reign of King James I (1603–1625) and his son Charles I (1625–1649) that racing was really encouraged. Races became more widespread when horses were brought to town markets to be sold throughout Britain. One of these town markets was set up at Newmarket, a small town in Suffolk, England. Newmarket soon became the Mecca of Thoroughbred

breeding and later was the setting of many Dick Francis novels on the subject of horse racing.

TRAKEHNER

Alternative Name(s): None
Region of Origin: Lithuania

History: The Trakehner is one of the oldest European warmblood breeds, with a history dating back more than 400 years. The breed is based on a local horse (then found in East Prussia, which is now in present-day Lithuania) called the Schweiken. The Schweiken was well known for his endurance and versatility and was considered the perfect breed for cavalry horses.

In 1732, King Friedrich Wilhelm I of Prussia founded the Royal Trakehner Stud Administration on the marshlands of East Prussia. He selected seven horses from various stud farms and took them to the new royal stud in Trakehnen to create speedy and beautiful coach horses. By 1787, breeding changed to produce lighter cavalry horses. Detailed documentation and precise selection of breeding animals were cutting-edge breeding tactics at the time, and they paid off for the Prussians. The stud farm was famed throughout Europe for its versatile, athletic, and beautiful horses. From 1817 to 1837, English Thoroughbreds and Arabian stallions were mixed into the breed.

Physical Description: At first glance, grace and presence distinguish the Trakehner. Although it is a solid breed, the Trakehner is lighter than any of the other warmbloods. The horse's head is refined with a dished nose and large eyes. Long legs and a long, well-set neck add to the refined appearance. The body is solid, yet with pleasantly rounded hindquarters. The Trakehner stands from 15.3 to 16.3 hh and in all solid colors. The additional bloodlines of the Arabian

BREED IN BRIEF

WALKALOOSA: UNITED STATES

The Walkaloosa is essentially a gaited Appaloosa. Many of the old-style Appaloosas, which were prized by North America's native Nez Perce tribe from the eighteenth century on, possessed a type of comfortable amble, which cowboys and settlers referred to as the "Indian Shuffle."

The modern-day Walkaloosa appears in all Appaloosa coat patterns and has an extra gait aside from the trot, which can include the running walk, the single foot, the fox-trot, the rack, or the pace. Walkaloosas are popular trail horses, highly valued for their comfortable ride. Walkaloosas are used for pleasure, pack, and trail riding.

WELARA: GREAT BRITAIN

The Welara pony was initially developed in Sussex, England, in the 1920s by Lady Judith Wentworth, a famous Arabian breeder who owned the stud Crabbet Park. Wentworth crossed her Arabian stallion, Skowronek, to her Welsh Pony mares to create a beautiful yet versatile pony. No breed registry was established for this new pony, however, until 1981, two years after an American breeder bought a half-Welsh, half-Arabian colt that performed amazingly in competition. With the registry came a name for the new breed, the Welara.

Today, the breed is used for gymkhana, dressage, and jumping. The Welera stands 11.2 to 15 hh and is found in all colors and patterns, except spotted.

and Thoroughbred refined the Trakehner, giving the horse the lovely floating trot and elegant canter that is his hallmark.

Uses: Today the Trakehner is prized all over the world for his jumping and dressage talents.

Of Note: The breed was nearly decimated after World War II, when breeders fled to West Germany after 1945 to escape capture from the Soviet army. Few horses survived the flight, but dedicated breeders worked to increase their numbers. The breed is thriving today.

WELSH PONY AND WELSH COB

Region of Origin: United Kingdom, Wales

History: Wales, which is one of the four countries that make up the United Kingdom, has a varied landscape that helped shape its history. Valleys produced the coal that gave the south international acclaim; and the high peaks, which cover a quarter of the land, intimidated marauding raiders. The environment of these wild landscapes also helped to develop the world-renowned Welsh Ponies and Cobs' incredible characteristics of.

Trakehner

Welsh Mountain Pony (Section A)

The Celtic people, who immigrated to the area in prehistoric times, were well known for their incredible horsemanship; in fact, they worshipped Epona, the goddess of the horse. The Celts hitched these tough little ponies to their chariots and drove them into battle. The warriors were so ferocious that the Romans called these native people "Furor Celticus" when they invaded the British Isles in AD 43.

After the Romans left in AD 410, the wild high moorland of the Cambrian Mountain range effectively kept the land isolated from any other conquerors for hundreds of years and allowed the Welsh to retain their language and ancient Celtic customs. The Welsh ponies continue to thrive and lend their use to hill farmers and shepherds, landowners and deliverymen.

In 1901, farmers, landowners, and breed enthusiasts in Wales established the Welsh Pony and Cob Society. Their aim was to record pedigrees and to ensure the survival of the Welsh breeds. Since the original wild pony had evolved into different-looking animals, the studbook was divided into four sections (A, B, C, and D) to establish a standard for each type of pony or horse.

Section A: Welsh Mountain Pony

Physical Description: The Welsh Mountain Pony is the smallest of all four sections, standing no taller than 12 hh, and he is the only one of the four to be found living semi-feral in his natural environment. The Welsh Mountain Pony is a small, spirited pony, possessing strong bones and tough hooves. The pony's attractive,

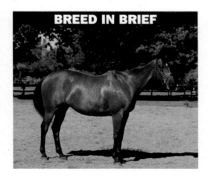

BREED IN BRIEF

WIELKOPOLSKI: POLAND

Poland's Wielkopolski is a warmblood that was established in 1964 from two warmblood breeds that are now extinct: the Poznan and the Masuren. These breeds had Arabian, Thoroughbred, and East Prussian Trakehner bloodlines.

The Wielkopolski is a strong, large-muscled horse, with powerful legs and elegant movement. The Wielkopolskis are an athletic breed, used for eventing, dressage, and show jumping, They are also used for riding and harness. The horses have good constitutions and endurance.

The Wielkopolskis stand 15.2 to 16.2 hh; most of them are bay or chestnut in color. Some of them have black, brown, or gray coats.

WURTEMBURGER: GERMANY

The Wurtemburger is one of Germany's warmbloods. The breed was developed in the seventeenth century at the then-famous Marbach Stud in Germany. Breeders put local mixed-breed mares to Arabians, Spanish horses, Barbs, and later Anglo-Normans. The resulting horse was good for light draft and farm work, as well as for riding.

During the twentieth century, as demand for the draft horse fell, breeders decided to create a sportier Wurtemburger. In the 1960s, they introduced Trakehner blood. The modern Wurttemburger is used for jumping, eventing, and dressage. The breed is well proportioned, with a straight yet handsome head, and stands 16.2 to 17.2 hh. The horse is found in all solid colors.

Welsh Pony (Section B)

Welsh Pony of Cob Type (Section C)

dainty head is set off by small pointed ears and large expressive eyes. The neck is carried high and crested, and it moves in a quick movement that rounds out from the shoulders. All colors but pinto are found, including roan and palomino.

Uses: The Welsh Mountain Pony was used for shepherding. Today, he is used for lead line classes, first riding classes, riding for the disabled, and driving.

Section B: Welsh Pony

Physical Description: The Welsh Pony is larger than the Mountain variety but stands no higher than 13.2 hh. All colors except pinto are found, including roan and palomino. Section B ponies share the same description as Section A ponies. However, the Section B pony is taller and lighter, with an emphasis toward riding suitability with long strides and easily ridden gaits.

Uses: In the past, shepherds used the old Section B ponies to help them herd their sheep, and hill farmers

used them for transportation. Today, the new Section B is a fancy child's show pony, found in hunter and show pony classes, as well as in driving (both competitive carriage and showing).

Section C: Welsh Pony of Cob Type

Physical Description: Although the Welsh Pony of Cob Type is a strong and sturdy pony, he is anything but plain. Possessing a refined head similar to the Welsh Mountain Pony, his ears are larger and set wider apart, and his neck is arched and carried high out of strong shoulders. The breed's muscled quarters, courtesy of Welsh Cob blood, give a powerful, chunky appearance; and this pony moves with short, smart action. All colors are found, including roan and palomino, but not pinto. The breed is intelligent with a kind temperament. These ponies stand no higher than 13.2 hh.

Uses: This pony is an extremely valuable animal for driving. In the past, the Welsh Pony of Cob Type

The Welsh Cob (Section D)

was used in slate mines and farms, as well as for delivering goods and hauling military equipment. Today, the Section C pony is ridden by both adults and children in trekking; harness; and hunter, jumper, and under saddle classes.

Section D: Welsh Cob

Physical Description: The Welsh Mountain Pony (Section A) is often mistaken as the foundation animal for The Welsh Cob (Section D), but this isn't exactly true. They may have initially come from the same genetic pool, but the Welsh Mountain Pony and the

Welsh Cob have each been around for at least 2,000 years. The description of the Section D type is much the same as that of the Section C, but the Section D is larger and of stockier build. The Welsh Cob's legs are also lightly feathered. He stands 13.2 hh and taller, with no height limit. Aside from pinto, all colors are found, including roan and palomino.

Uses: In the past, the Welsh Cob was used for the cavalry, farming, and transportation. Today, the breed is used by adults and children for hunting, showing under saddle, and eventing. The Welsh Cob is also a successful competitive carriage horse.

NATIVE AMERICAN INFLUENCE

When Europeans first set foot on the shores of the New World, no horses existed there and so they formed no part of Native American culture. But Native Americans, especially the Nez Perce, soon became skilled horsemen and breeders, who put their marks on the breeds (*see, e.g., pages 60 and 67*). The circa 1908 photo here shows a group of Dakota Indians and their distinctive-looking horses.

Activities with Horses

Section 3

From recreational riding events and fun games to competitive sports that involve big money, horses can fulfill many different roles—and myriad dreams. The horse made the transition from the role of work animal to that of a recreational companion, and as a result there are a great number of examples that truly show how the horse enhances the human world's leisure time.

The chapters in this section talk about and show some of the best-known riding disciplines and activities for the nonworking horse. Many of these activities were born from jobs that the horse performed at earlier points in history.

Chapter 5

English, Western, and Driving Competition

The phrase *English riding* is an umbrella term for riding styles that originated centuries ago in Europe. Much of English riding has its roots in the hunt fields of England, as well as in classical riding's haute école (high school) in Austria, France, and Germany. Western riding divisions have their roots in the ranch lands of the American West. More than a century ago, the performance and show divisions started at county fairs as a way to show which rancher had the most skills and the best cow horses

Driving horses used to be the main mode of transportation for people until the automobile became affordable for the masses. Today the sport of combined driving challenges the agility, endurance, and bravery of drivers and their teams.

ENGLISH COMPETITION

English competition includes the disciplines of show hunter, show jumping, eventing, dressage, combined driving, and saddle seat. Although all of these fall under the English riding category, they are very different from one another, having evolved from specific tasks that horses have performed over the centuries.

Show Hunter

The show hunter division, with its sleek mounts and spotless turnout, has its origins in traditional field hunting. Good field hunters carry their riders calmly over field and fence, and this ability forms the basis of

what is judged in the hunter show ring. A field hunter would be an unlikely candidate for the show ring, however, because a show hunter also has to have a highly stylized way of going.

Show hunters are judged on their ability to quietly jump a course of natural-looking obstacles, at the perfect pace and in the correct number of strides. They must do so with expression and impeccable style, which means that the knees are even and tucked well up under the chin and the back is rounded over the jump. Hunters are often more relaxed than are event horses or jumpers. The ideal show hunter covers the ground with long, sweeping, elegant strides. Because the horse (rather than the rider) is judged in this division, his appearance is a major consideration. Tack is conservative and uniform, and grooming is flawless. Hunter classes are offered not only at jumping competitions, but also at breed shows, such as Quarter Horse, Arabian and Paint.

Show Jumping

Jumping evolved from the horse and rider's need to conquer obstacles they encountered as they traveled from place to place. People enjoyed jumping so much that they turned this activity into a sport. Show jumping evolved in the middle of the 1800s. Before then, competitive jumping consisted of a single jump made higher and higher, similar to today's puissance (high jump) class.

In show jumping, the horse must jump over a series of colored fences that are set up in a challenging

Hunter courses are made up of natural-looking obstacles, such as this fence at the 2008 WEF Challenge Cup.

Eventing competitors Mr Pracatan and Andrew Hoy jump a drop fence into the water at the 2005 Badminton Horse Trials.

pattern, and he must complete the round cleanly. Riders collect penalties for knockdowns and refusals, and those with clear rounds move into the jump-off, where they tackle a shorter, higher course against the clock. A good jumper is a very powerful animal. He is more spirited in his jump and way of going than is an event horse or a hunter. Not only are good jumpers energetic and fast, but they are also agile and responsive, which are important elements in a jump-off. Show jumpers should be built uphill, meaning that their power should come from the haunches, which helps them overcome the higher fences. Jumpers also should be compact or shorter coupled, which also gives them more strength when jumping.

Eventing

Eventing, which is also known as combined training or horse trials, is like the triathlon of all the equestrian sports. It requires obedience, strength, stamina, and courage to complete the three phases: dressage, cross-country jumping, and stadium jumping. Formerly called the *militaire*, eventing was once utilized to test a cavalry horse's courage and endurance under pressure. The dressage test demonstrates submission and obedience, the cross-country course shows ability and courage while enduring miles of solid obstacles, and the final jumping test proves that the horse is conditioned well enough to jump a technical course of fences.

A good event horse must have many traits that will help him excel at all three phases. He'll need good movement and obedience for the dressage phase, crafty jumping skills, and endurance and speed to tackle the fixed obstacles in the cross-country course. An event horse must also have strong, sound conformation to help him complete the stadium-jumping phase. The Thoroughbred, a hotblooded horse, used to be the most popular breed in eventing; today, in the highest competition, there are more warmbloods and cross-bred sport horses in eventing than ever before.

Dressage

Two thousand years ago, dressage was developed as a way to make cavalry horses nimbler and more responsive on the battlefield. Dressage battle movements—such as the airs above the ground movements, which are still shown today by the famed Spanish Riding School in Vienna, Austria (*see opposite*)—

Horse talk

ENGLISH, WESTERN, DRIVING ACTIVITIES

Learning these terms will help you understand and talk about English, western, and driving activities.

■ **Combined driving:** English sport in which a horse and driver team compete in three events: dressage, cross-country marathon, and cones.

■ **Cones:** The third combined driving phase, testing precision and concentration of a horse and driver through an obstacle course marked by cones.

■ **Cow work:** A horse shows his ranching abilities by moving a cow throughout the arena in various exercises.

■ **Cutting:** A western sport in which horse and rider separate a cow from the herd.

■ **Dressage:** A 2,000-year-old method of English training and competition in which the horse learns to carry his rider in balance, relaxation, and harmony.

■ **Dry work:** A compulsory reining pattern in the western working cow horse competition.

■ **Eventing:** The triathlon of all English equestrian sports: dressage, cross-country jumping, and stadium jumping.

■ **Field hunter:** A horse used primarily for foxhunting.

■ **Hunter over fences:** A horse that jumps a course of natural-looking obstacles, with good pace and form.

■ **Reining:** The western version of dressage; it has its origins in ranch-horse work.

■ **Saddle seat horse:** A horse that is shown in a flashy high-stepping trot and a smooth rocking-horse canter.

■ **Team penning:** Western sport in which riders separate numbered cattle from the herd and drive them into a square pen.

■ **Warmblood:** A sporthorse that is an amalgamation of selected coldblooded and hotblooded breeds.

■ **Western pleasure:** A western class judged on pleasant demeanor and smooth gaits.

were demonstrated in displays in front of emperors, kings, and queens to show them the might and power of their cavalries. These displays eventually evolved into competitions, and the sport's modern Grand Prix test upholds the traditions and methods of the dressage forefathers.

All breeds of horses can benefit from learning some dressage, as the principles help a horse learn to use his body to carry his rider in balance and harmony. A competitive dressage horse should possess a healthy conformation, which will allow him to remain sound throughout the many years of training required to be successful. He should have strong and well-muscled hindquarters, a neck that comes high out of the withers, shoulders with good angles, and a lovely freedom of movement. He should also have a

A Lipizzan jumps into the airs above the ground movement at a dressage performance by the Spanish Riding School in Vienna.

English, Western, and Driving Competition | **159**

Sidesaddle

The sight of a woman sitting astride a horse was thought to be extremely vulgar in medieval times—and also probably impossible to do in the long skirts in fashion at the time. The solution was for women to sit sideways. Over the centuries, the sidesaddle went through many changes, becoming safer and easier to use as time went on. For instance, the first sidesaddle had no setup to secure the legs. Then in 1580, Catherine de Medici designed a sidesaddle that had pommels to hold the legs of the rider in position. Over the following centuries, the saddle went through other changes. Even a western sidesaddle  was designed. Sidesaddles proved to be very popular in the United States. In 1897, the Sears, Roebuck and Company department store chain offered sidesaddles in its catalog.

Today, particularly in America, sidesaddle riding is extremely diverse. Sidesaddles are popular in gaited horse shows, saddle seat, western, and even jumping competitions. In fact, all disciplines can be ridden in a sidesaddle. Breed organizations for horses (such as Morgans, Arabians, and Paso Finos) offer classes specifically for sidesaddle riding. Unless the rules ban sidesaddles, you can even compete against astride riders in any class. Costumes vary from traditional English hunt attire to southern antebellum dresses.

Riders use their right legs to help stay in their sidesaddles by keeping equal pressure against the pommel through the right thigh and by pressing the right ankle against the horse's shoulder. A rider should sit squarely in the saddle with equal weight distributed through the seat, shoulders, and hips while facing forward.

A significant difference between aside and astride riding is the loss of the use of the right leg for cuing. Because the leg isn't gripping the right side, riders have to be more reliant on weight aids and voice cues. Some riders use a dressage whip to cue on the right side. A horse new to this style of riding will adjust to the sidesaddle change quickly; it is critical, though, that the saddle fits correctly.

willing nature, which is often described as *rideability*. The U.S. Dressage Federation states, "Dressage competition at various levels of achievement is the ultimate test of the training program to determine whether the desired harmony between horse and rider has been achieved." Showing is simply the icing on the cake.

Saddle Seat

A truly American form of English riding, saddle seat was developed to show off the flashy gaits of American show breeds, such as the Saddlebred and the Morgan.

All saddle seat riding is done on the flat, meaning there is no jumping involved. With head held high, neck upright and arched, ears pricked forward, and a show-off attitude, the saddle seat horse moves with a comfortable, high-stepping trot and a smooth rocking-horse canter. The saddle seat show classes were devised as a way to exhibit the beauty and movement of this type of horse. The horses with the most spectacular and high-stepping gaits are shown in the park horse division, while horses that have less flashy movement, yet still fulfill the other requirements of the saddle seat horse, are shown in pleasure classes.

The American Saddlebreds, with their smooth way of going and well-mannered disposition, are considered the quintessential saddle seat horse. Saddlebreds are shown in three-gaited (walk, trot, and canter) and five-gaited (walk, trot, slow gait, rack, and canter) classes. Morgans and Arabians, among other breeds,

An American Saddlebred shows the saddle seat high-stepping trot and proud form that earned rider and horse a blue ribbon.

have saddle seat divisions, as do gaited breeds, such as the Tennessee Walker.

WESTERN COMPETITION

Western competition includes western pleasure, working cow horse, reining, cutting, team penning, and team sorting events. The western disciplines primarily originated from ranching, where a good cow horse was a cowboy's legs. Although some of the western disciplines, such as western pleasure, are a far cry from a cowboy's original needs, others, such as cutting and team penning, are close to the way cow horses work today.

Western Pleasure

Western pleasure style stems from the ideal of what a western riding horse should be: a pleasure under the saddle. However, this very stylized type of horse isn't exactly what you'd ride on the trail. Like the English show hunter, the western pleasure horse is judged on his style and way of going. The ideal western pleasure horse will have a relaxed but collected cadence to his walk, jog, and lope, along with a pleasant and quiet demeanor. The western pleasure horse must also halt, stand quietly in the lineup, and back up willingly—all while on a loose rein. In some classes, the judge may ask for extended gaits. Not all horses can be truly competitive in pleasure; the horses that succeed have the correct conformation (born with a level topline), as well as a calm disposition and the ability to sustain the slow, controlled movement required of the discipline. Winning horses will have a flowing, balanced way of going and a willing attitude, which gives the appearance of being fit and a pleasure to ride.

Working Cow Horse

With beef in great demand for the first time in the nineteen century, a cow horse had to be nimble, quick, and obedient to get the job done, and he needed enough endurance to take his cowboy home at day's end. Traditional ranching methods are still practiced, and working cow horse competitions are increasingly popular. Known also as reined cow horse competition, this event could be considered a counterpart to English riding's combined trials because it has two phases, or sessions, in which the horse competes and completes two or three types of work.

The first competition consists of dry work, in which the horse and rider perform a compulsory set reining pattern. The content in this pattern is similar

A western pleasure class horse and rider circle the show arena in a relaxed, collected walk.

to that found in reining competitions, although there is less emphasis on slides and more on hard stops (see below). The second session, the cow work, utilizes a real cow. To begin, the competitor must take the single cow to one end of the arena and hold it, a move known as boxing. Then the competitor must take the cow along the fence (fencing). Finally the competitor must make the cow go into the center of the pen and hold it in a tight circle (circling). Some events add herd work that includes a cutting phase, in which the horse must cut out a cow from the herd and keep it away within a certain allotted time frame. The ideal reined cow horse is obedient and highly responsive to cues and possesses great cow sense, meaning the horse can read the cow's body language and respond to it with perfect speed, coordination, and balance.

Reining

The reining event, which won recognition from the Fédération Equestre Internationale (FEI) in 2000 and is governed by that organization, is one of the world's fastest-growing equestrian sports. It was added to the World Equestrian Games competition in 2002, which were held in Spain that year. Considered to be the western version of dressage, reining has its origins in ranch horse work. America's ranches in the nineteenth and twentieth centuries required the cowboys' ranch horses to be willing, agile, and brave around cattle, as well as extra responsive so that the horse would immediately react to the lightest of cues from the cowboy.

In competition, all work is done at a lope, and horses compete in a compulsory pattern of slow and fast circles,

At WEG 2006, Francois Gautier and Snow Gun execute the sliding stop, one of the most exciting maneuvers in the reining pattern.

rollbacks, flying lead changes, spins on the spot, a run-down to a sliding stop, backups, and a quiet pause, all while demonstrating tremendous agility and willingness. Judges award and deduct points for each maneuver, and they judge the patterns on precision, smoothness, finesse, and degree of difficulty. There is also a freestyle division, in which competitors are allowed to design their own patterns, set to music; they often perform in costume.

Cutting

The sport of cutting truly is like a roller coaster ride because the horse is the one in charge. Cutting began on the plains of the American West, where a rancher needed a good horse that could help separate cattle from the herd so the rancher could administer medications or move them to another spot. Local competitions soon sprang up, so cowboys could test their horses against others. In 1946, cutting became a true competitive event when the National Cutting Horse Association was created.

In cutting, the horse is judged on how well he cuts, or separates, an individual cow from its herd and holds it away within a specific time. The event starts with a small herd of cattle at one end of the arena. The horse and rider go into the pen and choose a cow to separate from the herd. Once the cow is cut from the herd, the rider puts his or her hands down, dropping rein contact, leaving the horse to do the work of keeping the cow from rejoining the herd. Cutting horses are bold, quick, agile, and sometimes aggressive with the cattle. A good cutting horse can intimidate the cow: going eye-to-eye, pinning back his ears, and sometimes even dropping to his knees to put pressure on the cow. The competing horse and rider have two-and-a-half minutes to cut at least two or more cows from the herd. Competitors are judged on a scale ranging from sixty to eighty, with seventy points being average.

Team Penning

Team penning began on a ranch in Ventura, California, in 1942; today it is a worldwide western sport, with participants throughout Canada, Europe, and Australia. Like other cow-related sports, team penning has its origins in cutting cattle from the herd to vaccinate, move, or castrate them. In team penning, a team of three riders has either sixty or ninety seconds to separate three specific cattle from a herd of thirty yearling cattle and drive them into a square pen at the arena's other end.

The cattle are identified by numbers—0 through 9—on collars or painted on their backs. When the announcer calls a random number, the riders go into action, identifying the three cattle bearing that number, cutting them from the herd, then driving them to the small pen. If the team lets more than five cows of other numbers cross the start line, the team will be disqualified; if any cows with the wrong numbers get in the pen, the team must take them out and send them back to the other side of the starting line before time is up.

Team Sorting

Known also as ranch sorting, this event is similar to team penning, except that it involves a pair of riders moving certain cattle from one pen into another, while keeping the rest of the herd from getting loose. It's usually performed in penning's off-season. The event starts with two round pens connected to each other and ten calves in each pen, numbered from 0 to 9. The clock starts once the judge raises the flag, and the riders start to sort out the cattle in numerical order, beginning with the random numbers they were assigned before the competition began. For example, if the number 6 is called, the riders must first sort out number 6, then 7, 8, 9, 0, 1, and so on. If a calf gets sorted out of order, the team is disqualified. Although there is a time limit, teams are judged on the number of cattle sorted, rather than on the time it took to sort them. As in other cattle events, horses that are brave, agile, and responsive to rider cues are the best mounts for the sport.

COMBINED DRIVING

It used to be said that, if a horse didn't jump and wasn't good at dressage, he could become a driving horse as a last resort. But driving has recently experienced the type of revival that dressage did in the 1980s. People are catching on to the fact that there's something very romantic and even exhilarating about driving. And one of the many benefits is that most breeds can drive because size doesn't matter. As in three-day eventing, the horse and driver team competes in three events: dressage, cross-country marathon, and cones.

Each phase tests the training and ability of the horse and the driver, who is called the whip. Dressage—the first phase—tests the horse's balance, flexibility, and submission, as well as the whip's ability to memorize and execute the test correctly. In the second phase, the marathon tests the horse's endurance and stamina over distance, as well as the team's ability to use their skills to negotiate obstacles on the course. Cones—the last phase—test the precision and concentration of both horse and whip, as they maneuver between sets of cones on a course without knocking any down.

A team of four negotiates the marathon course during the 2007 combined driving competition in Aachen, Germany.

Chapter 6

Rodeo and Other Horse Sports

odeo involves many different rough stock events, as well as timed events born out of the work of the vaqueros (Mexican cowboys) and the cowboys of the West. Although rodeo comprises some of the more controversial sports involving the use of horses and cattle, it remains part of America's Western heritage.

There are also several other kinds of activities and horse sports equestrians can participate in. These competitive activities test a horse and rider's endurance, gymnastic abilities, trail knowledge, quickness, and agility.

RODEO

Since the 1870s, equine welfare advocates and animal rights activists have taken issue with rodeo, saying that the events are cruel to animals. In response, the American rodeo industry, especially the Professional Rodeo Cowboys Association, which governs and sanctions American rodeos, has made strides to regulate rodeos and improve the lives of rough stock. Despite these changes, several animal rights groups in North America still oppose rodeo. Internationally, rodeo is banned in the United Kingdom and the Netherlands. Of course, western riding is less prevalent in Europe.

Barrel Racing

A crowd-pleasing event, barrel racing is the only event in rodeo dominated by women. Competitors gallop around three barrels (usually three 55-gallon metal or plastic drums) set in a triangle. The fastest horse with all barrels still standing at the end wins. Barrel racing tests the horse's athletic ability, as well as the rider's skill as she or he guides the horse through a cloverleaf pattern around the three barrels. Blazing speed is the draw for riders who love this adrenaline-filled sport, as most winning runs are completed in seventeen seconds or less.

Competitors need more than speed; they and their horses must have the ability to make tight turns and lead changes on the fly. If rider and horse get too exuberant and run off pattern, then they are disqualified. Time penalties are incurred when a barrel is knocked over, which usually means that the pair is knocked out of the money.

Team Roping

Team roping was developed out of the traditional ranching procedure used to secure a steer: two cowboys roped the front and hind ends of a steer and stretched the rope between their horses so that the animal could be branded or vaccinated. Known also as heading and heeling, team roping is the only rodeo event in which male and female riders can compete together in the arena.

At the beginning of the run, the steer is given a head start. Then one horse and rider pair (the header) attempts to lasso the steer's horns, while the other

Horse talk

RODEO AND OTHER HORSE SPORTS

Learning the definitions of these terms will help you to talk horse sports.

■ **Bronc:** Unbroken horse used in rodeo competitions.

■ **Bucking chute:** Area where a cowboy mounts the bucking horse or bull.

■ **Canter:** A three-beat gait.

■ **Chukka (also called chukker):** Polo matches are divided into six chukkas, which are seven and a half minutes each.

■ **Dallying:** Winding the lariat or rope around a western saddle's horn.

■ **Gymkhana:** An event for all ages of riders and breeds of horses that challenges skills through games, such as pole bending and egg-and-spoon race.

■ **Longeline:** A long line attached to the horse and handled by a longeur on the ground. The horse moves in a large circle around the longeur.

■ **Mark out:** A rodeo rider must touch the bronc with a spur at the shoulder before the horse's front legs touch the ground.

■ **Pickup crew:** A crew, mounted on horses, that releases the bronc's flank strap after a cowboy's ride is completed.

■ **Rough stock:** Denotes untrained animals ridden in rodeo events, such as saddle bronc, bareback bronc, and bull riding.

■ **Surcingle:** A leather roller that attaches around the horse. In vaulting, the surcingle has rings that riders utilize as they do their exercises.

■ **Vaulting:** A form of gymnastics on horseback.

horse and rider pair (the heeler) tries to lasso the steer's two hind legs.

The heeler finishes the run by quickly stopping his or her horse while simultaneously dallying the rope around the saddle horn. Once the animal has been captured, the riders face each other and lightly pull the steer between them, so that it loses its balance and lies down.

A team roping run is usually completed in less than fifteen seconds (although occasional runs are done in less than five). The team that performs its job the quickest wins the event.

Penalties can be added to the team's total elapsed time. A team is penalized if it starts before the steer has traveled the length of its allotted head start (called breaking the barrier or breaking out) or if the heeler can only rope one hind foot. If either roper misses the target, the team receives no score for the run (considered a no-time).

Rider and horse round a barrel in a fast, tight turn during a barrel racing competition.

A bareback bronc sends a rider flying back as he attempts to stay on for at least eight seconds.

Piggin' string clenched in his teeth, a tie-down roper drops his lariat over a fleeing calf. The horse readies for an instant stop.

Tie-Down Roping

Tie-down roping (formerly calf roping), the oldest timed event in rodeo, is based on the ranch's need to restrain calves for branding, castrating, or vaccinating. In this event, the calf is released from a chute and runs down the arena. The rider pursues and ropes the calf around the neck mid-run; the horse instantly stops and sets back, drawing the rope taut. The rider jumps off, runs to the calf, throws it down, and ties three of its four legs together with a short rope called a piggin' string. The shortest time wins. Winning runs can time within five to seven seconds. If the calf falls or the horse pulls it off its feet, the rider must wait until the calf rises to finish the run. A great calf-roping horse will have good speed and a good stop; he will instantly hit the slack of the lariat and back up to keep tension on the rope.

Saddle Bronc

Saddle bronc is one of the two rough stock events that involve horses. A saddle bronc is either a horse that has been bred for the event (who exhibits strength, agility, and bucking ability) or a horse that has proven to be unsuitable for normal saddle riding. The event's origins are in the Old West's society's method of breaking horses to saddle—when a cowboy would saddle a horse, mount up, and ride the buck out.

In saddle bronc riding, the horse is first put into a bucking chute; there, the rider gets on, using a special type of hornless saddle with stirrups that move freely from the rigging. When the chute opens and the horse begins bucking, the rider must stay on the bronc for eight seconds, holding on to only the braided rope that is attached to the bucking halter. Finding a rhythm with the bucks, the rider has to "mark out" the horse by touching him with a spur at the shoulder before the horse's front legs touch the ground. The scores are determined by how well riders mark out their mounts, whether they stay on for the full eight seconds, and how wildly the horses buck.

Bareback Bronc

Bareback bronc is similar to saddle bronc in that the rider must mark out the bucking horse and stay on him for eight seconds. However, instead of a saddle, the bareback bronc rider only uses a special rigging to hold on to. Both the bareback bronc and the saddle bronc are made to buck higher and more aggressively by a flank strap, a sheepskin-covered leather strap placed toward the back of the abdomen. Animal rights advocates claim this strap is used in a way that induces pain to increase bucking; however, rodeo advocates counter that a horse in pain cannot perform his job and will actually stop bucking. Once the rider has come off the horse, a pickup crew releases the flank strap.

OTHER ACTIVITIES

Other activities include endurance riding, competitive trail riding, vaulting, gymkhana/mounted games, fox hunting, polo, ride and tie, and jousting. Many of these are very ancient and traditional activities.

Endurance

The marathon runners of the horse world are found in the endurance division. Recognized by the Fédération Equestre Internationale (FEI), endurance riding events can cover 50 to 100 miles or, on limited-distance rides, 25 to 35 miles. A rider must finish the course on a fit and healthy horse within a maximum time limit.

At required rest stops throughout the course (called vet gates or vet checks), veterinarians test the health of the horses, checking pulse and respiration while also checking for soundness and dehydration. During the stops, the horses' heart rates must return to a lower resting phase, about sixty-four beats per minute. Time continues to accrue throughout the checks, so a rider

should make sure that he or she has well-conditioned horse, one that will recover quickly at each vet gate and be released sooner to continue the race. Horses found to be lame or in distress are pulled from the race.

In addition to a first-place reward, competitors can win a top ten award (given to the first ten horse and rider pairs to finish) and the best condition (BC) award. The BC award is given to the pair that finishes in the top ten with the horse in the best condition at the end of the race. Condition is determined by how much weight the horse carried and how well he recovered in the vet gates. So, if a pair did not win the race but carried more weight than the winner and received the same vet scores as the winner, that pair could receive the BC award.

Like marathon runners, good endurance horses have a long, lean look, rather than the bulky, muscular build common in sprinters. Those long, slow-twitch muscles allow the horse to keep moving, mile after mile. A comfortable stride is nice for the rider, but an endurance horse must possess strong legs and tough hooves that hold up on rocky paths and uneven footing. Breeds with good-sized hearts and lungs, such as the Arabian, are most successful in endurance riding.

Competitive Trail Riding

For riders who love riding the trail but don't want to race to the finish, competitive trail riding is a great choice. It is a judged trail ride in which competitors are asked to travel a natural trail for a set distance

Four competitors make their way through the desert, competing in a challenging endurance race held in Dubai.

Frenchmen Nicolas Andreani performs freestyle vaulting on a cantering horse at the 2010 WEG.

A young girl competes in a timed obstacle race at a National Pony Club gymkhana with mounted games.

(usually 10 to 15 miles). Throughout the course, the horse and rider teams must stop to negotiate different obstacles, usually with varying degrees of difficulty. Higher scores are awarded for completing the more difficult obstacles.

The horses are evaluated on their performance, manners, condition, soundness, and trail ability; the riders on their equitation and horsemanship. Riders are usually asked to complete the ride at various gaits, although the trot is used most often. Competitors set their own pace in the specified gait for their division. The judges observe competitors at various points along the trail. The horse's pulse and respiration are checked periodically for recovery ability and conditioning, during mandatory holds/lunch stops. These holds/stops generally last between 10 and 20 minutes, although they can be longer depending on ride management. When all the riders have completed the final checkout, scores are tallied, an award ceremony is held, and all riders receive their scorecards.

Vaulting

Often considered a form of gymnastics on horseback, the sport of vaulting is also recognized by the FEI, and it has small pockets of passionate competitors worldwide. In vaulting, the horse is not ridden, but rather used as an apparatus. A very steady, usually larger breed of horse is put on a longeline at the canter. The horse wears a surcingle (or a roller) with

special handles and a thick back pad. The handler, or longeur, will be in control of the horse from the ground. As the horse circles, the vaulter will demonstrate various moves and positions on the back of the horse. There are six compulsory exercises (basic seat, flag, mill, scissors, stand, and flank), in addition to the mount and dismount, and a freestyle performance. Each exercise is scored on a scale of 0 to 10. Horses also receive a score. They are judged on the quality of their gaits. When done at the highest levels, vaulting is a beautiful art.

Gymkhana/Mounted Games

Usually, young riders participate in a gymkhana, which comprises a host of different games on horseback. Pony clubs and 4-H groups often hold mounted games for their members. Gymkhanas, also called play days or O-Mok-See (for the event's Blackfeet origins in America), have roots all over the world, from India to Europe. The word itself, *gymkhana,* is of Indian origin.

Gymkhana classes are timed speed events that can include barrel racing, keyhole racing, stake racing, pole bending, and figure eight competitions. Other gymkhana games, such as egg-and-spoon races and ride-a-buck competitions, also test riders' skills and horsemanship. There are even costume classes and opportunities for competitors to dress their horses and ponies in fun and original outfits.

Members of the 160-year-old Ledbury Hunt follow an articial scent. England banned the hunting of foxes by hounds in 2005.

Fox Hunting

Fox hunting certainly has its fill of proponents, as well as detractors, in contemporary society. The sport of fox hunting—which often consists of tracking and chasing a fox (or perhaps a coyote or a bobcat) with members of a hunt group on horseback and a pack of hounds—is rife with debate. Many of today's fox hunts, however, do not involve hunting a real animal at all. Instead, a scent is dragged over the terrain and that's what the hounds follow.

Fox hunting was developed centuries ago as a way to control the vermin population in rural communities, at a time when foxes were overrunning the countryside. In the hunts, the hounds killed the foxes. Today, for welfare reasons, the game is hardly ever killed—in fact, the chance a hunter will even catch sight of the game is slim. Usually, the hounds are called off the scent before game is sighted.

In the United States, it's more of a fox chase—a day out in the countryside, jumping natural obstacles and enjoying the thrill of the pursuit. Because these chases follow strict protocol and a host of rules, hunts are usually private or by invitation only. As a rule, foxes are not chased once they have gone to ground. Fox hunters usually accept stewardship of the land and make an effort to preserve fox populations and territories as much as possible.

Polo

The action-packed game of polo is arguably the oldest equestrian sport in recorded history. Records go back as far as 2,500 years ago, when ancient Persians created a game called Chaughan. The Chinese played a similar game on horseback that dates back several thousand years. Most accounts note that ancient civilizations learned about the sport from Central Asian nomads. British tea planters in India saw the game played in the early 1800s in Manipur, but it was not until the mid-nineteenth century that the British cavalry created a rule book for the sport. In 1866, British tea planters in India formed the first polo club in the world, the Calcutta Polo Club.

Today's polo is played between two teams of four, on a large, soccer pitch-type field complete with goalposts. The team scoring the most goals in a six-period game wins. Each period, called a chukka, is seven and a half minutes long. Time-outs are only allowed for penalties or for a player or a horse's injury, and no substitutions are allowed unless a player or horse must be replaced.

Polo horses must be very fit to endure the constant galloping, stopping, and reversing required during the games. They are traditionally referred to as ponies, although full-sized horses are used. In the sport, the pony is credited with being responsible for at least 80 percent of a player's effectiveness. Occasionally, an exceptional pony will play in two chukkas with one or more chukkas in between to rest. Because each rider needs multiple horses to play for the duration of the game, polo is typically a sport for the wealthy, however, arena polo is played in a smaller, enclosed space and requires fewer horses, making the sport more affordable. There is also a game called polo-crosse which is a hybrid of polo and lacrosse. Only one horse is required per rider.

Ride and Tie

Most popular in the western states, the thirty-year-old sport of ride and tie is the ultimate activity for the fitness rider. Ride and tie requires long-distance running, endurance riding, and keen strategy. Three team members are needed for ride and tie competitions: one horse, one rider, and two riders/runners. One member begins the competition mounted, and the other starts on foot (either running or walking). The first rider travels as far as he thinks the runner can reach; then he or she dismounts, finds a safe place to tie the horse, and sets off on foot. When the first runner reaches the horse, he or she mounts up, rides as far as he or she thinks the second runner can go, then dismounts, ties the horse, and begins to run again.

There are no set rules regarding how far each runner must go; it's all based on the strengths and weaknesses of the team. Races are usually 20 to 40 miles long, although shortened races are also held for fun. The field can include anywhere from 20 to 100 teams, so the first three miles are often dusty, crowded, and exhilarating. After the start, the trail narrows, and runners and horses find their own pace. All three members of a team must finish before their time is tallied. Most teams like to finish together, but it is up to each team to decide whether or not to do so.

Jousting

Jousting first began during the Middle Ages as a civilized way for the aristocracy to demonstrate the knights' might. In jousting competitions, knights fought each other with lances. But these war games had a price. Knights were often terribly injured and disfigured; some even killed, including King Henry II of France. With such common casualties, a new display of might was eventually introduced. Enter the Running of the Rings in the seventeenth century. This sport of collecting three rings on a shortened lance

Players enjoy an unusual game of snow polo. Riders have played the sport of polo across the world for thousands of years.

A colorfully costumed jouster and horse bring the centuries-old sport to life in a modern-day tournament.

was a way to show control and a steady hand, determining the best soldier around. Jousting still exists today; in fact, in the 1960s, American revivalists of the sport passed a bill making jousting (the three-ring kind) the state sport of Maryland.

Modern-day jousting is a family sport, set up for novice, amateur, semipro, and pro levels. The jousting track is about 80 yards long. Three arches with rings hanging from metal arms are set equal distances over the track. The rings have inside diameters measuring from ¼ inch to 1¾ inches. Riders gallop down the track in a half-seat or galloping position and try to spear the rings. The fastest competitor to get all rings wins.

It is not a reenactment sport, although costumes can be worn during the opening ceremonies. Most competitors opt for comfortable riding clothes, such as riding boots and breeches. Usually, English-type saddles are used. All breeds are suitable for jousting, but a speedy, small horse with an easygoing temperament is most desirable.

A New Horse

Section 4

H

aving a horse of their own—that's a dream come true for many horse-crazy boys and girls. Adults, too, who never lost the desire to have a horse can realize this wonderful dream.

It's important that a rider, especially a novice, choose the right horse. The last thing a green rider needs is a green horse. That combination won't benefit rider or horse. Experienced riders also have to make sure that they choose the right horse, one that will allow them to pursue their goals, whether that be perfecting their riding skills or competing. As important as choosing the right horse is choosing the right equipment. From saddles to protective leg wear, a horse needs to be appropriately and safely outfitted. The rider, too, needs safe and appropriate riding apparel.

Chapter 7

Choosing the Right Horse

Every rider has a mental image of his or her ideal horse. The characteristics of this dream horse may include a spirited nature, fancy gaits, and a proud carriage—the perfect description of the horses that trot over the pages of glossy calendars. Very rarely will attributes like easygoing nature, sturdy gaits, plain looks, and well trained make the list of desired characteristics; if they do, they are somewhere at the bottom. Sadly, too many people put an emphasis on a horse's exterior beauty rather than on his capabilities. They hanker after the spirited horses of their childhood dreams.

But a horse with a wild spirit can be a bad choice, especially for the novice rider. To safely learn the craft of riding, a beginner needs a horse with very particular skills. Spirit and beauty have to take a back seat to those skills.

Like novices, experienced riders need horses with the right set of skills to help them reach their riding goals. If a rider wants to compete in dressage, he or she should look for a horse with those skills, not one that would do better in eventing. A horse meant to carry riders of different skill levels should be more of a generalist or an all-around horse.

If you going to buy a horse, begin by honestly evaluating your own skills. Then think about the kind of horse that will help you to improve those skills and reach your goals. Research the different horse breeds and types, then research the places to find them. Some avenues are better than others. Take your time. Don't rush out to find the wrong horse, when the right horse can be yours with some extra time and planning.

A HORSE FOR THE NOVICE RIDER

A well-trained horse is indispensable for advancing a novice rider's skills. If you are a novice, you will find that a well-trained horse will provide you with a solid and generous background so you can concentrate on your position, balance, and timing. You can develop these skills without worrying about your safety or whether the horse is likely to misunderstand your cues.

If you're riding a horse that's ill-suited for your present abilities, you won't be learning—you'll be in survival mode the whole time. Instead of honing your skills, you will be fighting just to stay on the horse. A rider can't learn how to relax and follow the horse's movement if the horse is unschooled or has gaits that are difficult to ride.

Fear plays a big role in a rider's early days. Novice riders quickly begin to understand the strength of their horses and how fast they can move away when they perceive danger. Not having the experience to interpret what is going on in the horse's mind can lead to a crisis in confidence with a horse not suited to a novice. Whether or not the rider is injured in a fall, the fear factor will slowly break down the success of the partnership.

The Wrong Horse

One of the biggest mistakes that novices make is purchasing a green (untrained or in early training) horse, thinking rider and horse will learn together. This is a disaster in the making. A young horse needs guidance and consistency, which someone new to the saddle just can't provide. A beginning rider must have a horse that knows his job and is patient and forgiving—characteristics unlikely to be found in a green horse. In any riding pair, someone must be the leader or the teacher. This could be a well-schooled horse with a novice rider or a green horse with an experienced rider.

The bottom line is that a horse needs support from his rider, and he can't learn what the rider doesn't know. Horse and rider can certainly learn together, but when they do, they're likely to learn all the wrong things. The horse learns all the rider's problems, including fear, hesitation, and insecurity.

On the other hand, is it possible for a novice rider to buy a horse that knows too much? The answer is yes. A highly trained horse—one so sensitive that the slight movement and weight shift on the rider's part means something specific—can easily get confused by mixed or unintentional rider signals. Overbuying is common among riders who are looking for push-button horses and instant success. If a well-trained horse becomes dull to your aids, by the time you get good enough to start to learn more, you may discover that the horse's skills in certain areas are no longer easy to access; the horse may even require expensive retraining. To become a good rider, you must take the time to find a horse that will help you learn.

The Right Horse

Ideally, you should find a sound horse that has a good disposition and a level of training just slightly more advanced than your skill level. Then you can catch up to the horse without confusing him. Eventually, when your skills have come far enough, you can even teach the horse more skills.

Consult your veterinarian for advice before purchasing a horse, but evaluating the following aspects will help you narrow down your decision.

Health: Horses with chronic ailments—such as navicular disease or respiratory disorders—are often offered up for sale very cheaply. These horses are unsuitable mounts because their illnesses often prevent them from being ridden.

HORSE CHARACTERISTICS

Learning the following terms will help you talk horses and horse characteristics:

■ **Crib biting:** This vice, which is also called cribbing, is when the horse puts his front teeth on a surface, such as the fence or the stall door, and sucks in air. Some studies link this behavior to gastric ulcers.

■ **Dishing:** The hooves paddle outward as the horse moves, rather than in a straight line.

■ **Moving close behind:** The hind legs travel close together in a narrow stance.

■ **Schoolmaster:** A very experienced horse, usually a specialist in a discipline or movement.

■ **Stall walking:** A vice in which the horse paces his stall. Usually caused by boredom or loneliness.

■ **Stall weaving:** Similar to stall walking; however, in this vice, the horse bobs back and forth, swinging his neck from side to side. Usually caused by boredom or loneliness.

■ **Upside down neck:** When the bottom neck muscle is stronger than the top muscle.

Movement/gait quality: Horses should have true, honest gaits. Once they strike a canter, they should be cantering solidly. Once they trot, they should be moving so the rider can feel the rhythm.

Size: All too often, a rider will purchase a horse that is too small or too big for that particular rider, which makes it very difficult to ride properly. If the horse is too big, the rider won't be able to sit the gaits or apply his or her leg aids easily. If the horse is too small, the rider won't be able to put his or her legs in the right position.

To provide optimum performance ability, a rider needs to choose an appropriately sized mount, as this rider has with her pony.

Temperament: The horse should like people. Everyone wants a companion and not a sour old thing that dreads your presence. And the horse should be kind and forgiving. In other words, he should be mentally suited to deal with a novice rider. The horse shouldn't hold it against you if you do something wrong, such as using conflicting aids, like using your legs and the reins at the same time. Conflicting aids may cause a less-than-tolerant horse to buck out of frustration. A patient, schoolmaster-type horse is ideal for the beginning rider.

Training: A good novice horse must know his job and know it well. He shouldn't be too quick in his responses, thereby giving his rider time to make mistakes and learn from them. In addition, the discipline a rider wants to compete in should be the discipline the horse is trained in, whether it be dressage, jumping, or barrel racing. Not only is it difficult to learn a certain discipline on a horse with none of those skills, but also horses trained to certain disciplines, such as barrel racing, are too energetic for more refined disciplines, such as dressage.

In addition, a novice's horse shouldn't require harsh equipment or training devices. If the present owner is using strong hardware often, it's an indication that the horse does not have solid training and the owner must rely on equipment for control.

The Trade-Offs

There are always trade-offs and compromises with the purchase of any horse. Because a novice rider's horse must have certain attributes, you need to make certain

KEEPING UP A HORSE'S SKILLS

If you're a novice rider with an advanced horse, one way to ensure that he won't lose skills while you're learning is to ride him under a trainer's supervision and take lots of lessons. Consider putting your horse in training, as well. Your trainer will be able to keep the horse's skills up until you reach his level. Yes, this method is expensive, but then so was your advanced horse.

The flashy appearance of this pinto is appealing, but coat color shouldn't be the deciding factor when you buy.

that the horse you choose has them and let other, less important desired qualities go. It will take much longer to find your horse, perhaps years longer, and cost much more if your list of necessary qualities goes on to specify that the horse must also be a mare, six years old, 16.2 hands high, and black with four white socks and a blaze. Some traits that you think are nonnegotiable really are irrelevant to the ultimate goal: finding the best horse for your needs.

Age: A horse doesn't have to be young to be a good performer. A horse can perform well into his teens. In fact, recent studies have shown that horses are living longer, healthier lives and can continue working into their twenties. Don't narrow your search too much by insisting on a six-year-old.

Appearance: People tend to want flashy colors and markings, and they stay away from the plain brown horses, but specifying color and markings will drastically reduce the number of equine candidates. There is an old saying in the horse world: "You don't ride the head." In other words, beauty isn't a function of a good saddle horse. The best horse for you may not win any beauty pageants, but he could be the smartest and most generous horse in the barn.

Breed: Sticking to a specific breed may also extend your search time, particularly if you hanker after a rare breed (which will cost more, too). Although some

STEPPING-STONE HORSES

You should approach your riding education as a constantly changing journey. People often want one horse for their entire riding careers, and although it is admirable to want to keep your horse with you forever, this can take a toll on your riding skills. If you want to learn the basics and are happy in that state and have found a horse that you love and cherish, by all means, keep him forever. There's nothing quite like the deep friendship and trust that a rider and a horse can develop over many years of working together and enjoying one another's company. But if your dreams are to hone your skills, and then show or move on to another discipline, you'll have to think of your horse as a stepping-stone. You'll ride him only until you've both learned as much as possible. Your old horse will still have a lot to give, and if you can't keep more than one horse, you can find good people looking for a horse like him.

breeds may have certain characteristics that aren't suitable for a novice, such as a hotter temperament, this isn't a hard and fast rule. There are wonderful, kind starter horses in all breeds.

Movement quality: You may have to forget about that beautiful, prancing trot and the rocking-horse canter that you've always dreamed of. Good, workmanlike movement may take the place of that floating action. This, however, is actually a good thing, especially for a novice rider. A horse shouldn't move bigger than what you can sit and control.

Soundness: The novice's horse may show some wear and tear caused by teaching people over the years. He may, for example, have joint issues such as arthritis. But this doesn't mean he won't be perfectly usable over the next few years with a little tender loving care from you. Soundness isn't an absolute term; it's a relative one. Sometimes the difference between unsound and sound is a matter of the workload. A horse that can no longer stay sound while jumping five-foot courses may be an ideal two-foot horse for a rider who wants to learn to jump.

HORSES FOR FAMILIES AND EXPERIENCED RIDERS

For riders who aren't novices, the horse market is a much wider one. Knowing what kind of horse you are looking for before you begin shopping will help narrow the margins. Horses fall into certain training categories, including all-around horse, green horse, competition-ready, and schoolmaster. Consider carefully which type of horse fits your needs. It's a waste of time, for example, to look at a schoolmaster when you want to make your mark on a competition horse.

A rider looking for a horse the whole family can enjoy will probably want to steer clear of a green/young horse. Dad, learning to ride, won't appreciate the steering and balance challenges inherent in an unschooled mount. In turn, a green horse won't appreciate having his mouth tugged on and will grow to dread the riding sessions. An all-around horse would be a better choice for a family with members at different skill levels.

The All-Around Horse

The all-around horse is the jack-of-all-trades of the horse world. He's a willing companion who's ready to do just about anything, such as jumping, trail riding,

These turned-in toes show poor conformation. Check hooves carefully when buying a horse.

and dressage. An all-around horse is the sort of horse that can be ridden by any member of the family, within reason, and has the ability to compete at a local show. But he won't be a superstar. When considering an all-round horse, look for the following traits.

Conformation: Because conformation can vary widely with an all-around horse, you need to follow a general guideline, rather than a specific one, when you are assessing a horse's conformation. The horse's build should make you feel secure in the saddle. He should have a decent front, which means having plenty of shoulder and an ample neck. His hooves should be all one size, with a good heel and foot angle. Dishing badly (a circular action in front) and moving close behind (when the back legs don't follow the same track as the front) both show conformational weaknesses. All-around horses can be found in almost every breed.

Safety: To find out if safety is one of the horse's traits, put him to the test. Ride the horse by anything that's different, such as a garbage can lying on its side, a flapping tarp, or a pile of colored jump poles. Any horse might shy at something new, but if the horse's reaction is too quick, then he may not be as safe as you require.

Another way to evaluate a horse's reactions is to test his trail safety. Ask to take the horse out on the trail, both alone and with another horse. The bottom line: you should feel safe and secure on the horse, on the trail and off.

Temperament: A rider with one of these versatile mounts shouldn't have to be one step ahead, as is often the case when you're dealing with a young horse or a competition horse. Competition horses frequently have a quirk because they have that edge to them. You don't want quirks with an all-around horse. How can you tell if a horse has a good temperament? Go into his stall and groom him. Watch how he handles the

attention. He should be happy with you in his stall and not try to protect his territory. Next, take him out and walk him outside. He shouldn't barge off; instead, he should walk quietly beside you. Tack and untack the horse as if he were your own, and evaluate his reactions. Make sure you can pick his feet up and clean them and that he doesn't kick out.

Be prepared to take some time to find your horse, as good all-around horses don't come on the market every day. People often hang on to an all-around horse because he can do a bit of everything. When you are buying an all-around horse, you are buying a horse for his temperament and suitability. Keep in mind that there are many levels of all-around horses, and one with certain talents is often priced higher.

The Green Horse

The green horse is a clean slate, untouched by another rider's training. Green horses usually need experienced riders and handlers to help integrate them into Humansville. For the right rider, a green horse is a great project. The successful training of a green horse is an achievement to be proud of, as it reflects well

A rider needs to consider his or her own level of riding before deciding what type of horse to purchase.

ATTACHMENT TO GREEN HORSES

Don't get too emotionally attached to your green horse. It's very easy to get a vision about your goals when choosing a new project. But as you go through the horse's training, you may find that he's not the jumper you hoped he'd become. Perhaps he's showing more talent for dressage. Understand that buying a young horse is not much more than a gamble, and you may end up selling him and starting again.

on horsemanship skills acquired through hard work, practice, and training. But a green horse is nearer to his instinctive behavior than is a trained horse, and with the wrong rider, he can be easily upset and ruined by incorrect training methods.

Assessing a green horse is different from assessing an educated one. A horse may have only just been started under saddle or perhaps hasn't been ridden at all, but there are still ways to tell if a horse has the temperament you desire. It's best to check him out thoroughly before you buy rather than struggle with him and go through the heartache of having to cope with a bad temperament or making the decision to sell him.

Training acceptance: If the horse hasn't been ridden, you can assess what the horse will be like under saddle by hand-walking him. If he's nice and light on the lead, then he'll probably be nice and light under saddle. If he's a thug to lead, then he's probably going to be a thug to ride. If he's inattentive while leading, then he's probably going to be inattentive under saddle. If he's new to going under saddle, make sure he is happy to work. He shouldn't toss his head, hump his back, swish his tail, wrinkle his nose, or pin his ears. While moving forward, he shouldn't resist your aids or hesitate when you put your leg on.

Conformation: It's often said that it's difficult to evaluate conformation in a young horse, but that's not necessarily the case. Normally you can tell whether a horse is balanced in his conformation, regardless of the age. Some horses may look gawky at certain stages of their development, but horses that have nice conformation normally have good conformation from the beginning. If a horse has a short neck as a yearling or a two-year-old, chances are he will still have that short neck when he's five.

Avoid a horse with a *ewe* neck (which means his neck is heavier at the underside than on the top) or a horse with a short neck, as short-necked horses are often too strong to ride. The horse should have balanced front and back ends. He shouldn't look as if the back end belongs to a bigger horse and the front to a smaller horse. This will affect how easy or difficult he'll find the work and how easy he'll be to ride.

Don't be blinded by a horse's beauty. Don't fall in love with a pretty face and miss the conformational faults. Remember the type of horse you want, as well. If you want to do dressage, look for a good mover that's built uphill. If you want to spin and slide in reining patterns, you should buy a horse with the aptitude and conformation to do that. We all can fall into the trap of asking our horses to excel at something they are not built to do.

Ability to stay sound: On the ground, turn the horse around on his forehand to see how he crosses his back legs. If he resists moving in one way, this can alert you to a weakness on that hind leg. While you're riding, check to see if the horse will turn. If he resists turning in one way, that will demonstrate weakness on the same side of his body. Ask to see the horse trotted in hand. Is the rhythm even? If it's not, this weakness may hold the horse back as you advance in his training. The horse should stop and go when you ask him to, not five strides later, even if he has not been under saddle long. Starting and stopping isn't just about training; it's about balance and conformation. A bad conformation will affect how he's able to balance the rider.

Temperament: You can evaluate the horse's temperament through touch. See if the horse will let you touch him all over. A horse that won't let you touch him, especially where your leg and tack would touch—the back, barrel, head, ears—may be difficult to work with.

The Competition Horse

No matter what the discipline, competition horses are first and foremost athletes, which is why they are often called sport horses. The high-performance horse is a brave, big-hearted, honest, cool-headed, and athletic animal; and above all, one bursting with talent and enthusiasm. This type of equine athlete is the perfect choice for riders who have high aspirations, a confident nature, and a desire to reach their competition goals. When considering a competition horse, look at the following traits.

An athletic horse and rider compete in an eventing competition. Buy a horse talented in the discipline you plan to pursue.

Adaptable: As a competitive rider, you're going to be traveling to lots of places, with different stables and arenas, and it is vital that your horse be willing to adjust. Your life will be much easier if your horse doesn't come unglued every time you leave your farm. You should be able to load him and tie him to the side of the trailer without worrying that he is going to pull back and run off.

To check a horse's comfort levels, ask to take him to another place, even back to your home stable. This is a great way to find out how adaptable the horse is. The seller may not consent, however, particularly if there are many people scheduled to look at the horse. Others in the equestrian community may know the horse. Ask your trainer or other people if they've seen the horse at a competition and what they have observed.

Conformation: When looking for a competitive mount, make sure the horse's conformation is relevant to your sport. Different conformation is predisposed to different disciplines. Regardless of the sport, an equine athlete should have well-shaped and healthy feet, and a good depth of girth for plenty of lung room. He shouldn't be overmuscled in the front because it will

be hard for him to come back on his haunches, which is important for any sport. The set of the neck depends upon the sport. Western pleasure requires a level profile so these horses have a low-set neck. Dressage riders prefer a high-set neck, which makes it easier for a horse to move in a correct outline. The neck should be long enough to give the rider a good length of rein.

An equine athlete has to have enough bone to cope with his body mass. The horse should be built uphill and back on his haunches, with his hock set well underneath him. He shouldn't have too long or too sloping a pastern. But then again not too upright, which would give a jarring action. He should have a nice flat knee. If he's back at the knee, his tendons could be put under too much stress. An athlete should have a nice strong forearm, a sloping shoulder, and a shorter back, which allows the horse to move well and be comfortable to ride.

Personality match: Some of the best horse and rider partnerships in the world are the result of darn good matches. The horse should suit your personality because you'll get along with him better. If you have a burning desire to achieve, then try to find an outgoing horse that likes to go, rather than a timid horse that looks to you for confidence. Quiz the seller on the horse's personality. Do you like what you hear? Imagine yourself at a competition, and try picturing how a partnership with that horse will develop. Do you think that personality will suit you?

Talented in his discipline: If it's a dressage horse you're after, he should move well and have the ability to do lateral work. If it's a jumper you want, he should have a good technique over fences, with a brave yet careful jump. Watch the seller ride the horse first, and ask to see the horse put through all the movements or over all the types of jumps the seller says the horse is familiar with. If he seems reluctant or doesn't have a clue as to what is being asked of him, he may not have the talent the seller claims. Yet don't discount a horse just because he isn't well schooled. Maybe the owner isn't experienced enough or the horse needs a tune-up. If you are an advanced rider and have the confidence and know-how, you may be able to train him yourself. When trying the horse out yourself, push him a little and see if he will let you school him.

Willingness to learn: Even if the horse is already trained to a certain level, as a competitive rider you're going to want to put your stamp on him, whether to improve his way of going, move him up a level, or introduce new jumping challenges. He has to be willing to change to your riding system. Does he handle the change easily, or does he get anxious?

You may need to travel far and wide to find your new competition partner, particularly if you're looking for an advanced horse. Top-class horses aren't often found around the corner. Be prepared to look all over the country and possibly even abroad. In fact, there are sellers who specialize in importing competition horses from other countries; ask a trainer, instructor, or fellow competitor if he or she knows of horses for sale. Competitions are a good place to find horses too. If a horse has caught your eye, go up to the rider and ask if the horse is for sale. If he isn't, ask if there is another horse like him.

To get the right horse, you will, of course, need to do a lot of research on the horses you're thinking of buying. If the horse has competed a lot, make sure to check out his show record. Ask to see photographs, test scores, ribbons, and even competition videos.

It is also important to understand that a high-level competition horse is much more expensive to ride and maintain than an ordinary horse. You need a lot more help to reach your goals, so you may be spending money on training and more lessons, not to mention the expenses associated with registration and entry fees to competitions and the extra care and maintenance of the horse to keep him performing at his peak.

The Schoolmaster

A schoolmaster is a horse that has the experience and the ability to help riders learn and perfect their competition skills. The rider who benefits from a schoolmaster is one who is just learning what's involved in a certain discipline, such as dressage movements, jumping, or cross-country. If you are looking for a horse with whom you can learn dressage, find a horse that has competed in that discipline. Don't make the mistake of assuming you can retrain a schoolmaster to be a competition horse; a schoolmaster is unlikely to be advanceable or retrainable. Learn what you can from him, then move on to a competition horse. No matter what discipline you are training for, all schoolmaster horses can be evaluated by looking at the following traits.

Age: A schoolmaster horse is unlikely to be younger than twelve years old. For this type of horse, age is not a major factor. More experience means an older horse, and therefore the horse may be stiffer and need extra time to warm up. However, if the

An English rider and her horse train in the early morning. Whatever horse you buy, take every opportunity to establish a bond.

horse is fit and well, he may offer you many years of learning and enjoyment.

Conformation: The most important conformational points when considering a schoolmaster are those that will affect soundness and may make the horse uncomfortable to ride. If the horse has a stiff back, he'll be bouncy and jarring in his movements, which will affect his usefulness as a schoolmaster. A horse with a long back, upright pasterns, or hocks set out behind will create an uncomfortable ride. A horse with poorly formed hocks, such as sickle hocks or tied-in hocks, or one that moves close behind, shows a weak conformation, which will create soundness problems as the years go by.

Good equitation: A schoolmaster's purpose is to help you perfect your riding skills. If you have to distort your body to ride the horse's movements, or if you are unable to maintain the right position for your discipline, you won't progress as you should.

Reactions: A schoolmaster shouldn't be too quick in his reactions, yet there must be a balance between willingness and quickness of reaction. No matter what the horse has to offer, if you don't feel secure with him or you're frightened of his reactions, you won't learn—particularly if you are a nervous rider.

Soundness: Relative soundness is an issue that you have to consider because many times a horse becomes a schoolmaster because he's no longer sound enough for competition. He may need medications or supplements to help him along, but he can still serve a useful purpose as a schoolmaster. Accept that you may have to invest some money, especially in veterinary services, so his life is comfortable. Ask the seller if there are any medications or if there is special care that the horse needs. It may be a bad idea to make a radical change or remove the horse from his health regimen, so make sure you can afford his special upkeep.

Specialty skills: Schoolmasters are often specialists in certain skills but are not so good in others. The horse may be incredible at lead changes but not so good at sliding stops. Perhaps the horse is a very good show jumper, but maybe he has a weakness over certain jumps or he's limited in his scope. Make sure the horse you are looking at is good at what you want to learn.

Training: A good schoolmaster should be able to understand correct aids; even if his response is only modest. You should not have to exaggerate your aids, or use incorrect cues. When you try out the horse, make

sure your aids are as correct as possible. Better yet, bring your instructor along to try the horse first. If he responds to unorthodox cues, you won't learn properly. Conversely, you don't want the horse to respond to any sloppy aid. A good schoolmaster will ignore incorrect aids, only obeying correct ones. Don't expect an easy, simple ride if you are riding incorrectly.

Willingness: The horse may not be the best mover in the world, but if he accepts the rider's correct aids and keeps thinking forward no matter what, then the rider will be able to learn.

Make sure the horse is indeed a schoolmaster. Many horses are passed off in the marketplace as being schoolmasters when they don't really qualify. Often, a horse labeled as a schoolmaster is really a dull and nonreactive horse or a horse whose training has gone wrong. This type of horse may have a distorted view of what the training should be. To ride such horses is usually a horrible experience because they are often resistant, tense, and confused and, as a result, often have impaired gaits and are awkward and uncomfortable to ride.

FINDING YOUR HORSE

Locating a suitable horse can be an overwhelming endeavor. There are no shops to visit to compare prices or quality, and there are no weekend super sales. Worst of all, the horse market can be just like a secret underground, with only the insiders aware of the tricks of the trade. Knowing where to look can help crack the mysteries of the market.

Finding a good beginner's horse can be especially challenging. Good first horses are golden; people who own them tend to keep them forever, or pass them along to younger family members or a friend. Often they are available only through word-of-mouth, so visit stables in your area where people take lessons, and call up trainers to inquire about horses for sale.

Where to Look

The best way to find your horse is to enlist the help of your trainer. He or she knows your abilities best and is often privy to word-of-mouth sales. Trainers are usually paid for this kind of assistance. A trainer will either charge you an hourly rate or up to 20 percent of the horse's purchase price. It may seem expensive, but finding the right horse will save you money in the long run. You won't waste time and gas looking at unsuitable horses, you won't spend money on prepurchase exams (see page 191) on lots of horses, and you'll most likely be paired with the right horse much faster than you would looking on your own. There are many ways of finding a horse that meets your needs; which avenues you choose to explore depend on the time and money you are willing to spend.

Word of mouth: Putting the word out that you are on the hunt for a horse frequently pays dividends, as people may point you in the right direction. You may also be told of sellers to avoid. If a seller is doing something disreputable, it is almost always widely known. Nobody can do anything bad for too long without someone's finding out about it. So ask around, and get more than one person's opinion. Check out the horse market just as you would the car market. Who did you buy your car from? Were you happy? Did you get what you wanted?

Breeders: Breeders often have lots of horses in different stages of training for you to choose from. Breeders usually work hard to keep their customers happy, since the best advertisement for their business is a satisfied, successful client.

Adoption: Charities and horse rescues can be wonderful places to find a horse. You'll be giving a horse a loving home and, in return, will be getting a suitable companion. For a minimal fee, rescues adopt out horses that they've rescued, rehabilitated, and retrained to suitable riders.

Classified advertisements: Whether they are in local papers, club journals, or larger horse publications, classifieds can lead to terrific buys; however, the volume of choices can prove problematic. You may have to make tons of phone calls and wade through a

This young rider and her pony seem to have a good partnership. Families tend to hold on to a good beginner's horse for their kids.

lot of ambiguous and possibly inaccurate descriptions. Shrewd shoppers must be able to translate these ads and ask smart questions before they see the horse.

Friends: Your friend has just bought the most beautiful horse that is absolutely perfect for her. And exactly what you were looking for, too. This horse is obviously not the only one of his kind out there. More horses like this can be found with a little simple question: Where did you get your horse? If the horse came from a breeding farm, then the breeder may have other horses from the same stallion or mare. Ask your friend to take you over and introduce you. If the horse was located at an instructor's stable, call the instructor and inquire about other horses like your friend's horse. If you don't know the person who owns the horse you're admiring—perhaps you've seen the horse at a show or at your stable—strike up a conversation. After all, people love to talk about their horses, and chances are you'll get more information than you need.

Internet: The Internet is perhaps the most modern way of approaching horse shopping. Many sellers are now advertising on the web and your new horse may be just a click away. You can narrow your choices through search engines or online databases and make inquiries via e-mail.

Professionals: Well-known trainers and instructors tend to have a good network and usually know where to find great horses. These professionals often hear about horses that aren't yet on the market. If you don't have the time to look for a horse yourself, you can hire a trainer or instructor to look for you. As

mentioned earlier, expect to pay that person a certain percentage of the cost of the horse or an hourly fee.

Do bargains exist? Yes, they do, and you don't necessarily have to be extremely lucky or in the right place at the right time to get a great deal. Look for someone who is moving up to a more experienced horse. Many times owners are simply interested in finding good homes for their horses, rather than making a profit. People will often approach trainers and ask them to help rehome their horses. A way to attract a good bargain to yourself is to have a great network of people or a trainer with a network. Just bear in mind that, if it's a bargain that you need, the search may take you some time to complete.

The Assessment

You've narrowed the search, and now your horse may be within your reach; it is time to go see the horse and make an assessment. Before you do so, decide on a plan of action: how you'll access the horse on the ground, which questions you'll ask, and how you'll go about your test ride. Take a friend or a trainer with you on the shopping expedition so you can have a second—and probably more objective—opinion.

The trouble with shopping on your own is that you can easily fall in love with the horse you're considering, even if the horse might not be the best fit for your needs. People tend to shop with their emotions. They look at a horse through rose-colored glasses and don't see some of the potential problems. After all, it's difficult to walk away from a horse that has a sweet face or kind expression. So have a trainer, a horsey friend, or someone who knows your riding abilities come with you to help keep your search on track.

Your companion may point out problems, such as a club foot, that you have not noticed and advise you to end a deal before you spend your prepurchase exam money. Your companion can screen the horse and decide whether it's worth it for you to continue. A trainer will not be afraid to ask searching questions. Your trainer knows your skills well and can help direct you while you are riding the horse. Another person will ask questions that you may forget to ask.

FIRST IMPRESSIONS

Once you have arrived at the stable, take time to run through your questions with the seller again. If you receive answers that are different from the ones you

AUCTIONS

Auctions are a buyer-beware situation. Horses that are auctioned off, rather than being sold through the general market, usually wind up at auctions for reasons. For the most part, these are mystery horses, and not much is known about them. You often won't have much chance to get to know the horses, ride them, or talk to their owners. It may be difficult to find out information about the horse, which may not have been previously vetted; and the stress of bidding can be overwhelming for the buyer. Unless it's a reputable breeding farm where its reputation is at stake or the sale is in conjunction with a competition, stay away from auctions.

When you assess a horse, check out the condition of his particular stall for signs of vices, such as cribbing or stall walking.

received over the phone, if the seller avoids answering your questions, or if you begin to suspect that something is not right, walk away.

If everything seems fine, then you can proceed to checking over the horse. Before you ride, observe the horse. Look at his stall for signs of vices, such as kicking marks on the wall and crib bites. Although these marks may be signs of a previous occupant, point them out and question the seller about any vices the horse may have. Watch the horse in his stable. Does he weave? Or is he content? Examine the bedding to see if there is a circular pattern, a sure sign of a stall walker (a horse pacing around his stall—not a good sign).

Then, spend time with the horse before you mount up. Go through all the steps you'd usually do—pick up his feet, groom him, and tack him up. Then take him for a walk in-hand. Although the horse will be

slightly different on the ground than under saddle, this walk will give you an indication of his manners. For instance, does the horse walk with you or does he barge ahead?

THE TEST RIDE

If the screening has gone smoothly so far, ask the seller to ride the horse while you observe. By doing so, you may discover right away that this horse isn't for you. If you don't like what the horse is doing and you're apprehensive, stop right there. Chances are the rest of the trial isn't going to improve. Walk away then and there. The longer you remain, the greater the chance that you'll talk yourself into buying him (or the seller will talk you into it). Go with your first instinct—this horse is not for you. If, however, the horse looks good and you're still interested, take note of how he's being

This horse is unhappy in the bridle, which may be caused by poor training or riding or by physical problems.

ridden, then try to duplicate it. You'll get the best out of the horse that way, because he is acclimated to that style of riding. Don't be afraid to question the seller about the horse's tack. And if the horse is being shown to you in a double bridle or training equipment, such as a martingale or training fork, ask to see him ridden in a snaffle or without the equipment. Sometimes riders use these for control. If a horse can't be ridden without them, this is a tip-off that he is hard to handle.

When you're trying the horse, take some time to get comfortable on him and let him get comfortable with you. Begin your assessment by walking him on a loose rein, changing directions at the walk and halt, and asking him to stand still. Evaluate how the horse handles this quiet work. If all goes well, it's time to ride the horse. Begin with a walk, trot, and canter. Then, change direction, making circles and straight lines, riding the horse much as the seller did. You may feel a little self-conscious with everyone watching you, but try to forget about the spectators and concentrate on your ride.

Next, ride the horse as if he were already yours. Make him go the way you think he ought to and see how he responds to your aids. Do something wrong, such as changing direction at the last second, to see whether the horse lets you. If the horse does allow it, do it the correct way, and see whether you get a better reaction. Don't skimp on what you're asking him to do; really put him through his paces. Walk, trot, canter in both directions; move him on to a faster tempo (speed) and bring him back to a slower tempo. Does he move on willingly? Does he come back or do you feel as though you are blasting off over the horizon?

Try the horse out in the discipline for which he is trained. If he's an upper-level dressage horse, ride him in half pass or ask for a flying change. If he's a jumper, pop him over a fence or two. Take a normal-size fence, one you'd usually do. If the horse is more advanced than you are, don't jump a bigger fence than you're used to. If facilities aren't available to try the horse, the seller should offer you another option, such as hauling the horse to another place for you.

If possible, have a friend record your ride, so you can watch it later. Reviewing your ride may aid you in making your final decision. This is very helpful if you've ridden several horses.

After you've had your initial test ride, take the horse out on the trail or on the field with another horse and rider. Let the other ride on ahead without you. How does the horse you're riding react to being left behind? Does he try to catch up to the other horse? Can you handle that response? Then let your horse ride ahead. Does he mind leaving the other horse? Does he stop in his tracks to wait? Then ride off on your own for a little while. Does he like being ridden out on his own? Note his response around cars. Is he calm? Is he fearful? Can you handle him?

Untack, groom, and put the horse away as if he belongs to you, and watch how he handles your requests. Then go home and think about everything you saw and felt before you make a decision.

Review all the aspects of the horse, good and bad. Discuss the horse with the person you brought along with you and review your video. Above all, don't let anyone talk you into purchasing a horse you think isn't right for you. In the same vein, don't barge ahead with a horse that more knowledgeable people think is wrong for you.

Don'ts and Dos Checklist

Don't shop with emotions instead of intellect. It's easy to fall in love at first sight. You need to be practical and unemotional when looking at prospects.

Don't consider inappropriate horses. Stick to your basic idea of what kind of horse you would like. Don't look at western pleasure horses when you want to own a dressage horse, because the training is completely different. Yes, it's possible to retrain a horse, but there are no guarantees.

Don't get caught up in the hype. Again, stick to your idea of the right horse for you. It's easy to get swept away in the glitz and glamour of the horse world.

Don't buy the first horse you look at. The novice buyer in particular should kick a lot of tires. It's important to look at a lot of horses to see what's out there. Study the type of horse you are interested in; do the research. It's rare to find the perfect horse the first time out. Expect to spend several months searching.

Don't buy a horse fresh off a lay-up (a break due to an injury). The lay-up may hide a chronic problem. Choose a horse that has been working consistently.

Don't get on a horse that you haven't seen ridden. Make sure you can ride the horse. If you don't feel comfortable on him, don't think anything is going to change.

Don't buy a horse that you can't ride well, that you can't put on the bit, or that doesn't respond to your aids.

Don't buy a horse that is green if you are a novice. It's this simple: green horse + green rider = disaster. You may end up spending a lot more money for training your young horse than you ever dreamed possible.

Don't let anyone flatter you or pressure you into buying a horse.

Don't be embarrassed to ask seemingly stupid questions. Never buy a horse that makes you leery in any way.

Do make sure the horse is not on medications. If he is, find out why and what medications. If they are too expensive for you to continue or you're not comfortable with the reasons they are being used, then you may have to pass on the horse. You can always discuss these issues with your veterinarian.

Do examine the horse's stall to make sure he doesn't have stall vices and water hasn't been withheld from him. An old trick is to dehydrate a horse to make him quieter and easier to ride.

Do think hard about buying a horse that has potential but no track record. Only buy this type of horse if you are a gambler. That is what equine potential is: a gamble. You may have to fork out a great deal of training money to get that potential to shine.

Do bring along your trainer or another knowledgeable person when you go to see a horse.

Do buy a horse trained in your discipline of choice.

Do try to ride the horse on three separate occasions and at different times of days to get an overall assessment.

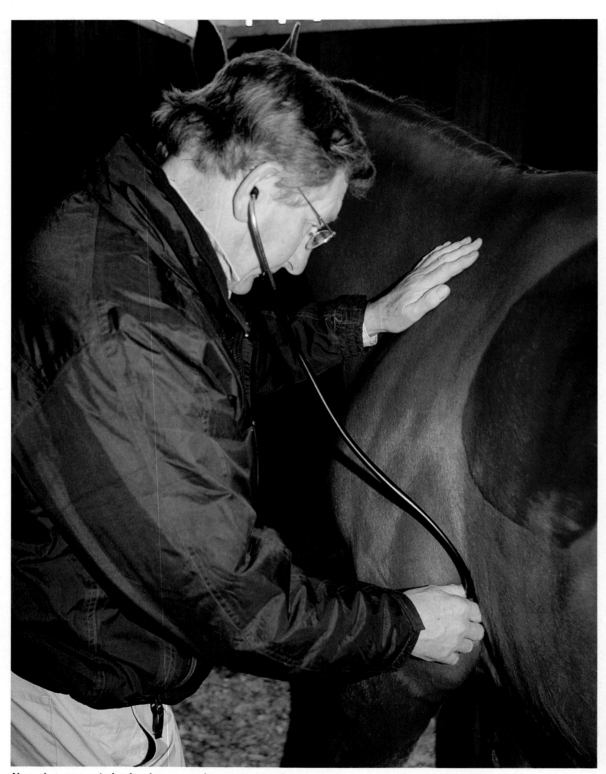

Always have your veterinarian do a prepurchase exam to make sure the horse you decide to buy is healthy.

You've finally found the horse that may be the one. Yet how can you know for sure? The answer may be to take the horse for a trial period. This is a great way for the two of you to get to know each other and for you to see how he reacts to a variety of situations. It is a sensitive topic for most sellers, so take the time to talk to the seller and explain why you think a trial will be best for all parties involved.

Understand that the seller may be less than enthusiastic about the idea, or unwilling to entertain it at all. To allow you a tryout, the seller has to take the horse off the market, risk the horse's health, and lose control of the horse during this period.

With a little creative negotiation, however, a tryout may be arranged. You can offer to take out insurance on the horse and name the seller as sole beneficiary of the policy so he or she is covered in case something happens to the horse. Or you can construct a clear, concise contract agreeing to pay all veterinarian bills should something happen during the proposed trial. These few precautions may allay the fears of a hesitant seller. Whatever you decide, set out the terms of your trial agreement in writing. If something goes wrong, and you have such an agreement, you'll be able to support your claim. If the seller is reluctant to let the horse off the property, a good compromise is to try the horse for two weeks at the seller's stable. This way the seller still has control over everything.

PURCHASING YOUR HORSE

If after countless hours of research, telephone calls, stable visits, and trials you have found a horse you want to purchase, there are still a few tasks to complete before any money is exchanged. For one, you should have the horse vetted to ensure that he is sound and healthy. This is commonly called a prepurchase exam, and it is designed to screen the horse for active or potential problems. If the horse passes the exam, you'll need to carefully go over the contract terms (you may even want a lawyer to examine them) and negotiate a final price.

The Prepurchase Exam

The reason to have a prepurchase exam done is to have a veterinarian thoroughly examine the horse to identify any medical problems he may have and then to assess the relevance and significance of those problems. The exam will also tell you whether the horse can do the job that you want him to do.

No matter the cost of the horse, you should never skip a prepurchase exam. Even if the horse costs only $1,000, it's worth spending $200 for the exam because you'll save money in the long run by not acquiring a horse with a problem. Cheap horses are usually cheap for a reason and should be thoroughly examined so you can make an educated decision.

No prepurchase exam is watertight, because a horse is a living animal, and there are no guarantees that a health problem won't be waiting around the corner. However, the exam will point out issues that will help you decide whether the animal is worth chancing. Generally speaking, veterinarians don't look for reasons to fail the horse; they look at the problems the animal may have and then see whether he is appropriate for the job despite those problems. The vast majority of horses, just like the vast majority of humans, can do a job despite their health issues. A prepurchase exam can include any of the following and should include a report of any conditions or issues:

- **Physical exam:** an examination of the heart and lungs, eyes, head, teeth, and throat.
- **Movement observation:** an observation of the horse standing still, walking, and trotting in a straight line, turning and backing, and being longed at the trot.
- **Flexion tests:** tests in which the veterinarian flexes the horse's joints, then watches the animal trot off to see if the flexion caused soreness or lameness.
- **Exam under saddle:** the veterinarian watches you mount, walk, trot, canter, and possibly gallop on straight lines and circles.
- **Blood test:** a test to check for medications.
- **X-rays:** x-rays are not black and white; they are shades of gray. A horse may have had the issue shown in the x-ray for a long time without experiencing any problems. Consult with your veterinarian before getting x-rays done.

Price Negotiation

You don't always have to pay the asking price, but this differs in every situation. For a sound, well-working, or competing horse, it's fair to bargain a bit. Big price drops are probably not going to happen unless it's a distress sale or the horse is up for a quick sell. If you're unsure as to whether the horse is set at the right price, ask your trainer or a knowledgeable friend's advice.

Chapter 8

Choosing the Right Equipment

One of the most important things you can do for your horse is to make sure his tack fits properly. His wearing a poorly fitting saddle is comparable to your wearing poorly fitting shoes: too tight and you've got squished toes; too loose and blisters abound. A horse suffers the same fate when a small saddle pinches his withers or a big saddle rolls around on his back. The rest of the tack must fit well, too. An ill-fitting bridle won't allow the bit to sit correctly in your horse's mouth, causing discomfort and training problems.

Your comfort and safety are also important. You must also get the right clothing, boots, and helmet. Different disciplines call for different apparel. Boots and helmets are a necessity for safety, and the correct riding attire can mean the difference between a comfortable ride and an uncomfortable ride.

SADDLES

Finding the right saddle is no easy feat. It must fit you as well as your horse. A well-fitting saddle will make the world of difference for your riding and for your horse's comfort. Because of the way they are crafted, English and western saddles are fitted differently, so it's important to know this as you shop. There is no such thing as a one-size-fits-all saddle. The girth or cinch (cinch is the western term for the girth) and saddle pad may seem like simple accessories, but the size and the kind of material used are also crucial.

English Saddles

English saddles generally come in tree sizes of 30, 31, 32, and so on (denoting the width of the tree in centimeters from point to point), or narrow, medium, and wide:

- An average Thoroughbred wears a medium or 31.
- A Saddlebred generally wears a narrow or 30.
- A warmblood wears a wide or 32.

Keep in mind that these are generalizations, meant to give you an idea of what size is likely to fit your horse. Sizes also vary among brands. A wide-size saddle from one English saddle brand may not be the same wide saddle of another brand; flocking (stuffing) varies. It's better to have a saddle that is too wide than one that is too tight. A thick pad can make up one size. A saddler can also restuff a saddle to improve the fit.

The best way to find the proper saddle size for your horse is to purchase a flexible curve ruler (an inexpensive drafter's tool sold in art supply shops or online) to trace your horse's withers. Stretch the ruler over your horse's withers, starting at the center of the curve across the withers and down two fingers behind the point of the shoulder. Set the ruler down on a large sheet of sturdy cardboard, and trace the inside shape.

Cut out the shape, and take it with you to the tack store. Set this template inside the pommel of the saddle (pretend the cutout is your horse's shoulders and withers) to find a close fit. If you're purchasing a saddle through a catalog or online, send the template in with your order.

An English jumping saddle, such as the one above, includes a knee roll and skirt on either side of the pommel.

The cutout should not fit perfectly inside the pommel like a puzzle piece. You should be able to fit two to four fingers vertically between the cutout and the pommel. The bottom half should fit well inside the panel.

FIND THE FIT

Never purchase a saddle without trying it on your horse first. Find a tack shop, a catalog, or an online store that offers a trial saddle period. If one does, be sure to take extremely good care of the saddle, because it may not be returnable if it is scratched or worn. Ask the company about its policy regarding how to attach stirrups and girths for a trial, so they won't damage the saddle while you're trying it out.

Once you have the saddle home, stand your horse squarely on a level area, and place the saddle on his back without a pad or girth. Place the saddle far forward, and slide it back into position to rest just behind the shoulder muscle, not on the shoulder blade.

To check for proper contact, put your hand on the top of the saddle and apply slight pressure. You won't be using a girth so you'll need to hold the saddle down with your hand. Slide your other hand underneath the sweat flap next to your horse's back. Push with your

An owner fits four fingers underneath the saddle's pommel, testing the fit of the saddle.

fingers along the panel to see if your fingers can slide underneath the panel at any point. If they can, the saddle is not making good contact with your horse's back, and you will have to find a better-fitting saddle.

To check the pommel clearance over the withers, place two to four stacked fingers under the pommel (four would be for significant withers), but keep in mind that is it normal for the clearance to shrink after the saddle breaks in. Regardless of your horse's wither size, if you can fit more than four fingers under a new saddle, it's likely too narrow; only one finger, and it's too wide.

The saddle must be properly balanced. Check to see if the flat area of the seat is parallel to the ground. The tip of the cantle should be slightly above the pommel (the cantle height will vary by saddle style). The pommel should never be higher than the cantle. If it is, your balance will be off because the saddle will push you behind the horse's center of gravity. If the lowest part is toward the cantle, the saddle is too narrow and will pinch your horse. If the lowest point is toward the pommel, the tree is too wide and will press down painfully against his withers.

The gullet should clear the horse's spine, not rest on it. You should be able to look through the saddle, all the way down the length of your horse's back.

If these aspects check out, add a pad and recheck the fit. Don't use a pad that is too thick—it will skew the results. A thin, quilted pad is best for a properly fitted saddle unless your horse requires a therapeutic pad.

TEST RIDE

Girth up and mount. Check your fit by placing your hand flat behind your seat (fold your thumb under). You should have only four fingers' space from your seat to the cantle. Many people go by the sweat-pattern markings on the horse's back to see if the saddle fits him correctly. Don't rely on sweat patterns to find your English saddle fit. Sweat patterns can be erratic with a new saddle or a newly flocked saddle. The flocking, or stuffing, needs a chance to break in to the shape of your horse's back. If after two weeks the sweat patterns are still uneven, have a saddler look at the saddle.

Western Saddles

Western saddles are less complicated than English saddles to fit. Since these saddles are made without stuffed panels, the main fitting concern lies within the tree. The western saddle tree is made up of the cantle, the bars, and the swells. The front of the tree, called the fork, is joined to the back of the tree, called the cantle, by bars. These bars run along the horse's back and denote the shape of the saddle and the width of the gullet. There are three standard bar widths (sizes) in western saddle trees: the Quarter Horse bar fits 80 percent of all western horses, the Semi-Quarter Horse bar fits higher-withered horses, and the Full Quarter Horse bar fits flat or mutton-withered horses. It's better to have a saddle that is too wide than too narrow. A thick pad can temporarily fill up some of the excess space; however, there is no substitution for a proper fit.

FIND THE FIT

To find the perfect fit, place the saddle on the horse without a pad and cinch it. Check that you are able to get two to three fingers clearance between the withers and the top of the gullet.

Now add a pad and make sure that the skirting, the rounded leather panels under the tree, follows

The presence of a horn on the pommel and fenders above the stirrups are two distinct characteristics of a western saddle.

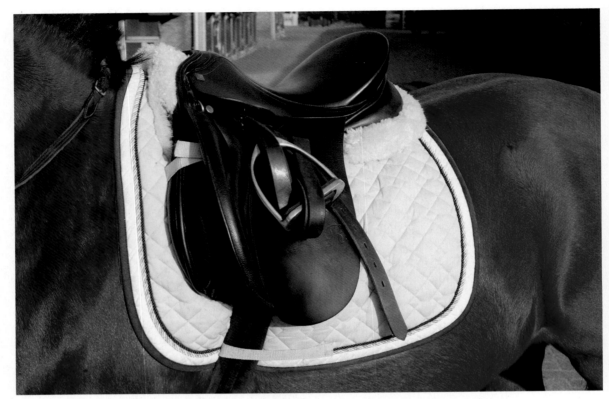

A saddle pad can help make minor adjustments to saddle fit, but it can't be expected to resolve major fit issues.

the contours of the back of the horse. You want the bars of the saddle to make contact with the horse's back. If you see space under the back of the saddle, or the skirt is lifting up, then the skirt does not fit properly and you will have to look for a better-fitting saddle.

Because western saddles, unlike English ones, do not have stuffed panels, sweat patterns on western saddles are useful to look at when assessing a saddle's fit. Ride the horse, and then take off the saddle and check the pattern of your horse's sweat marks on it. If you see dry spots, the saddle may be bridging, or not fitting properly. You can try a different pad to improve the fit or a different saddle altogether.

TEST RIDE

The saddle may fit the horse, but it should fit you too. The most popular seat size for western saddles is 15 or 16 inches. Choose a seat size that allows adequate clearance for your legs and your stomach. You should comfortably settle into the deepest part of the seat.

Saddle Pads

A saddle pad seems like a fairly harmless device, but it can cause injury if it's not fitting properly or not made from appropriate materials. Pads of synthetic material can generate too much heat and friction, causing bumps along the horse's back called collagenolytic granuloma, which look like large mosquito bites and are caused by damage to the collagen fibers in the skin. The fibers dissolve, and the horse's natural inflammatory response causes the bumps. (These can also be caused from the friction of an ill-fitting saddle.) These are common along the dorsa and the girth area. All-natural materials such as cotton, cotton felt, or genuine sheepskin are better options than synthetic materials.

The bumps can go away if corrected early or if a veterinarian can inject them with a corticosteroid. If you start to see these bumps, you don't want to ignore them; they can cause hair loss or become infected.

Gel pads are designed to better distribute weight across the horse's back. Jumpers, in particular, benefit from a gel pad because the small saddles don't allow for much weight distribution. Always make sure, when you tack the horse up, that you pull the pad up into the gullet of the saddle. Otherwise the pad will squash down on the withers and cause problems.

Girth and Cinch

Although there are many girth and cinch styles available, what is more important than style is the material used. Ill-fitting girths can cause irritation, which can predispose the horse to girth galls (sores that develop due to friction) and fungal infections such as ringworm. Neoprene girths and cinches are good for short rides, but if you are riding all day in hot weather, they can cause thermal injury (hair loss and chafing). Fleece covers may seem like a way to reduce heat and friction, but they can be slippery and make the girth hard to tighten. A girth with elastic ends is a good idea as it provides an even pressure, but there should be elastic on both sides or not at all. Be aware that after time the elastic can stretch, causing the girth to be uneven.

ENGLISH GIRTH FIT

To measure your horse for an English girth, place your saddle and pad on his back. Choose the pad you usually use because a thicker sheepskin pad may require a slightly longer girth than a thinner quilted pad. Hold a dressmaker's measuring tape on the middle holes of the billets on one side. Pass the tape under your horse's belly (make sure it is positioned where the girth will sit) to someone on the other side. Have your friend hold the tape to the middle holes of the billets on the other side. The resulting measurement will give you a good idea of your horse's girth size. Girth lengths run in two-inch increments, and sizes range from 32 inches to 52 inches. Round up to the bigger girth if your horse is between sizes. To find the fit for a shorter dressage girth, subtract 20 inches from the total. So a horse that wears a 50-inch-long girth will wear a 30-inch dressage girth.

WESTERN CINCH FIT

To measure your horse for a western cinch, put the saddle and pad on your horse. Measure with the tape from the off (right) rigging ring to the near (left) rigging ring, and then subtract 16 inches from the total. This is because there should be about 8 inches of latigo (the strap that secures the cinch) between the saddle and the cinch buckle. So if your horse's measurement is 48 inches, you'll need a 32-inch cinch.

These girths are (*left to right*) an English chafeless girth, a contour girth, another chafeless girth, and a cord girth.

Tie the cinch by sliding the latigo through the ring on the cinch. Run the latigo through the ring on the saddle (toward you) and back down to the cinch ring. Repeat to take up any excess, and then tie with a knot by looping the latigo through the saddle ring, back over itself, and then back up through the ring. Insert the strap end downward. This should look like a neck tie knot. Tighten by pulling up the inside strap.

BRIDLES

Before you purchase a bridle for your horse, determine the fit. As with saddles, bridle sizes are estimations. Your pony may have a large head and need a standard-size bridle, or your Quarter Horse may have a dainty head and need a cob-size bridle. Whichever size your horse wears, make sure the bridle leaves you enough leather to adjust up or down one hole.

There are various types and sizes of bridles. Here is a general size chart for ponies and horses.

English:
- Pony—several sizes are available.
- Cob—Arabians or horses with petite heads.
- Horse—an average size intended for the typical Thoroughbred and Quarter Horse.
- Warmblood—large horses, draft crosses.

Western:
- Arabian or pony—small sizes for horses with petite heads
- Standard—average size intended for Paints and Quarter Horses
- Extra large—large-headed cross-breeds, draft crosses.

Different types of bridles and bits adorn the wall of this tack room. Choose the tack that best fits your horse.

The Right Bit

Choosing a bit is probably one of the most perplexing tasks to undertake. It is also one of the most important. There are many styles to choose from, and which one is right for your horse can be difficult to determine.

The biggest mistake people make when considering a bit is failing to look inside the horse's mouth before purchasing a bit. The horse may have a shallow mouth or a deep mouth (lengthwise), which would determine the bit to use. If you're unsure of what you are looking at, ask a veterinarian or an equine dentist for advice.

Research has shown that pressure on the horse's palate is one of the most common causes of bit problems. Horses that toss their heads, lean, or place their tongues over the bit are often trying to relieve the pressure on their palates. Many ergonomic bits have been developed that answer this problem by contouring to the shape of the horse's mouth.

Although there is very little evidence that a horse likes one material over another, the material should help the horse salivate. You don't want a dry mouth because there will be friction. Many bits are copper because that metal helps with salivation. Size is determined by inches, and the average size worn is 5. The best way to find out your horse's size is to measure with a piece of string. Tie a pencil to the piece of string. Slide the free end of the string through your horse's mouth to the other side and make a knot on the end. The pencil should rest at one corner of your horse's mouth and the knot at the other. Remove the string and measure with a ruler, rounding up to the nearest quarter inch. Figure one size up if you're choosing a loose-ring snaffle The rings on a small bit can pinch if they don't clear the edges of your horse's lips.

Types of Bits

There are hundreds of bits to choose from. They run from classic choices to innovative, ergonomic designs. Bits can be made from synthetic material, metal, or a mixture of metals. They fall into two basic categories: a direct pressure bit (snaffle) and a leverage bit (curb). Differing from these is the gag bit, a hybrid of snaffle and curb. You can also ride a horse without a bit. These bitless bridles are called hackamores (English) or bosals (western). These bridles work by putting pressure on the nose. The reins are attached to a specialized noseband.

SNAFFLE BITS

The snaffle bit puts pressure on the tongue, bars, and corner of the mouth through the bit rings, located on each side of the mouthpiece. The mouthpiece can be jointed, linked, or solid; the cheek rings also vary. The snaffle is used in English and western riding. Styles of snaffle bits can be made up of the bit rings and mouthpieces below.

Eggbutt: The rings on an eggbutt are flattened on one side and merge into hinges. The mouthpiece widens out at the sides creating a very mild bit. The rings' shape prevents rubbing and pinching of the bit.

D-ring: As the name implies, these rings are shaped like the letter *D*. As with the eggbutt, the rings merge into hinges. The shape prevents the mouth from being pinched by the hinges.

Full-cheek: Long metal pieces extend out from the mouthpiece. These prevent the bit from sliding through the horse's mouth. They also put gentle pressure against the side of the horse's face to keep the head straight.

Loose ring: The ring is not fixed; rather, it is able to slide freely through the holes in each end of the

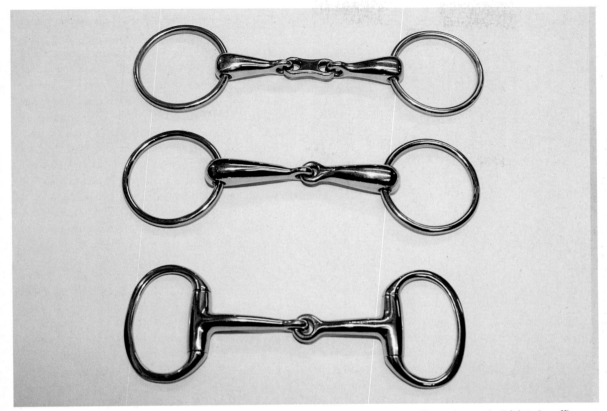

The bits above are (*top to bottom*) a loose ring French link snaffle, a loose ring jointed snaffle, and an eggbutt jointed snaffle.

mouthpiece. This helps the horse to relax his jaw; it also prevents him from grabbing the bit rings.

French link: This is a two-part mouthpiece linked with a lozenge of metal. It lies over the horse's tongue better than the jointed mouthpiece.

Jointed: This is a two-part mouthpiece linked in the middle.

Mullen mouth: This is a straight bar mouthpiece. It is very gentle against the roof of the horse's mouth.

Twisted: As the name implies, this mouthpiece is twisted; sometimes it can even be made up of thin twisted wire. This is a harsh bit because the twist of metal creates friction in the horse's mouth.

CURB BITS

The curb bit puts pressure on the mouth, poll, and chin groove using leverage by way of the shanks that extend from each side of the mouthpiece and the curb strap or chain, which controls the lever action of the bit. Mouthpieces can be smooth, jointed, or have a port (a bump in the mouthpiece of various heights depending

on the style of the curb). Curbs are used in English and western riding. Curbs include the following:

Kimberwicke: This is a mild curb with a curb chain. It has offset D-rings, which act as shanks; the reins are attached near the mouthpiece. The reins can be fitted into the top slot or bottom slot of the D-ring for less or more control respectively. It is used with one set of reins. They are commonly used on ponies who can get the better of their smaller rider. It is mainly an English bit.

Pelham: This is a leverage-type bit that allows the use of two sets of reins: one attaches at rings at the

Both of these western bits have a port in the mouthpiece, which puts pressure on the roof of the mouth.

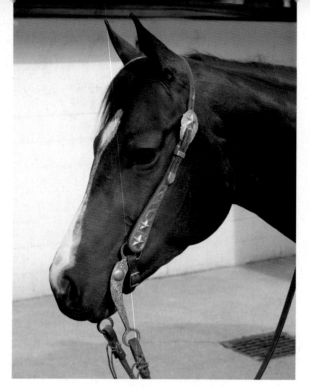

This western curb bit and bridle is used for showing and is much more decorative than the everyday bit and bridle.

mouthpiece (which can be jointed or smooth or have a port) and the second at the shanks. The rider can choose between the action of a snaffle or the curb as he or she rides. It is used in English and western riding (called a cowboy Pelham).

GAG BIT

The gag bit resembles either a snaffle or a curb but with no curb chain. The rider uses two reins and can choose between the gag action (which puts pressure on the poll and the sides of the lips at the same time) and a regular snaffle action. Sometimes this bit requires a special gag bridle. The gag bit is generally used with strong horses that pull on the reins. It is commonly seen in polo, show jumping, and eventing. It is never used in dressage or in the hunter arena.

Nosebands

Nosebands or cavessons are standard issue on English bridles, but western riders may also use them during training sessions (never in competition). The types are: flash (an ordinary noseband with a second attached to the first, which buckles under the bit); drop (one single noseband than is adjusted under the bit); and figure 8 (crisscrosses around the horse's jaw and under the bit).

These types of nosebands help encourage a horse to keep his mouth closed when riding. Although not usually seen in western or hunter competition, flashes, drops, and figure 8s are allowed in jumping, eventing, and dressage competitions.

You can adjust your noseband quite snugly (use your fingers only to tighten; tools can overtighten and cause pain) without causing discomfort to your horse; however, if the noseband is positioned too low on his nose, you can pinch his nostrils. Standard rule of thumb: two fingers below his cheekbone for the first noseband, and positioned directly under the bit for the drop, or second noseband. Make sure the buckle and keeper are positioned toward the back of the horse's nose or on top. The buckle should not rub on the lips.

TACK CARE

Examine your saddle and tack regularly for signs of wear and tear. A well-fitting western saddle is virtually maintenance free. Clean all styles of tack with a good leather cleaner, preferably every time you ride. At the very least, wipe the dust, sweat, and dirt away with a soft cloth. Dust

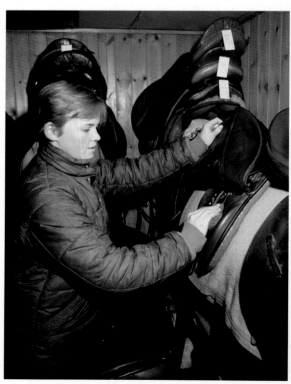

A rider wipes the dust, sweat, and dirt from her saddle in order to keep the leather in good condition.

Fitting the Bit

How you fit the bit depends on what kind you decide to use. Follow the steps below to determine the proper fit of your horse's bit.

English Bridle with Snaffle or Pelham Bit

1. The bit hinges should clear the corners of the horse's lips. Move the hinges around to make sure they don't pinch.

2. Adjust the cheekpieces so that the bit wrinkles the corners of the mouth of your horse in one or two folds.

3. The noseband should fit two fingers below the cheekbones. You should have enough leather to be able to tighten the noseband so that only one finger fits under it.

4. The top of a flash noseband should also fit two fingers below the cheekbones. You should be able to adjust the bottom half of the noseband quite tightly below the rings of the snaffle.

5. You should be able to fit at least four fingers sideways inside the throatlatch.

6. If using a Pelham, straighten out the curb chain and attach it to the opposite hook. The curb chain is the right length when the chain engages as the shanks reach a 45-degree angle. You can test this by picking up the curb rein and pulling the bit back.

English Double Bridle with Bradoon and Weymouth Bit

With the double bridle, two bits are used: the bradoon (a snaffle with smaller rings) and the Weymouth (a type of curb used with the double bridle). The bradoon should be the same size that you would normally use in a snaffle, but the Weymouth should be a little bit wider. A bradoon and Weymouth bit is pictured above.

1. The Bridle and noseband should fit as described above.

2. Put the snaffle in the horse's mouth first and then guide the curb in.

3. Adjust the cheekpieces so that the bradoon wrinkles the corners of your horse's mouth in one or two folds. The curb should hang just below the snaffle.

4. Straighten the curb chain and attach it to the opposite hook. You'll know the curb chain is the right length when the chain engages as the Weymouth's shanks reach a 45-degree angle. To test this, pick up the curb rein and pull the bit back.

Western Bridle with Snaffle Bit

1. Adjust the cheekpieces so that the bit wrinkles the corners of the mouth in one small fold.

2. Attach a leather or nylon curb strap below the reins.

3. Adjust the curb strap so it is loose enough to hang down against the chin, but not so loose that your horse can get it into his mouth. You should be able to fit two fingers stacked between the chin and the curb strap when the reins are loose. (Western trainers use the curb strap in order to help keep the snaffle bit from sliding through the mouth of a horse while they are teaching the basics of turning.)

Western Bridle with Curb Bit

1. The curb hangs lower in the mouth than the snaffle. Adjust the cheekpieces so the bit wrinkles the corners of the mouth in one fold.

2. Attach a curb strap (can include a chain).

3. You should be able to fit two fingers stacked between the chin and the strap when the reins are loose.

A woman longes a horse with side reins, a helpful aid at any stage of training as they mimic proper rein contact.

and sweat can cause the leather to become slippery. Don't oil your English tack too frequently—once or twice a year is a good rule of thumb for oiling.

Today's English saddles are built for comfort with softer, thinner leathers; oil causes the fibers of the leather to separate, and pressure will break down the leather. Some manufacturers tell you to never oil their tack. Western saddles, due to their hardier nature, can be oiled more often. But remember that oiling a lighter colored saddle may cause it to darken. Use a light brush for suede seats and the fleece underneath the saddle.

Don't wait to clean leg wraps, saddle pads, rub rags, and stable towels until they are so dirty they can stand by themselves. Dirty equipment can chafe your horse's skin and back. Wash these items with a mild fragrance- and bleach-free detergent about once a week.

TRAINING AIDS

Training aids may be one of the most argued about topics among horsepeople. Some say they should never be used, while others claim they have a place. There is truth in both opinions. Some gadgets, such as side reins, can

be very helpful. The right aid can help retrain a horse that has been ill trained. Aids, used under the guidance of more experienced horsepeople, can assist an amateur rider or a green horse to get the correct feel.

Training aids used incorrectly and by someone who is inexperienced can be damaging to a horse, both mentally and physically, and at times downright cruel. Aids put on incorrectly can force a horse to compensate elsewhere in his body, causing muscles, ligaments, and tendons to give out prematurely. Incorrectly adjusted aids can trigger a claustrophobic response and can cause your horse to fight against the aid.

If you're having trouble with your horse and are thinking of buying training aids, understand that these aids are to be used temporarily; eventually, you should only be using a simple bit and reins.

Side Reins

Side reins are full leather, elastic, or half-leather/half-elastic reins with rubber "doughnuts" or rubber inserts. They resemble regular reins but have a buckle at one end that enables you to attach them to the girth, cinch, or surcingle, and a snap or buckle on the other end, which you attach to the bit. Side reins are usually used while longeing the horse, taking the place of a steady pair of hands helping the horse to understand what "on the bit" means. The side reins help the horse to remain on the bit, rounding his back and softening his trot.

When teaching a green horse to use side reins, first attach one side rein, wait for him to relax, then put on the other rein. The reins should hang loose at first, barely contacting the bit, until your horse becomes adapted to the bit pressure.

As your schooling session progresses, slowly shorten the reins. Your horse should be able to bring his nose slightly in front of vertical (drop an imaginary line from his forehead to the ground). Adjusting the side reins too tightly can cause your horse to rear or fight the reins.

Draw Reins

Draw reins are either full-leather or nylon-sliding reins designed to help retrain horses that have learned to go with an inverted neck carriage. Draw reins also help horses that lift their heads high and use their conformation (high set on necks) against their riders.

Draw reins are slid through the snaffle bit rings and attached to the girth, either under the horse's chest or on the sides of the girth by the horse's elbow. They are used in conjunction with your regular rein. Hold your normal rein around all four fingers and pick up your draw rein with your second and third fingers until your

This gelding is tacked up with draw reins. Draw reins should be used cautiously and only by an experienced rider.

horse gives to the pressure. Draw reins can give you tremendous torque. Tightly pulled-in draw reins can cause your horse to overbend or overflex, dump on his forehand, or compensate by bearing too much weight on his front legs.

Draw reins used under the horse's chest can be particularly damaging. Draw reins should not be used every day because they can become a crutch for the rider who will never learn how to put a horse on the bit. Draw reins do not duplicate "one-handed" riding in western. If you choose to use draw reins, take a lesson with a knowledgeable instructor first.

The German running martingale can be a better solution. This device is a draw rein that hooks on the snaffle rein. You can adjust it to a certain length, but once it is mounted, you cannot tighten it.

Running Martingales and Training Forks

Running martingales (for English riders) and training forks (for western riders) are basically the same aid. They have either two rings attached to a V-shaped piece of leather connected to a breastplate, which extends up from your horse's neck, or two rings incorporated into the breastplate located near your horse's shoulder.

Martingales and training forks keep your horse from getting his head too high. Aside from the initial adjustment, the rider has no control over this device while riding. The horse discovers the aid only if he lifts his head.

To adjust this type of aid, slide each rein through a ring in the martingale or fork, and run the reins up over your horse's neck. The reins should be able to come back to the snaffle in a straight line. Martingales and training forks adjusted too low can have the same effect as improperly adjusted draw reins.

Standing Martingales

A standing martingale is a solid piece of leather that hooks to a breastplate and attaches to the noseband. Standing martingales can help keep a horse from throwing his head, but your horse should be able to lift his head normally while wearing a standing martingale. The martingale comes into play when your horse raises his head too high. Adjust the martingale while your horse stands normally, loosely at first, then tighten it as needed. Never force your horse's head down while he is wearing a standing martingale. A tight martingale can cause your horse to rear over backward.

This horse wears a running martingale, which helps prevent him from carrying his head too high.

Artificial Aids

Spurs are used in both English and western riding. There are many choices that range from a small blunt English spur to a large western rowel. Spurs are powerful so you must be careful how you use them. Used incorrectly, they dull the response to leg aids and lead to a horse's kicking at the spur. Control your leg, so that you don't spur the horse unintentionally. Keeping heels down, with toes pointed forward, ask with your leg(s) for the horse to move forward, or sideways, then quickly follow with a spur prick. Then use the leg again (alone) to see if he promptly moves away from your leg.

Whips are for training, not for punishing. When you use a whip or crop, be sure to use your leg aid first. If your horse doesn't respond, then pair the leg aid with your whip or crop, lightly at first, using just a little flick of your wrist. Increase the pressure a tiny bit if you need to. Try very hard not to use the whip frequently as this can make your horse dull to it and to other aids.

Crops can be used in on the shoulder or behind the leg. The long dressage or schooling whip is designed to be used behind the leg, without moving the hand. Simply flick your hand and the whip will curl around your leg and tap the horse's flank. Whips and crops are usually carried in the inside hand, although they can be carried in whichever hand is needed for schooling purposes. For instance, if the outside hind leg is lazy, carry the whip in the outside hand to encourage the horse to step through with his weaker hind leg.

Always check your rule book regarding whips and spurs when competing.

BOOTS AND WRAPS

Injuries can set a horse back in his training for days, weeks, or even months. What is most frustrating is when that injury is from the horse himself. During training, a horse can easily take a misstep and tread on his heel, knock one leg against the other and cut a tendon, or even exacerbate an already existing injury. Since wrapping a horse up in bubble wrap isn't an option, over the years riders have tried to avoid these interference injuries by using boots and leg wraps.

Boots and wraps protect a horse's legs from knocks, which can come from the horse himself or from a jump pole. For this purpose, they work quite well. Although people also use boots for support, this isn't what they are made for, and they don't work very well as support. Tight wraps or boots limit flexion of the joints during the swing phase, but they have not shown to restrict sinking of the fetlock for more than a few strides. People who use boots on their horses are those who anticipate that their horses may be subject to traumatic injury—eventers, for example. Boots are also appropriate for young horses not yet well coordinated.

However, boots and wraps can actually cause damage if they are not used correctly and with some caution. Tendon core overheating, which predisposes the horse to later tendon damage, can be caused by these devices. The overheating is caused by heat produced in the tendon that cannot dissipate—which may cause more heat buildup. Remove the boot or wrap immediately after you finish exercising, and get rid of excess heat as quickly as possible with ice or cold water.

If you plan to use boots or leg wraps during a competition, you should check the rule book carefully. If they aren't allowed, you can still warm your horse up with wraps and boots; just be sure to remove them before you enter your class.

Types of Boots

There are many boots out there in the market with new and improved versions coming out every year, so how do you make your choice? The main thing to keep in mind is your horse's needs. A young horse learning to balance himself and his rider will need bell and interference boots. A seasoned dressage horse that knows his job may only need simple fleece-lined interference boots. Below is a list of the most common boot types.

BELL OR OVERREACH BOOTS

Bell boots, also called overreach boots, have evolved through the years more than have any other type of boot. They are also the most controversial. Some disciplines, such as dressage, say that the boots slap against the hooves and cause the horse to snatch his legs up instead of bring them forward. However, bell boots are

ADVICE ON TRAINING AIDS AND GADGETS

The first and foremost advice to remember is that less is best. Use the training aid to show your horse correct training—not to force him to submit. Your ultimate goal is to get your horse to accept the bit and to have a strong basic foundation in training.

- Horses are hind-end creatures. Their "engines" are in the back; ride forward, or no training aid will help you.

- When you drop the reins, the horse should stretch his head and neck down. If he jigs and sticks his neck up instead, you've skipped some basics along the way and perhaps let yourself become dependent upon training gadgets.

- Any training aid should be treated as an explanation to your horse and not as force.

- Anything forced is not proper. If you create force, you can end up with unsoundness. Stop and reassess the situation if you've turned to using force.

- If you are uncertain about how to use any of the aids, seek help from a qualified professional.

very useful for protecting a horse from an overreach injury. (An overreach is when the horse reaches too far forward with his hind leg and steps on the back of his front leg.) Bell boots cover the whole hoof and bulb of the heel and are normally used only on the front legs. They can, however, be used for the hind hooves during trailering in case a horse should step on himself.

The newer bell boots are composed of neoprene and are made to stay in place and not spin around the hoof as the horse moves. The more traditional rubber bell boots are fastened with buckles or Velcro straps or are designed to be pulled over the hooves. There is also a petal-type bell boot made of pieces of rubber joined together on a plastic strap. One of the benefits of this type is that the petals can be replaced if they are damaged. Bell boots are also beneficial for turnout. If you use bell-boots frequently, make sure the top of the boot isn't rubbing the skin.

FETLOCK BOOTS

Fetlock boots are designed to cover only the fetlock joint. Many horses, particularly young unmuscled horses, travel very close behind, meaning their stance is very narrow, and they are very likely to knock their hind fetlock joints together. While this may not cut

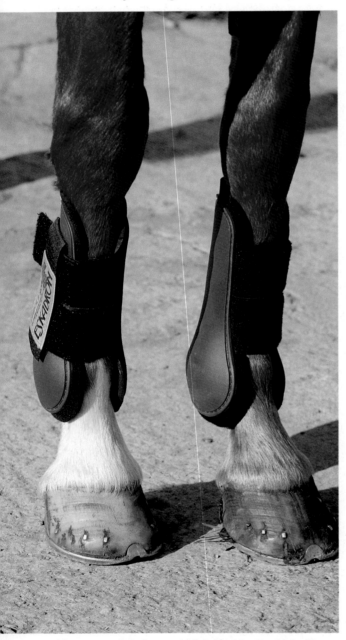

A horse wears interference (or brushing) boots, which will protect his legs if he hits one hoof or leg against another.

the fetlock, knocking can bruise the joint and cause chronic swelling, which will worsen if the joint isn't protected.

The boots are made of various sorts of material and usually have extra padding over the joint. There are also many variations available. A rubber ring or sausage boot is a good choice if only one fetlock is being hit. A sausage boot looks much like a strip of rubber hosing and it fastens around the hind pastern. It also protects the coronary band and is used when the horse is in the stall and during turnout, but not when riding.

INTERFERENCE OR BRUSHING BOOTS

Interference boots, also called brushing boots, are probably the most diverse of all the boots. They come in many shapes and sizes with different cutouts, lined with fleece or felt, and are made of either neoprene, vinyl, or leather. Interference boots can come in several different lengths. The tallest boots normally cover from the bottom of the knee to the end of the fetlock, but shorter boots are also available.

Dressage riders are usually big fans of interference boots because the boots protect during lateral work. The boots are also good for turnout because they won't accidentally unravel and tangle around a horse's leg like a polo wrap will. (*See Polo Wraps, opposite.*)

OPEN FRONT JUMPING BOOTS

Open front jumping boots are favored heavily among jumpers. This is because the boot covers and protects the tendons and ligaments and the back of the leg in case of an overreach. These boots leave the cannon bone exposed so a horse can feel the jump poles if he hits them. Jumping boots are usually made of leather, neoprene, or vinyl; they are lined with felt or fleece and have straps that cross in front of the cannon bones.

SHIPPING BOOTS

There have been great strides in the development of shipping boots in the last few years. Until recently, riders had to make do with polo wraps and bell boots. The trouble with this combination is that polo wraps can unwrap during transit.

The more modern shipping boots are padded with shock-absorbing synthetic material usually fastened with Velcro from the coronet band to over the knee in

POLO WRAPS

Polo wraps were first used by polo riders (hence the name) to protect their horses against mallet knocks, interference from hooves, and knocks from other horses. Today, they are used by riders of all disciplines and even with their drawbacks are still very popular. Polo wraps are made of a stretchy type of cotton or fleece-type material and usually fastened with Velcro. They take more maintenance than boots because wraps absorb sweat and moisture, so regular washing is a necessity. Polo wraps were once thought to offer support to tendons, but this has now been disproved. You also must make sure the Velcro is attaching well and that the wrap material still has stretch and give. If not, your wrap can unravel, usually when you are riding. Polo wraps are not good during the cross-country phase of eventing or on a trail ride as the wraps can absorb water. They are good for arena schooling such as dressage.

the front leg and from the coronet band to up over the hock in the hind leg. Sometimes a plastic dish called a scuff plate lines the back of the wrap and helps protect the hoof.

SKID BOOTS

Cutting and reining horses, due to the nature of their work, need a different type of boot than any other horses; the skid boot fills the bill. Skid boots are designed with a cup to fit under the rear fetlock to prevent chafing during western work such as sliding stops, spins, and cutting calves. The boots are usually made of very heavy rawhide leather and have straps that buckle. These boots need to be kept scrupulously clean as a buildup of sand and dirt can make chafing worse. Some trainers prefer to leave the hair on the horse's rear fetlocks and pasterns long and only use skid boots if the hair isn't long enough.

SPLINT AND SPORTS MEDICINE BOOTS

A splint boot's job is to protect the tendons, ligaments, splint bone, and inside fetlock of the lower leg (both front and hind legs). The boots have a rigid plate built inside the boot that lies against the inside of the leg, protecting it from an accidental cut from a hoof. They are usually made of neoprene with suede outside panels.

Sports medicine boots are touted to actually extend the working lives of horses that may have otherwise faced long rehabilitation and to prevent injuries to working horses. This is mainly because they work to absorb shock from interference and hoof concussion. They do that by encasing the leg from the knee down and around the fetlock in high-density neoprene that can take a hit and dissipate the energy.

Putting on Boots and Wraps

Finding the right fit for boots and wraps is as important as choosing the right type. One manufacturer's version of a size large may differ from another's; ask the manufacturer or your tack shop dealer what size your horse needs. If you are buying an interference or a splint boot and the material doesn't meet in front of the leg when you are fastening the straps, the boot may

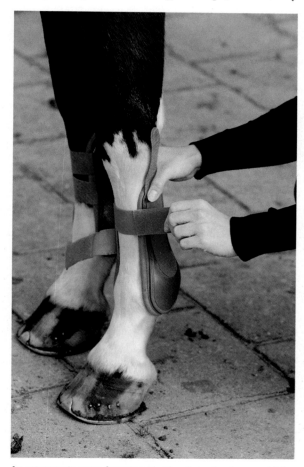

An owner puts open-front jumping boots on her horse, which protect the tendons and ligaments in case of overreach.

be too small. On the other hand, if you are overlapping material to fasten the straps, the boot may be too big. Overlapping, however, could be the style of the boot, so ask questions before purchasing.

Boots are usually marked left or right. If they aren't marked, the boot should sit on the leg so that the ends of the Velcro or buckles are on the outside and face the back so the horse won't accidentally kick the tapes or buckles open.

Before you start, make sure there isn't any dirt or shavings inside the boot or on the horse, then place the boot above the leg and slide it down into position. Fasten the middle strap first to hold the boot in place, then fasten the other straps. The boot should be snug enough that you can't twist the boot around the leg or slide it down easily. When you remove the boots, let them dry before you put them away.

Leg wraps are much more difficult to put on than boots. To make things easier, roll the wrap so that the Velcro pad is turned under first with the hooks on top. This way the Velcro will be in the right place when you've finished wrapping.

Make sure your horse's leg is clean and free of shavings before you begin to wrap. Hold the rolled wrap at the outside-middle of the top of the leg, and wrap at a downward angle while keeping even pressure—slightly tug each time you make a new wrap around the leg. Wrap around the fetlock and diagonally back up the leg.

Before the wrap runs out, take a straight turn or two and then fasten the Velcro. If the wraps become heavily soiled, launder them before your next use. If not, let them air dry, reroll them, and put them away.

HUMAN RIDING APPAREL

No matter what discipline you choose, comfort is the key with riding clothing. Relaxed fit jeans with some give and stretch are great for western riders. English riders wear close-fitting breeches, which prevent fabric bunching around the thighs and slide easily into tall riding boots.

Riding helmets have come a long way in the past few years. Modern helmets are no longer heavy, hot, or cumbersome; they are made of cutting-edge and lightweight material in a variety of styles, not just the traditional velvet. Fashions include western leather, a breathable and open design, and even a sporty high-tech look.

Clothing and Boots

Breeches for English riders can have a full-leather seat insert (preferred by dressage riders) to combat slipping, or be made of a simple, breathable cotton/Lycra blend, similar to leggings. Shirts can run the gamut from a polo shirt, blouse, T-shirt, or moisture-wicking sports shirt.

Women should wear a type of sports bra designed for high-intensity exercise. Men may choose to wear a jock strap or, even better, a dance belt, which is a specialized undergarment worn by male dancers to support the genitals. Both undergarments keep the male anatomy up and out of the way of the front of the saddle.

All disciplines should wear a boot with a heel, which prevents the foot from slipping through the stirrup. The classic cowboy boot is popular with western riders at competitions, but there is also a lace-up style ankle boot called a roper that many western riders use for everyday riding. English riders usually wear a tall boot, either with laces (called a field boot) or without laces (sometimes called a dress boot). Short paddock

FITTING YOUR HELMET

Follow these steps to properly fit your helmet:

1. Put the helmet on without using the strap. Move the helmet around. If you can feel the skin on your head moving, or the movement makes your eyebrows move, the fit is good. If the helmet only moves your hair, the helmet is too loose.

2. The helmet should feel firm and snug but not tight. Your head will expand a bit when it gets hot and the padding will begin to mold to your head after wearing it a few times.

3. Take the helmet off and put it back on three times. You'll find that the helmet will begin to feel molded to your head and you'll get a better idea of fit.

Note: The chin strap's job is to make sure the helmet doesn't come off the way it was put on. It shouldn't hold the helmet to your head; the proper fit should keep the helmet from shifting. The strap is properly adjusted when you can talk comfortably.

A properly outfitted rider wears light, breathable breeches, tall leather half chaps, paddock boots, and a properly fitted helmet.

A western rider wears relaxed-fit jeans, heeled boots, and a cowboy hat for a relaxed, afternoon walk.

boots are also popular, with laces, zippers, or elastic inserts. Half chaps are often worn over the calf to act like a tall boot. Half chaps are made of suede or leather and secured with Velcro or a zipper.

Helmets

No matter what it looks like, the helmet's job is to provide a stopping distance between your head and the ground and is absolutely necessary when riding horses. A fall on your head causes the brain to ricochet inside your skull, which causes bruising. The body immediately produces histamine, which causes swelling—just like when you hit your thumb with a hammer. If the swelling continues, the brain tissue will crush neighboring healthy cells. This is why you may feel fine after hitting your head but not so good an hour later. With this in mind, designers have built the helmet with a bubble-wrap type safety.

Helmet padding has been injected with millions of gas bubbles set in ten or more layers. When you fall, the bubbles burst, taking much of the impact. The idea is that the bubbles' bursting slows the impact, much like diving into a deep swimming pool, to your head and help keep bruising to a minimum. Your helmet also offers protection against hooves, rocks, branches, and other dangers.

An ill-fitting helmet will fall forward and leave the back of your skull vulnerable. Modern helmets are designed with a throat strap to help support the fit of the helmet. Helmets won't deteriorate over time

HELMET DO'S AND DON'TS

Don't wear your helmet without the strap or with it hanging loose.

Don't wear your helmet on the back of your head.

Do be suspicious of a loaner helmet. You can't know how many times it's been in a fall. Buy your own.

Don't have a "lucky helmet." If you've had a few falls in it, change it—even if you won last time wearing it.

Don't wear a bike helmet. They are better than nothing, but a horse-riding helmet is specially designed for a rider's needs.

Horse talk

EQUIPMENT AND CLOTHING

Learning the following terms will help you to talk about horse and rider equipment and clothing.

■ **Billets:** Part of the English saddle. The girth attaches to these straps.

■ **Cantle:** The back of the seat on a saddle.

■ **Cinch:** A western girth.

■ **Collagenolytic granuloma:** Bumps on the skin caused from synthetic saddle pads.

■ **Coronary or Coronet band:** Found directly above the hoof wall and protected by hair and thick skin. This is the source of growth for the hoof wall.

■ **Dorsa:** Muscles in the horse's back.

■ **Double bridle:** A bridle that includes two bits: a curb and a snaffle.

■ **Fetlock:** The joint on the horse's legs above the pastern.

■ **Girth galls:** Sores at the girth caused by chafing.

■ **Girth:** Tack that holds an English saddle on the horse.

■ **Gullet:** The underneath of an English saddle, which allows clearance for the spine.

■ **Latigo:** Strap that attaches the cinch to the Western saddle.

■ **Noseband:** Part of the bridle that goes around the horse's nose.

■ **Overreach injury:** When the horse's hind hoof steps on the back of the front hoof. This is a common injury for jumpers.

■ **Pastern:** Part of the horse's leg that lies between the hoof and the fetlock. This part of the equine anatomy is vital to shock absorption.

■ **Pelham:** A single bit that acts like a double bridle. Used with two sets of reins or one set of reins attached to a special connector.

■ **Pommel:** The front of the English saddle.

■ **Ringworm:** A fungal infection of the skin.

■ **Saddle tree:** Rigid foundation of the saddle.

■ **Snaffle:** A direct pressure bit.

■ **Surcingle:** A padded device that takes the place of the saddle during longeing.

■ **Swells:** The front of the western saddle that holds the horn. Also called the pommel or saddle fork.

or with constant wearing. The padding, however, will begin to break down if you store your helmet in a hot area such as the back of your car.

Falls will also lessen the effectiveness of the helmet. If you've fallen on your head and lost consciousness, chances are that you've maxed out your helmet's bubbles. Throw the helmet away and buy another. If you've landed on your shoulder or your hip before you hit your head, that part of you took the maximum impact, so you may have only popped through one layer and diminished your helmet 10 or 20 percent. Is only 80 percent protection okay with you? Replacing your helmet is going to be up to you and your needs.

You won't be able to see if your helmet has been affected on the inside, but outside is a different story. After impact, a painted helmet will show concentric rings on the top and striated lines on the sides, while a velvet helmet will have crushed pile.

Horsekeeping

Section 5

O wning a horse is not like caring for a dog or a cat; nor is it like overseeing other types of livestock. The horse requires a different level of care and knowledge. You must have a good grasp of his needs before bringing him home.

That begins with knowing how to safely transport him to his new home. Although many horse owners choose not to purchase a horse trailer, relying instead upon friends and professional companies for transportation, it's a good idea to know the ins and outs of trailering so you can make good traveling decisions.

You must also figure out where and how you will stable your horse and what types of food and supplements he will need to stay healthy. Another aspect of keeping him healthy is learning how to groom him.

Chapter 9

This bumper-pull trailer is hitched to a truck that can more than handle the size and weight of the fully loaded trailer.

can handle. Some of the newer, larger SUVs have the specs to perform like a ½-ton truck and could be considered to pull a two-horse trailer without a dressing room. To be able to safely accommodate this lighter trailer, your SUV must be able to handle at least a 5,000-pound (fully loaded) trailer with a 500-pound minimum tongue weight. Anything more than 13,000 pounds needs a bigger tow vehicle, such as the Ford F-450, an extension of the company's super-duty line.

LOADING THE HORSE

Before loading your horse, make sure that you have outfitted him in protective travel gear, such as shipping boots. You also need to think about the time of year and precautions you should take to make sure your horse does not get chilled or overheated.

A horse that will walk into a trailer easily is worth his weight in gold. However, going into a darkened cavelike box goes against every instinct that the horse has. Additionally, once a horse has a bad loading or a bad hauling experience, he may become reluctant to go into a trailer again.

If you are a horse-hauling novice, work with a skilled horseperson to load your new horse. Even if you have had some experience, an extra person on the ground can make loading an unfamiliar horse easier.

Travel Gear and Weather Precautions

Outfit the horse in travel gear by placing shipping boots or wraps on the legs to protect from bumps or scrapes (see chapter 8 for boot and wrap information). Some people put a protective head bumper on the horse's poll and use a tail wrap as well. The horse should also wear a leather halter rather than a nylon one so that in the case of a severe accident, the halter will break free.

Although horses have a harder time dealing with heat than with cold, they should still be protected from cold while trailering. If the temperature is below freezing and a horse is being transported in an open stock-type trailer, the chill factor created by going 60 to 70 miles per hour can severely injure the horse. Avoid transporting your horse in an open stock trailer in such cold weather. If your horse is wearing a blanket, change it as the weather changes. If temperatures increase during the day, the heavy blanket your horse needed in the morning may now be too hot, causing your horse to sweat and lose fluids.

Regardless of the duration of the trip, your horse should be protected from extreme heat as well as from extreme cold. One good option is to travel when weather conditions are better. During hot weather that means traveling in the evening, at night, or in the early morning. Avoid high noon.

Loading Techniques

Do your loading in a quiet area, with few distractions around. And set aside as much time as possible for loading; the more you hurry, the less your horse is likely to cooperate. Have the trailer open and ready, with the ramp down or doors open. Have your helper assist in guiding the horse in, as well as in closing up the trailer. Horses generally know that they are going to be loaded once they see a trailer, but allow your horse some time to inspect the trailer if he's wary.

Horses gain confidence from their handlers, so be confident and relaxed during loading. Calmly lead your horse as far as you can, keeping your own attention forward. Reward the horse's every positive step by releasing the pressure momentarily on the lead rope, but take up the pressure again to make him go forward. Some horses may hesitate, but then decide to go in, so be prepared to exit out the escape door (the small "people only" door in the front of the trailer) as soon as he is fully loaded.

There is a difference between a horse that has a fear of loading and a horse that is just being stubborn about doing so. If you're dealing with a stubborn horse that will walk into a trailer halfway and then back out, don't let him think he is calling the shots. Make it your choice to back him out, and do it swiftly. Once back

Safety equipment, such as the shipping boots and a blanket on this horse, help protect your horse when he's traveling in a trailer.

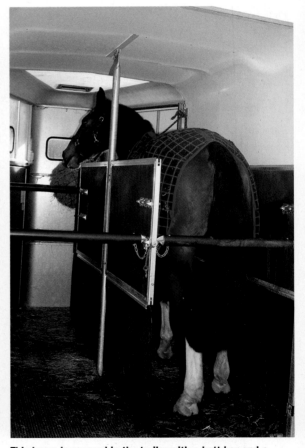

This horse is secured in the trailer with a butt bar and a trailer tie with a safety catch. He also has a hay net.

outside the trailer, create busy work of circles, backing up, and going forward. Cease the busy work when the horse makes an attempt to load—this way, he knows he can relax inside the trailer, but outside he must work.

If the horse is truly fearful, work with a professional trainer to make the horse more confident in loading and unloading.

Once the horse is standing in the trailer, attach him to a trailer tie that has a safety snap, while your helper puts up the butt bar and ramp or closes the doors. Make sure all latches are secure. As mentioned earlier, studies have shown that the horse benefits by having full freedom of his head and neck so that he has enough room while aboard to lower his head to cough, so it is fine to leave the horse untied. Keep the air in the trailer fresh by opening all the vents and maximizing your trailer's ability to exchange air by adding fans or putting screens on the windows.

Many people will put hay in the trailer to keep the horse calm and occupied while traveling. This is fine, although it's best to soak the hay in water beforehand to cut down on any dust or debris flying around the trailer. If you're going on a short haul and your horse is relaxed, leave the hay out altogether.

DRIVING WITH A TRAILER

While most people learn to drive a passenger vehicle through a formal driver's education course, the majority of drivers learn to haul horse trailers through experience. Considering how precious the cargo is, "learning by doing" hardly seems rational, but trailer-driving schools are fairly few. However, there are ways to safely learn without endangering your horse, and techniques available to help sharpen your driving skills.

Confidence comes with experience. If you're driving a trailer for the first time, take along an experienced

When driving with a trailer, be extra vigilant. Be sure to keep a safe-driving distance between you and other vehicles.

passenger who can help out in sticky situations and warn you of problems to come. And take time to practice with an empty trailer in a parking lot.

Hitching the Trailer

Hitching up is one of those jobs that looks easier than it is. Unless you have a friend to direct you as you back up, hitching can be very tricky. It is easier with a gooseneck trailer because you can see the ball while you are backing up to the trailer. Hitching up a tag-along (bumper pull) is more difficult because you can't see the hitch or the coupler on the trailer. Practice is needed.

You can create a visual guide by taping a strip of electrical tape on your truck-bed, right above where your truck hitch is located. As you back up, aim the strip of tape toward your trailer's hitch and you should land within inches. There are also products on the market that to help you to align your trailer.

All trailer hitches are not compatible, and the ball on your truck may be too small for a borrowed trailer's hitch. If so, the trailer can bounce off it. You will need to purchase a different ball from an auto supply store. If you are borrowing a trailer, ask the owner to show you exactly how to hook everything up.

Stopping and Going

Rapid driving can cause serious accidents, and quick braking can cause a trailer to jackknife. When driving a horse trailer, always drive your rig at the speed limit or

slightly under it, and anticipate that people will pull out in front of you. Allow plenty of space between you and the vehicle ahead. Although the space you leave varies depending on driving conditions, tow vehicle size, and gross weight, a good rule of thumb is to at least double the car-safety following distance to four seconds or one rig length for every 10 mph. To gauge the four-second distance, begin counting seconds as the car in front of you passes a focal point, such as a road sign or tree. You should count to the number four as you pass that same point. Of course, you may find that other drivers won't respect your braking distance. If a vehicle pulls into the space you've created, fall back to regain a safe distance. This cushion could save your life or your horse's life in the event of a sudden stop.

For your horse to have a comfortable and safe ride, make all changes in speed gradual. Keep to the slow lanes. Unless you have to maneuver around stopped traffic or a driver going so slow that it's dangerous to follow him or her, stay out of the fast/passing lanes.

There are three ways to slow down the trailer: stepping on the truck's brake pedal, engaging the manual trailer brake control, and downshifting your transmission. Downshift to a lower gear when you need to slow down on a steep hill. This lets the engine do the job rather than overtaxing either your truck or trailer brakes. Use your truck's brake pedal for normal braking. Use the manual trailer brake when you need to slow the trailer and not the truck, such as if your trailer begins to sway or if it comes off the hitch.

Always keep in mind how much longer it takes to stop a truck and trailer than it does a car. Practice

A woman unloads her horse from a walk-through trailer; it allows her to unload without having to back the horse out.

stopping in an empty parking lot, so you can get a feel for how long it takes to bring your rig to a halt.

Turning

Although a gooseneck trailer (*see Horse Talk, page 216*) has the advantage of being able to turn around on a small radius, that advantage can go sour when you are maneuvering around normal corners. The gooseneck is more likely to cut the corner, so plan to drive farther out before you make your turn. By contrast, a tag-a-long trailer will obediently follow around a turn in the tracks of its tow vehicle and will be less likely to catch on the curb or other street objects.

Practice turning in an empty parking lot. Make sure your mirrors are adjusted properly. You should be able to see alongside the entire trailer (on both sides) and beyond. Then set out a single cone, and practice turning left around it. Keep an eye on the cone in your mirror as you make your turn. If you find yourself driving too close to the cone, steer a little bit wider next time. Now practice turning right around the cone. This way it is much harder to keep track of the cone in your right mirror. Just remember to swing out wider.

Backing Up

Backing up is one of the most difficult trailering skills to learn. To back a trailer you must turn your tow vehicle in the opposite direction from the way you want your trailer to go. If you want to send your trailer left, you must turn the steering wheel to the right.

A method that works to help train your brain is to put one hand in the center of the bottom of the steering wheel. If you want your trailer to go right, move your hand to the right. If you need to make a sharp turn, turn the wheel and then accelerate. If you need to make a gradual turn, accelerate and turn the wheel as you go. Once you're in line with your target, straighten the truck wheels to line up the tow vehicle and trailer.

Again, empty parking lots are great places to practice your skills. Set up an alleyway of cones or boxes and try to back between them. Start out backing straight into the alley, and then practice backing into the alley while you are turning. First, try it with the bend in your rig on the driver's side so you can see what is going on in your mirrors. Give yourself plenty of room before you begin, and make a sweeping turn backward to line yourself up. Then, practice with the bend on the passenger side.

Further Trailering Suggestions

Be sure to close and lock all windows and doors. It is dangerous to let your horse hang his head out of a moving trailer.

Plan your route so that you don't get lost in places you can't get out of. Know the height and length of your trailer. Don't drive under a railroad underpass or fast food drive-through unless you know how tall your trailer is.

Make sure to take a wide enough swing into gas stations as you pull up to the tanks. Heavy poles protect the pumps, and you have to make sure you give them enough clearance, particularly with a gooseneck.

Post a notice in a visible place listing any numbers emergency personnel can call to get help taking care of the horses in case you're in an accident and are incapacitated. Don't expect most emergency personnel to know how to handle the horses.

Never change a tire on the road unless you are qualified. Proper lug torque and torque sequence are very important. If you do change a tire by yourself, have a qualified mechanic properly torque the wheel as soon as possible. The moment you notice a flat tire, maintain a safe speed as best as you can. The safety of you and your horse is more important than a wheel or a tire. Get to the side of the road as quickly as possible and call for help. Once your repair is complete, build up momentum on the side before pulling into the lane of traffic.

In straight-load-style trailers, always put the horse on the left side of the trailer. There is an increased incidence of trailer turnovers resulting from horses' being loaded on the right-hand side because, in the United States, the middle of the road is usually higher than the sides.

Chapter 10

Stabling

Have you decided where you'll keep your horse? At a boarding stable or at home? Not everyone can afford the luxury of horse property, nor should novices take on the incredible responsibilities that having a horse at home requires. A good alternative is keeping your horse at a boarding stable. Living space for your horse can be as simple as a pasture with many other horses, or as elaborate as a high-end, full-service equestrian center. There are also many options in between. The key is to find a place that meets your needs and the needs of your horse.

For some people, a horse at home is the ideal situation. It is a feeling like no other when you can look out the kitchen window and see your horses in the field. There are a number of benefits to keeping horses on your own property, such as easy access and daily contact. You do, however, have to be prepared to provide and maintain appropriate housing, fencing, pastures, arenas, and other facilities.

BOARDING A HORSE

If you're choosing to board, do your homework. Begin by gathering a list of boarding places in your area. A good source for listings is still the local tack or feed-store's bulletin board. Check online for sources as well. Sometimes you can discover a private ranch that is offering space for a horse; while others are regular public boarding barns.

When you have selected a prospective facility, give the management a call to set up an appointment. Have the manager take you for a formal tour of the facility and point out the amenities. Determine whether you like the size of the arenas, and find out where your horse's proposed stall would be. Check that the horses appear healthy and well fed and that the facility looks well maintained.

Once you have found the place to call home for your horse, you will sign a boarding contract. Evaluate this carefully. It should have information on deposits, services, duration of contract, and how the contract can be terminated.

Do not simply accept a handwritten note on a scrap of paper as a receipt for your deposit. If the boarding facility says that it does not provide a contract, then you should draw one up for yourself—it will protect both of you in the long run.

Some individuals thrive in a boarding-barn environment. There are people to ride with and plenty of activities going on. Other individuals, however, may be miserable in that situation. They don't like other people "in their business," or they prefer to spend time with their horses rather than socializing with other people. Different boarding barns have different types of atmospheres. Some establishments can be very friendly, while others are cliquish. By carefully researching facilities ahead of time, you can avoid the pitfalls of choosing the wrong stable.

Boarding Amenities

Think about what your riding goals are and what you want in terms of care and maintenance for your horse and decide whether a particular facility has the amenities you are seeking. Here are some of the questions you should ask and some other factors to consider when deciding on a facility.

Questions to ask management:
- Are there riding instructors at this facility who teach my riding discipline or one that I wish to learn?
- What type of feed do you provide for the horses that you board?
- What hours is the barn open?
- How often do you feed the horses?
- How often does your staff clean the horses' stalls?
- Do you have vaccination and deworming requirements for all horses on property? What are they?
- What does the price of board cover? Are extra services available, such as blanketing in the winter?
- Does the facility require insurance?
- Is there anyone on the premises who is available to handle emergencies?

Other points to consider:
- Is there storage space available for trailers?
- Does the facility have any space where extra feed can be stored?
- Are there tack lockers for boarders?
- Does the facility provide washracks?
- Do boarders have access to nearby trails?
- Is there a round pen on the grounds that is available for training?
- Are outdoor areas lighted?
- Who are the riding instructors and trainers associated with the facility?
- Is there a riding lounge or clubhouse, and what is it like?
- Do boarders have access to all arenas and amenities?
- Are there on-site laundry facilities available for the use of boarders?
- Is there an indoor arena, and what is it like?

Try to speak with present boarders to get a feel for whether they are happy with the facility and with its management. If you do so, be sure to talk to more than one or two people. The more people you speak with, the more accurate will be the picture you get of the true state of affairs at the facility.

Boarding your horse at a facility may be your best option if you lack the stable space yourself.

SETTING UP A HOME FACILITY

If you have enough space on your own land and are thinking about keeping your horse there, you should make a thorough survey of everything you have and everything you need. All must be in place before you bring your new horse home.

You may, for example, already have an existing structure on your property that you're thinking might work for your horse. But will it? What kind of shape is it really in? Old barns do have a certain charm, and building a new barn can be expensive, but old barns also can prove to be inadequate for housing a horse in comfort and safety. If your present barn does not make the grade, then you can choose to augment it or to replace it altogether.

Other aspects to consider are fencing, pastures, manure management, arenas, and pest control. One of the biggest responsibilities of keeping your horse at home is stable maintenance and fire safety. Whether your barn is old or new, you must keep it clean and keep it uncluttered.

Old Barns

Old barns were designed to shelter livestock such as cattle, which have different needs than horses do. For instance, old barns often have small stalls. If your stalls are too small, not only is there a risk of your horse's getting cast (trapped where he cannot rise from the ground), but your horse will also have a hard time keeping out of his manure. He'll be tracking it all through the stall, which means cleaning the stall will take longer. If your barn has a good open floor plan, though, you can create a better stall system with prefab stalls.

One of the biggest factors to consider when you are deciding if it's worthwhile to renovate the barn is ventilation, which is a health as well as a maintenance concern. Ammonia fumes from urine can damage the horse's respiratory tract. In addition, spores and particles from mold, hay, and dust can be harmful and can accumulate, particularly if the hay is stored above the horse's stalls. Keeping ammonia from taking control can be a difficult task.

Many old barns are built against banks and have low ceilings and few windows. As a result, they lack the number of air exchanges necessary to maintain proper ventilation and health for the horse. Trying to fix ventilation issues is a losing battle. Properly placed windows and doors and open soffits at the eaves will create good air exchange, but they aren't easy to add to an old barn. If your barn is difficult to maintain, it may be time to replace it.

Most old barns were also designed for human convenience over safety. In the past, for example, hay was usually stored overhead in a hayloft. Keeping the hay there made it handier for the owner, but it also made the barn a more dangerous place. Given the combustible nature of hay, convenient storage, whether overhead or in a stall next to the horse, is not be the safest option for horses.

So evaluate your old barn very carefully, then decide whether it is adequate as is, needs to be renovated, or needs to be replaced altogether. If you decide to replace the barn, you can choose from several options for a new structure.

STABLING

Learning these terms will help you talk stabling.

■ **Cast:** When a horse tries to roll and becomes stuck against one of the walls or corners or the door.

■ **Harrowing:** Harrowing smooths the arena's surface, usually by pulling a piece of equipment, called a harrow or a drag, behind a tractor.

■ **Hot-walker:** A machine that walks the horse around in a circle.

■ **Modular barn:** A pre-engineered barn that is assembled quickly and is easy to customize.

■ **Pole barn:** A barn that's open and has no internal supporting structure. Often used to store hay or equipment.

■ **Sacrifice area:** A small paddock with no grass, used during rainy or muddy days to protect pastures.

This old barn may be filled with character, but it may be too far gone to be worth renovating to accommodate horses.

Modular Barns

Many people choose to purchase a modular or pre-engineered barn. The benefit of a modular barn is not only the speed of building but also the opportunity to customize. Manufacturers will plan a barn for you based on your needs. These barns are available in many sizes, from a basic pasture shelter to a forty-stall-plus barn with attached indoor arena. Modular barns can be customized to create an easy-to-maintain barn, such as one with wider stall doors and aisles that allow you to strip a stall clean with a utility tractor's bucket rather than by fork and wheelbarrow. And if your horse chews wood, you can request that the stalls be lined with wood but topped with a metal bar.

There are many companies out there that manufacture barns. See as many barns as you can before you choose a company. Don't just go see the barns that have only recently been put up; go see the ones that have wear and tear to evaluate how they have held up. Ask the manufacturer to give you references.

Handmade Barns

There is a growing trend for Amish-made barns because of the handmade quality and work ethic of the builders. People living near Amish country, in particular, can often find a builder easily, either by word of mouth or by referral. The Amish, with their shunning of electricity and laborsaving devices, design barns with easy horsekeeping in mind. For example, they use pressure-treated wood to line the indoor wash stall so the wood will not rot; they also use tongue and groove when they build the barn walls so that as you wash down the wood the debris will flush out rather than sit in the grooves or spaces between the wood.

BUILDING A HOME ARENA

Many home riders make due with a sectioned-off piece of ground near the barn that they call their arena. But without a good base and proper footing, it's difficult to make the going consistent enough for your horse's health and safety. The job of the base is to make a level surface to support the footing above. If the base is properly installed, there shouldn't be low spots, deep spots, or areas that collect water. The ideal footing helps cushion soft tissue by allowing the horse's hoof to slide just a little bit as he sets it down and provides a firm surface for the hoof to bite into as the horse pushes off. Natural ground will never provide these benefits. As a result, your horse's performance may suffer because he won't be able to trust the footing, and he will remember, perhaps spooking at uneven spots.

Before you put one shovel in the ground, get organized. Think about where you want to put the arena. Right next to the barn may seem convenient, but if that is an area that collects water then think again. Drainage is a concern in any arena, so if possible you'll want to avoid land that's already wet.

Next, you'll need to get a soil consultant, an engineer who understands all the ins and outs of working with soil. You can begin by seeking free assistance through your local office of the Natural Resource Conservation Services (www.nrcs.usda.gov). NRCS staff

Easing the Arrival

Once your horse has arrived at his new home, he will be a bit anxious, and will whinny and even pull hard at his lead rope in his first moments off the trailer. He may spook at foreign sounds and sights. If he is too uncontrollable, have someone with experience lead him around while he calms down. Place him in his freshly bedded stall or paddock, and make sure he has hay and water. He may be too distracted to eat at first, but he will eventually settle down. You may choose to get him out for a walk about the property later, after he has quieted.

In the days that follow, be sure to allow him time to familiarize himself with his living area—not just his stall, but the grooming area and riding arena or trails. Turn him out if he has a lot of nervous energy. Spend time grooming him and allowing him to get used to your voice and mannerisms. Horses are very adaptable creatures, and as long as you develop a regular routine, your new horse will soon be content with his new life and schedule.

can also provide good information on your soils (such as the percentage of sand, silt, clay, and gravel in your soil) and will often come out to your site. The NRCS office may recommend a soil consultant in your area. You can also look for someone online or in the yellow pages (under engineering consultants), or ask other people for personal recommendations.

The next step is to obtain a permit. Any time you disturb at least 5,000 square feet of earth (which is not being plowed for agricultural use), you'll need at least a grading permit. You may also need an erosion-and-sediment-control permit. You may not think you are disturbing that much land, but you will disturb twice as much earth as the size of your arena.

Base of the Arena

You will never have a good arena without an adequate base. Your soil consultant will help you determine your base requirements. The base should be about six inches below the surface, if it's to be stable. You want the base to be firm, essentially a road base, because it will create a level surface for the footing above. If it isn't stable, consider bringing in some crushed stone to form a firm base. The stone will have to be rolled when installed to create maximum hardness and evenness and to make sure the stone stays in place.

Creating a Drain

Anyone who has ever had training sessions interrupted because of a flooded arena knows how frustrating that is. Sinking money into an arena that sits underwater a good part of the year is a waste. Your soil consultant can also help you avoid drainage issues. He or she will first determine what your soil is like from the surface down to three feet. Perhaps you'll have three feet of pure clay or maybe six inches of loam and then six inches of sand, followed by gravel. This is important to understand because you need to know how your soil handles water.

If you do have soil that doesn't drain well, you don't have to give up on your dream arena. There are steps you can take to limit flooding. You can install a French drain, which carries water away from the arena, or crown the arena, so that the footing isn't flat with the

English riders train in their home arena, which took a great deal of proper planning, design work, and time to build.

This horse's owner erected traditional wood fencing—not only sturdy but highly visible. It prevents accidents but is costly to maintain.

ground. It stands up above the ground and is level (so as not to create gullies), it is sort of like making a big sandbox. The water will now run off of the sides.

Arena Footing Options

There are many types of footing additives on the market, from shredded felt to crumb rubber to polymicrofibers, but most footing recipes begin with sand. And any old sand won't do. You want sand that will bind and hold together. Sand comes in a variety of shapes, some are sharp and angular and some are rounded.

Sand performs in different ways depending upon its characteristics. It can be slippery if it's round, like tiny ball bearings. The round grain will act like beach sand and be unsteady underfoot. Particularly if the sand is dry, you will find it difficult to move through. Because sharp sand binds together and holds better, it is the preferred footing for arenas.

Sand by itself works well, but you will need to be careful when watering. Too much water can flood the arena, making it unworkable for a long time. Wet sand is harder than dry sand, but very dry sand can be inefficient and create a dust problem.

Footing additives can help to improve cushion and traction. They also help sand to clump together better while also preventing compaction. Some additives may help prevent freezing and cut down on trenching along the arena's walls. Additives create footing that requires less harrowing and watering. When choosing an additive, make sure that it's manufactured for riding arenas.

Sometimes people use arenas for dumping grounds for manure and shavings. Although it can hold water well, manure creates a health issue for both humans and horses. The shavings also break down quickly and create a slippery surface.

FENCING

Horses are nomads by nature. Although wanderlust and curiosity may be fine in themselves, enclosing any creature with this type of nature is problematic. Doing so requires good fencing, which can be expensive to buy and maintain. There are many types of fencing to choose from, however. Do your homework before you buy, and determine what makes sense for you.

First, think about your specific needs: How many horses do you have to enclose? What are the horses' temperaments? Do you have any stallions or foals? Do they run in their pastures, or do they spend their time behaving like lawn ornaments? How big is your pasture or paddock? Do you have enough money now to pay for top-grade fencing that will last you years and be inexpensive to maintain, or do you need a quick, cheap solution for the time being?

Assembled below is a list of the most popular fencing along with a discussion of their pros and cons. Once you've chosen your fencing, find out who manufactures it and call up the companies. Ask lots of questions. Among them, what guarantees are included, whether there are any extra costs, and what maintenance costs. If you still have questions, consult your local extension agent.

Wood Fencing

One of the major drawbacks to wood fencing is its palatability to horses (many planked-fence owners find that any decorative points of their wood fencing are munched down to nubs). Wood is also prone to termites, splintering, and breakage. Regular painting is needed to keep this fencing looking good and to protect it from the sun, wind, and rain. Planks can pop out easily, allowing escape artists to explore the countryside. It's expensive to buy and expensive to maintain and easily climbable for trespassers, In fact, a Texas A&M study showed that the average wood fence ended up costing twice the original price in additional materials and labor over a fifteen-year lifetime, with fully 50 percent of the rails being replaced at least once during that time period. Yet wood fencing appeals to many people because it is both traditional and attractive. Wood fencing is also clearly visible to horses, marking their boundaries well, which is particularly important when a horse is galloping toward the fencing with a good head of steam.

Polyvinyl Chloride Planks

Polyvinyl chloride (PVC) planks have the look of estate fencing mixed with slats of an old-fashioned Hot Wheels toy racetrack. This type of fencing is good looking, visible, and horse friendly—maybe a little too much so. Horses that lean, searching for the greener grass on the other side of the fence, find the PVC planks comfortable, but their leaning can stretch the planks of vinyl fencing. If a plank of good quality is stretched, however, it will return to its original shape. One of bad quality can break into splinters.

You can discourage leaning by topping the planks with a strand of electric wire. The drawback of PVC fencing is that it is more expensive than standard

The owners of this horse farm chose PVC fencing, which is more expensive to buy than wood fencing but less costly to maintain.

wood fencing. You can consider it to be an investment, however, because the maintenance costs are low and the panels last.

High-Tensile Polymer Fencing

High-tensile polymer (HTP) fencing was designed specifically for horses. You can liken it to a large rubber band wrapped around your pasture. If a horse hits this fence, it is highly resistant. The entire length of the system shares the shock of the jolt then springs back to shape. If it doesn't spring back, you can tighten it

quickly and easily, making it as good as new. It comes in one or more bands, although the single-strand fence is the cheapest and easiest to install and maintain. The fencing's pros are many: long lived, little upkeep, and easy for anyone to repair. It's good in wooded areas as it can handle the impact of falling branches. The fence line can be contoured to follow the lay of the land with no corners needed.

Mesh Fencing

Diamond-weave mesh fencing is strong and impassable by hooves and legs, has give, is long lived with little maintenance, and is stallion worthy. You must make sure, however, that the fence is built so hooves and legs can't get stuck under the wire. Maintenance costs tend to be low with wire fencing, but be sure the fence you choose is galvanized, giving it a protective zinc coating, which stands up to the elements for years.

When installed, the fence should be stretched to a firm tension so it does not sag. Make sure, however, that it is not overstretched. If the tension is properly set, the woven design will provide some additional give if a horse or another animal runs into it. (Wood, pipe, electric, and welded-wire fencing do not have this type of

This high-tensile polymer fencing will absorb a horse's impact and then spring back into shape. It is cost effective and resilient.

give, which will result in either the horse's being injured by the ungiving fence or the fence's breaking, exposing sharp edges that could further injure the animal.)

Installation is not easy. You can put it up yourself, but you'll need to be savvy with heavy tools, such as chain saws and hand stretchers. You'll also have to enlist the help of several people due to the heavy weight of the rolls. Instructions are available online, but anyone who has never built anything before may become easily overwhelmed. It is better to hire a professional.

Electric Fencing

Electric fencing is more of a psychological barrier than a physical one because any horse that has touched the fence and discovered that it bites back will give it a wide berth. The downsides are that electric fencing used in a small enclosure, such as a run outside of a stall, can discourage a horse from rolling or lying down. In addition, some horses are afraid of electric fencing, and even the tick or hum of the generator can cause anxiety. This doesn't happen often, but it does happen.

Horses can still run through the fence, particularly if they can't see it. If you're planning to install high-tensile electric wire on your property, tie streamers along the wire so your horse can see the fence perimeter. This fencing is cheap, but you'll have to constantly be on the lookout for blown-down streamers. If your horse does manage to run through an electric wire fence, he could injure himself.

Great improvements have been made with electric fencing over the years, but with them costs have risen. One such improvement is poly tape, which is made of an open mesh woven material encasing several chargeable wires. The benefit is that it is highly visible, light, and easy to install.

The drawbacks are that due to weather stresses, electric fences may last only a couple of years. Then you have the additional cost of replacing the entire system. With some systems, stretched webbing can be tightened and torn webbing can be spliced, but even though the mesh may look fine, the wires inside can be damaged, causing shorts to the entire system that will be difficult to track down.

A good, easy-to-maintain improvement on the mesh tape is braided electric rope. It doesn't short out as easily and is ideal for horses because of the material's soft nature. It can be tightened if it sags due to weather and wear; some styles are flexible, similar to HTP fencing.

The poly-tape electric fencing shown here is easy to install and very visible, but lasts only a couple years.

Barbed Wire and High-Tension Wire

It's inexpensive, but only at the outset of cost. Numbers quickly add up when you weigh in the cost of vet bills. A horse snarled in barbed wire is a nightmare no one should ever have to experience. Barbed wire was originally designed to enclose sheep and cattle, not horses. Injuries with barbed wire are rampant, and the wounds that they cause are jagged and leave scars. Tensile wire tends not to snarl a horse, but the shearing injuries of limbs are very high. If you must use wire, add streamers and planks for visibility and perhaps a strand of electric wire for safety and peace of mind that your horse will never go near it. Never use wire in a small enclosure; when your horse rolls, his legs can get caught.

PASTURES

No one wants to see horses pick through a weedy, over-grazed pasture. A poor pasture, however, is more than just an eyesore. Overused pastures have a big environmental impact. During a rainstorm, water droplets hit the ground and begin to move downhill. If you have a thick pasture, the rain will slow down and the grass will allow it to soak in. An overgrazed pasture with bare spots, on the other hand, provides a smoother surface so water will pick up speed and velocity and will not have time to soak in. As it moves, the water picks up loose soil, manure, and fertilizer. If rainwater is allowed to rush through the manure pile, it can contaminate your groundwater and well.

If the water off your land hits a stream at a high velocity, it will erode the stream bank. All that silt and pollution is deadly to a stream. When manure and silt

To keep a pasture this healthy, you must periodically rotate your horses. Horses spot graze, creating bare spots that need reseeding.

run off the land, different nutrients make it impossible for that ecosystem to exist and everything starts to crash. Prevention and maintenance are the keys to healthy pastures. It's challenging to manage pastures because horses are spot grazers. They graze in their preferred lawn area and use the rough for elimination.

Not only do bare spots cause erosion, but they also allow weeds to get a toehold. In spring and fall, walk through your pasture and look for bare spots, throwing down grass seed as you go. And don't be afraid to mow. Allow the horses to graze the field down, remove them, and then come right along with the mower. Topping the grass off will make the areas of rough more palatable to horses and cut the weeds back before they go to seed. Lawns should not be below the length that is right for your area, about four inches or so. (If you need guidance, contact a turfgrass extension specialist.) When lawns are down, rotate your horses to a sacrifice area.

MAINTENANCE AND FIRE SAFETY

A clean barn is a healthier and safer place for horses, owners, and visitors. Keep the stalls and other areas mucked out, swept, and decluttered on a regular basis.

Make sure you only keep your horses and their related equipment and accessories in your horse barn. Storing hay and other machinery in a separate building, as well as not smoking near the barn, will help you to prevent accidents, including fires.

There is no such thing as a fireproof barn. A barn fire can happen anywhere because of the nature of the

Stable workers sweep the stall area to maintain healthy conditions and help prevent fires from starting.

building's use. You can, however, lower the likelihood of a fire. Fire is a chemical reaction that needs three elements to get going and to survive: oxygen, fuel, and heat (the fire triangle). Horse barns have all of these in spades. Barns have good ventilation to allow influx of healthy air (oxygen). Barns are filled with hay and bottles of flammable liquid such as alcohol, hoof paints, and creosotes (fuel). People have electricity installed for objects such as lights, heated water buckets, and stall fans (heat). Take away any of the three elements and a fire can't happen.

Routine Barn Maintenance

It's easy to accumulate loose hay, bedding, and baling twine, and allow the buildup of birds' nests and dust. When this happens, the barn becomes not only untidy but also unhealthy and more dangerous.

Organic debris can be fuel for a fire. Innocent-looking cobweb can even be a fire hazard. Chains of webs can create a pathway for the fire to travel very quickly from one end of the barn to the other. Bits of flaming cobweb act as tinder and can drop into another stall and allow the fire to spread. Cobwebs can also spark a fire by touching a hot light bulb. Keep a safety shield over light bulbs so that webs and other matter cannot come up against the bulbs and start a fire.

Preventing rodents is very difficult in a barn, no matter how many cats you employ, but you can keep your barn cleaner to hold the numbers down. Rodents not only create mess and carry diseases but also like to chew on the plastic around electrical cables. PVC and other plastics include salts in the production processes that are tasty to mice so encase all wires in metal cable, ridged cable, or EMT conduit. If you're ever in doubt, get a qualified electrician's advice.

This is why it's critical that you keep your barn as clean and as clutter free as a horse barn can be. Sweep up debris (such as hay or spilled feed) daily, and pull down any cobwebs as they develop.

Think also about fuel sources for a fire on the outside of the barn, particularly if your barn is situated close to a road. A lit cigarette flipped out of a car window can spark a fire in dry weather. Cutting down weeds and trimming the grass around the barn on a regular schedule will make a firebreak and stop the flames from spreading either to the barn or away from the barn.

Compost Temperature

It's true that manure piles can get hot, but there has never been evidence of piles igniting. The internal temperature of the pile should be around 140 degrees Fahrenheit, which means the pile is composting correctly. You can check your compost's temperature by inserting a garden thermometer into the pile. If you think it's getting too hot, you can turn the pile with a pitchfork or a tractor.

Barn Storage

There's just something about a barn that makes people want to store anything and everything inside it. As noted earlier, it's best to keep only equines and equine-related equipment and accessories inside the barn. House your tractor and your other gas-powered equipment that could ignite away from your horses. Machinery also creates clutter, which can block exits and pose tripping hazards for visitors and for emergency personnel.

People often keep minerals and solvents and rags for cleaning—all of them sources of fuel for a fire—in the tack room. If you use an oily rag, dispose of it in a metal trash can away from any storage or lay it out flat, away from the barn, to let the oil evaporate. Do not store flammables such as paint, gas, and oil in the barn.

Think seriously about how you manage your hay. Keeping loose hay off the floor is only one step in preventing a fire breakout with hay. Hay is nothing more than dried plant matter, and nature wants to continue breaking that matter down. You only need to watch steam rising off a manure or compost pile to see how powerful bacterial reactions are. If hay is not dried (cured) correctly before it is baled, bacteria will begin to do its work and mold will form. Mold break downs grass, creating heat, which can lead to spontaneous combustion.

One step you can take to prevent this reaction is to salt your hay. This sets up a saline environment, which eliminates or lowers the ability for mold to grow. You can buy 50-pound bags of salt used for water softeners at building-supply stores. When you store your hay, sprinkle a scant cup of salt over it and repeat between each bale. This is an old process that farmers and ranchers use in the Pacific Northwest, and it doesn't change the nutrition of the hay.

If you smell a musty or sooty odor, you probably have a suspect bale of hay. You can test your bales by taking their temperatures. Insert an outdoor thermometer or a probe cooking thermometer. If the thermometer reads 150 degrees Fahrenheit, you have cause for concern; so check the temperature again in four hours. If it reaches 175 degrees or higher, phone the fire department immediately. Do not try to sort the bales out yourself. A suspect bale will already have two legs of the fire triangle: heat and fuel. Pulling a bale out into the open will introduce the third leg: oxygen. You don't want a bale to burst into flames.

Convenience has often been the uppermost consideration when it comes to where hay will be stored. As close to the horse as possible seems easiest, so most of us either store hay in a hayloft or in a stall or hay area next to the horses. But given the flammable nature of hay, it is safer to move the hay to a location outside the horse barn, to a pole barn or other structure. There are, however, ways to prevent a fire if you don't have these alternative storage options. Keep air on the hay so that it remains dry. Stack it in rows of two bales with a half- or quarter-row space separating each stack so that there is air circulation between the bales.

Moving hay outside in a pole barn will cut down the fire issue. Doing so does make the maintenance issue higher. To cut down on trips back and forth between barns, you could stack several bales in a stall fitted with fire retardant treated wood (FRTW). This is made of a mixture of lumber and plywood that has a fire-retardant solution mixed in. It works by delaying the spread of flames and cuts down on smoke. All fire-retardant wood will be stamped with FRTW. If the wood doesn't have this stamp then it is not legitimate. Alternatively, you can paint over a stall's existing wood with fire-retardant paint and varnish. These materials have solutions in them that slow the spread of flames. All fire-retardant goods have a rating system in hours, which will tell you how long you have before the fire begins to burn out of control.

Electrical Devices

The number one cause of barn fires is electrical, so have your electricity inspected once a year. Don't overload fuses by using too many extension cords. Although unplugging electrical devices when they are not in use is good, a better idea is to turn off the electricity when

This electrical switch safety cover is a simple yet effective tool in disaster prevention against barn fires.

leaving the barn. If possible, have an outside switch that turns off all of the power to the barn.

Only code-approved, permanently installed heaters should be in the barn. Portable space heaters can make a night of foal watch bearable, but any portable heating or lighting device is dangerous. Wrap up warm instead.

Coil bucket heaters are a big convenience when heating up water, and they work very well. But this is another area where complacency can set in. Never walk away from the heater. It's too easy to get sidetracked on another task and forget it. The coil burns right through the bucket, then the barn burns down.

Disaster Planning

You have to be on autopilot in a disaster; you must know what to do in an emergency situation to prevent or minimize injury and property damage. This is why police and firefighters train extensively for their jobs. You don't want to be devising your disaster plan in the midst of a disaster, when you're in panic mode.

The first step in establishing a plan is to map out the layout of your barn on paper: where the feed and hay are kept, where each horse stays, and where the tack is. Next, note down exit routes from the barn and places where you can safely secure the horses. Simply turning them out of the barn or assuming they will leave on their own is out of the question. The idea that horses will run back or remain in a burning barn is sadly true. Horses are used to their stalls and used to their routines, so if a stable turns into a fire scene the horse won't leave it. He will stay in the barn and die.

Finally, include how wide the driveway is and where horses can be put and where the nearest source of water is—this can be a well, a pond, or a river. Include emergency telephone numbers (veterinarian, your numbers at work and at home) and a list of experienced local horsepeople who can help to evacuate your horses.

Distribute a copy of your plan to everyone who has anything to do with your barn or horses. Take a copy to your local fire station, and keep another in a cylinder, marked "in case of emergency," by your mailbox.

Post a large notice by your phone stating the location of your barn, including street names and any landmarks, which can be difficult to remember in an emergency. Write the following statement on the note to remind yourself to say it to the emergency operator: "I have a horse-barn fire with living animals, not a storage-barn fire."

Make sure you have the right equipment in your barn at all exits, such as a flashlight and a fire extinguisher (have extinguishers inspected yearly). Everyone who spends any time at your barn should know how to work the fire extinguisher. Leather halters and cotton lead ropes won't get hot or melt like nylon will. And always have them right by the horses at all times.

Your Safety Drill

To prepare yourself to act efficiently and effectively in case of a fire, go through the following drill until you can take the steps automatically. First, call the fire department. (Just pretend to do this in your drill.)

If it is safe to do so, remove the horses. Do so one at a time, beginning with the horse closest to the exit first. Remain calm and try to act as though everything is normal. Gently coax the horse out of the stall, and lead him to the safe location you've chosen ahead of time. Tie your horses together in that location.

If a fire is small and can be put out with a fire extinguisher, this can be your next step. However, if the fire is any bigger, or you are at all uncertain, leave it to the firefighters. Black smoke is full of arsenic and cyanide, and only three of four breaths of it becomes lethal. Most people and horses die from smoke inhalation rather than flames.

Practice your drill frequently. Time it, then study what you can do more effectively. Run through it with one person doing everything, then with two people, then with more. You can ask your local fire department to inspect your barn and tell you how to improve your disaster program. And ask them if they will wear their

Fire-Safety Equipment

Make sure that you have the following equipment in your barn and that you keep it clean and in working order.

Home-based smoke detectors: Be sure to keep these clean—they can become clogged with organic matter and fault—and test them periodically to make sure they are working. Heat detector alarms aren't appropriate for barns because there needs to be flames before the alarm kicks in. By contrast, projected beam detectors, called optical systems, recognize a change in the obscurity level in a building, and they work very well for barns and need little maintenance. However, they are much more expensive than smoke detectors.

Intercom system: An intercom system between the main house and the barn will help alert you to any changes. Keep it switched on at all times.

Lightening rods: These help avoid fires by allowing lightening to follow a path to the ground instead of striking the barn itself. Make sure your installer is certified by the Lightening Protection Institute and that their rods are approved by Underwriters Laboratories Inc.

Fire retardant treated wood (FRTW): FRTW can help keep fire from spreading and help stop roof trusses from collapsing. FRTW can stop flames from spreading for up to two hours.

Fire-retardant paints and varnishes: Like FRTW, these products can help slow the spread of flames.

For more information, download the booklet *Making Your Horse Barn Fire Safe* at http://www.humanesociety.org/assets/pdfs/Horse-Barn-Fire-Publication.pdf

fire gear when they lead your horses out of the barn. It's a good idea to get the horses used to that big red machine and the people dressed like Martians who are coming toward them.

Fire prevention goes back to neatness and awareness. Evaluate your barn and your habits, and note where you think a fire might start. Remove those fire hazards, and make sure everyone who enters your barn understands that you are serious about fire safety.

Chapter 11

Nutrition and Grooming

Nutrition is a dynamic process, and it's difficult to have one feeding program for an individual horse for his entire work life. You have to be able to make adjustments depending on the horse's work level, his environment, his age, and any injury he has received. Ninety-nine percent of nutrients are going to come from the hay, grass, and grain that you feed; only 1 percent from supplements.

A horse's digestion is closely tied to his overall health and well-being. Although people may be able to get away with a junk-food diet for a while, a horse cannot, and food-associated diseases such as colic and laminitis (founder) will become an issue without a good diet.

In every horse's diet there are specific requirements, and these are best evaluated clinically. Certainly, there are computer programs available that can evaluate the amount and ratio of protein, fat, and carbohydrates for a particular horse's work level and age category, but the best way to evaluate the horse is to take into account his weight, energy, disposition, and appearance. Veterinarians and nutritionists are best equipped to make those evaluations because they see so many horses.

Grooming is also an important aspect of keeping your horse healthy. Your should make certain that he is cleaned and brushed, and you should be sure that his hooves are picked out daily.

HAY

Forage (hay or pasture) includes plants classified as grasses or legumes. Forages can be made up of all grass, all legume, or a combination of both types grown together. Grasses can withstand a variety of growing conditions throughout the United States. Some grasses commonly used for hay include timothy, brome, fescue, coastal Bermuda, orchard grass, and Kentucky bluegrass. Horsekeepers also feed beet pulp, a forage byproduct left over from the sugar-beet industry, which is easy to digest and made of highly digestible fiber.

Legume hays are usually higher in protein, calcium, and energy than are grasses. Two types are clover and alfalfa. However, they require a warm climate and good growing conditions to produce the best nutrients. They are also higher in minerals, such as calcium, but have an less-than-ideal ratio of calcium to phosphorus. Because they are high in protein, they are ideal for growing horses or those put to hard work, but the calcium/phosphorus ratio has to be balanced out by other feeds to prevent bone abnormalities in growing horses.

You will see hays in other forms, such as pellets, cubes, alfalfa/molasses mix, and haylage (found in Europe). These processed versions usually have a more consistent quality and are good in places where certain hays are not available. However, these hay forms do not satisfy the horse's need to chew long-stem fiber, which hay provides.

A herd of horses shares water in the Nevada desert. Water is essential to a horse's digestion and overall health.

WATER

Water is the most essential nutrient for the horse. Lots of clean, fresh water should be available to the horse at all times. Be aware, however, that horses will change their drinking patterns during the seasons. In warmer, humid parts of the world during the hot summer months, a horse's water intake increases from 4 to 8 gallons of water per day during 65-degree Fahrenheit days; it increases from 20 to 40 gallons of water during 100-degree Fahrenheit days. During extremely cold weather, water intake may be compromised, and the horse will drink less than needed.

In the winter, to help your horse to maintain his core body temperature to keep it up to where it should be, you must make sure that his digestion is able to function the best that it can. That means making sure he has a readily available source of drinkable water—and that does not include snow. Horses need to have a certain amount of water for the digestion process. If your horse is being fed hay and then he has to eat snow, he won't be able to eat enough snow to compensate for the amount of water he needs for digestion and hydration.

SUPPLEMENTS

Before giving your horse supplements, talk to your veterinarian; he or she can advise you. Then do some research yourself on the basics of nutrition. You also need to examine your horse's basic feeding regimen before adding anything. You should have a firm grasp on all of this before rushing out to buy supplements. Too many supplements are, at the very least, a waste of money. Worse, they can compromise a horse's health by decreasing the intake of essential minerals. Too much zinc, for instance, can interfere with the body's use of other vitamins. Don't use any supplement for which you can't determine a specific need.

Preventive medicine is always the most difficult thing to understand—and that's basically what you are doing with supplements: you're feeding to prevent disease, weightloss, and loss of condition. You're doing everything you need to do to maintain your horse's top condition.

A rule of thumb is that if you choose to use a supplement or change a horse's feed, you should probably try it for thirty days, then reevaluate how the horse looks, how he's going, and how he feels. If you don't see a big difference—or at least a measurable difference—then

INGREDIENTS

Learn what the ingredients on the labels mean to help you make the best feeding choices for your horse.

■ **Alfalfa:** Alfalfa is high in protein and calcium.

■ **Antioxidants:** Vitamins and minerals that control free radicals—such as vitamins C and E, beta-carotene, and selenium.

■ **B-complex vitamins:** These vitamins are protein builders.

■ **Biotin (vitamin B):** Naturally present in grass, it helps build strong hooves. It is also called vitamins H and B7.

■ **Blackstrap molasses:** Made of sugar beets, this pure sugar is full of iron, calcium, and potassium.

■ **Chelated minerals:** Include iron, copper, magnesium, cobalt, zinc, and magnesium. They are essential for red blood cells, nervous system function, and protein metabolism.

■ **Copper:** Helps build red blood cells.

■ **Essential fatty acids:** Aid healthy cell functions; improve skin and cell membranes; increase oxygen consumption, energy, and metabolism; increase kidney, nerve, and immune system functions. EFAs are considered beneficial fats.

■ **Flax seed:** A good source of omega-3 fatty acids.

■ **Folic acid:** Helps maintain red blood cells.

■ **Garlic:** This is a natural antibiotic.

■ **Kelp:** Kelp is high in iodine, which helps the thyroid control metabolism, growth, and energy.

■ **Lecithin:** Keeps cholesterol soluble.

■ **Lysine:** Amino acid for healthy tissues.

■ **Niacin:** B vitamin that reduces fat in the blood.

■ **Omega-3:** An essential fatty acid.

■ **Probiotics or direct-fed microbials:** Microorganisms that help digestion.

■ **Rice bran:** Outer husk of rice that contains vitamins, minerals, fiber, and antioxidants. Full of natural vegetable fat.

■ **Selenium:** A trace mineral that aids in building healthy muscles and tissues. Too much selenium can be toxic.

■ **Spirulina:** Blue/green algae that is full of vitamins, minerals, and chlorophyll.

■ **Thiamin:** Metabolizes carbohydrates.

■ **Trace minerals:** Generic name for minerals (such as selenium) that are only needed in minute amounts.

■ **Vitamin B-12:** Metabolizes amino acids and fatty acids.

■ **Vitamin B-6:** Metabolizes amino acids and proteins.

■ **Vitamin C:** Helps ward off infections and stress, and helps collagen synthesis.

■ **Vitamin D:** Helps the body absorb calcium. The body makes its own vitamin D from sunlight.

■ **Vitamin E:** An antioxidant that is particularly essential for newborn foals and pregnant mares.

■ **Vitamin K:** Helps blood clot and liver function.

■ **Yucca:** Improves digestion.

■ **Zinc:** Helps build strong hoof walls.

the supplement is probably not necessary. However, you may feed products that give no obvious results, so it's a bit of a gamble—all the more reason not to feed without professional guidance. You can start and stop most supplements as needed without inducing digestion difficulties. Just don't combine supplements unless your veterinarian advises you to do so.

FEEDING THROUGH THE SEASONS

As seasons change, a horse's nutritional needs require some tweaking in order to avoid illness and weight loss. With a little forethought and common sense, you can maintain that all-important digestive health throughout spring, summer, fall, and winter.

Spring

During the spring, pasture quality tends to be different from that in the late fall or winter. Younger grass tends to be richer in calories and more nutritious.

People often feel it's important to slowly introduce their horses to spring pasture for digestive health. But pulling a horse off spring pasture after an hour can be difficult—horses surrounded by green grass are reluctant to come in—and unnecessary. Spring pasture has a mix of green and dormant plants. If you keep them on pasture, they will slowly acclimate as the pasture becomes greener. Don't wait to turn them out until the pasture is fully green; if you do this, they will just gorge themselves. There is some risk of colic and founder if this happens.

If your horse is completely new to pasture, turning him out for an hour each day; extending each day by an hour or so to gradually change him over is probably the safest method. If you're concerned, one solution is to put a grazing muzzle on your horse. New pasture growth high in fructan sugars may cause laminitis in some horses, especially ponies or those that are insulin-resistant. People who have good pastures can count themselves lucky because the nutrients are very high in living forage. You can supplement with hay, but you will probably notice that your horses will ignore the hay altogether.

One caveat: In the cooler parts of the country, where pastures are more standard, people can get fooled and assume that all the tall grass they see is full of nutrients. Tall, mature grass is actually low in protein, calories, vitamins, and all the other good stuff horses need (apart from fiber). The more mature the grass, the more stem it has

Round bales of hay can be used to provide cost-effective roughage in the winter or when pasture grass is scarce.

and the less nutrients. Two-thirds of the nutrients in grass are in the leaves, so you want a good leaf-to-stem ratio.

Summer

During summer months' extreme heat, you can feed more high-fat and soluble-fiber diets. Hay forages have a higher heat increment or heat of fermentation than more soluble starch-type feeds, such as grains or protein supplements. So in the summer months when the temperatures reach the 100-degree Fahrenheit mark, reduce the amount of dry forage and replace it with fats, oils, or concentrates, such as the newer low-starch, high-fat, and soluble-fiber products, which create less internal heat upon digestion. The heat of fermentation is lower, and the horse doesn't have to struggle to rid his body of excess internal heat. Overall, he will be able to regulate his body temperature better. Remove several pounds of dry hay per day during the summer, replacing the calories with low starch-type feeds or mixtures of oils (usually an omega-3 fatty acid source) as a topical dressing.

Fall

Fall equals relief in hotter regions, but in colder ones, fall means it's time to get ready for winter. For horses, this can mean packing on some extra weight to prepare for winter, when they have higher energy requirements. Have your horse enter cold weather carrying a few extra pounds by increasing his calories. You can add some corn or vegetable oil to the diet.

For a hard keeper (a horse that has difficulty gaining weight), grain with fat added can be a good option. Sugar-beet pulp is a nice fiber source high in soluble

Nokota horses feed from a shared trough in wintery North Dakota. A high-fiber diet is important for wintertime warmth.

carbohydrates. It is an easily fermentable fiber for horses, so you don't have to worry much about digestive problems. It's also a high-energy source low in phosphorus (unlike some grains), an extra benefit because the horse won't have a high-phosphorus excrement (which means less contamination to the environment). Some believe it is important to soak beet pulp in water before feeding to avoid choke, but studies have shown this isn't necessary. If you choose to soak, do so within an hour of feeding; otherwise, mold can develop.

Winter

Heat is produced inside the horse's digestive tract, via bacterial fermentation. In other words, horses have their own little central heating system. You can help keep that warmth flowing by providing the right materials. For the horse, this means fiber. When a horse digests long-stem fiber (hay), microbial fermentation occurs and heat is created. This is beneficial to the horse in times of cold weather because fiber is digested slowly, so the heat is sustained for a long time.

But what is really cold to a horse? The thermoneutral zone (TNZ) is the range of environmental temperature at which the animal uses minimal energy to maintain body temperature—the ideal temperature for comfort. The TNZ for a horse is lower than that for a human. If the outside temperature is in the single digits (Fahrenheit) or lower, the horse will need more fiber to stay warm. If it's in the 30s, then requirements aren't as great. You must also factor in wind and cold rain, which will increase a horse's energy requirements.

Other calories gained from fat and grains can still be used for warmth, although they will not produce sustained heat. A horse will use that feed for whatever purpose he needs to, whether he is running a race or keeping warm, but the fiber automatically generates heat.

GROOMING

Grooming your horse is good for his health and a great way to bond with him. Spending quiet time together without the pressure of riding and training can really deepen your relationship. Bathing, clipping, pulling manes, and caring for tails are all part of life with your horse. Here are a few tips on how to make those chores more efficient.

Giving a Bath

Baths are an occasional activity because overbathing can strip the horse's coat of natural oils. However, a bath is a welcome treatment on a hot day or after a long workout—and a necessity before a show.

Choose a bath site that won't be slippery when wet or get too muddy. If there is no place to tie the horse, ask a friend to hold him for you. On a warm, still day, a cold-water bath is fine. If the weather is cooler or there is a cold breeze, warm water is best. If you don't have a hot-water source, you can purchase a bucket-size immersible water heater to take off the chill (note: never leave it unattended).

If you're using the hose, start with a soft spray and wet the horse from the shoulder on back, leaving the mane, the tail, and the face for last. If your horse has never been bathed or if you're unsure whether he has, start with a gentle spray on the legs to get him used to the water. If you're using the bucket method, use a sponge to squeeze water over the horse's body. Even if you're using a hose, however, wet the horse's face with a sponge; water straight from the hose can be startling.

Next, squirt the shampoo on the sponge, and begin to wash the horse using a circular motion (similar to the way you use a curry comb). Finally, wet the mane and tail, and apply shampoo directly to the hair. Wash as you would your own hair. Rinse everything thoroughly with either the hose or a clean sponge. Apply conditioner, if desired, repeating the above steps.

Remove the excess water from the horse's body with a sweat scraper and polish your horse with a dry towel. Rub the legs dry with a towel.

Equipment needed: Be ready with a sponge, bucket, hose and nozzle, sweat scraper, towel, equine shampoo and conditioner.

A woman bathes a horse in a safe enclosure. Baths should only be given occasionally; frequent bathing strips out natural oils.

Detangling the tail by hand helps prevent hair loss and breakage, resulting in thicker, healthier hair.

Detangling a Tail

A horse's tail can easily wind itself into knots. The best tools for this job are your fingers, which you will use to pull the snarl loose. This can prove to be time consuming, but using a brush can rip hairs and make the tangle tighten. And broken hairs will leave the horse with a thin and sparse-looking tail.

To make the job easier, first apply a detangling or conditioning product (either a gel or spray). This will make the tail hair slippery and easier to loosen and, it will help prevent the tail from snarling again. Pull the snarl away from of the bulk of the tail. With your fingers, gently untangle the hair, working from the bottom up (working the snarl from the bottom up will be easier and quicker).

Once the snag is free, gently brush or comb the hair. While brushing the tail, hold the bulk in one hand as you work to avoid pulling hairs out from the top.

Equipment needed: your fingers and a brush or comb to finish, a detangling or conditioning product.

Grooming Kit Essentials

When you assemble your kit, you will, of course, start with the container. Boxes and bags come in many colors and shapes. You can opt for a traditional tray-style box, with a closing lid (keeps dust and dirt out); an open tote; or a zippable courier-style bag. Here's what you need to fill it.

• Currycomb: a circular rubber brush with short teeth. The old-fashioned square metal currycomb on a handle is generally used for cleaning brushes.

• Stiff brush: a long narrow brush, similar in shape to a scrub brush. The bristles will be long and in clumps.

• Soft/body brush: more specialized than the stiff brush. It comes in many shapes. It can be very small to accommodate the face or wide and flat for the large muscle groups of the body. The bristles will be short and close.

• Hoof pick: choices include those with brushes on one end to remove debris and those with different lengths and shapes of the metal pick.

• Mane/tail brush or comb: this can be fancy or simple. There are brushes similar to those found in the human beauty department, with wide teeth and coated bristles.

• Sweat scrapers: they come in straight metal or plastic, as well as in a bowed squeegee.

• Towel/sponge: any household towel or sponge.

• Hoof dressing applications: these include oil poured from a can or a thick paste applied with a brush.

• Fly repellent: a needed seasonal item. It's available in spray, pour-and-wipe, and long-lasting drops (similar to cat and dog flea drops).

• Mane/tail conditioners: these run the gamut from gels to sprays to wipe-on cloths. Ones with green stain remover are a good choice for owners with white or gray horses.

Pulling the horse's mane helps keep it more manageable by shortening and thinning the hair.

Pulling a Mane

Pulling the mane shortens and thins it, making it easier to maintain. This is necessary if you want to show; long manes can't be braided in the traditional style (for hunter, dressage, eventing, and show jumping competitions).

Secure your horse, and tie a hay net in front of him for distraction. Comb the mane free of tangles, beginning at the withers. With one hand, hold a small section of mane, and with the other hand, comb the mane toward the crest a few times. You'll be left holding a few long hairs. Wrap these hairs around the comb a couple of times, and pull downward in a quick motion. If the hair does not pull out, start over and try with fewer strands of hair. Comb the section through and repeat if you want the mane thinner or shorter. Now repeat the process with the next section.

Shoot for a length of around four inches. That will make a nice, manageable hunter braid if you want to show. If your horse's mane tends to stand up straight when it's shortened, try to pull the hairs from underneath instead of on top; doing so will help you avoid the Mohawk look. You can also gel the hair and braid the mane down to help train it to lie flat.

If your horse objects to having his mane pulled, don't force the issue; try using a clipper blade (off the machine) to pull it instead. Use the blade like a pulling comb. Back-comb and slice the hairs with the blade. You can also use a mane-pulling device, which has an internal blade system that cuts the mane in a way that leaves it looking more natural than scissors would.

The pulling process is the same with a comb, only you will cut the strands of hair instead of wrapping and pulling. You can also use scissors turned vertically to snip the ends of the hair. This works very well if your horse has a thin mane. Never use scissors horizontally because this will give the mane a chopped, blunt look.

Once you've pulled the mane, try to keep it at the desired length all the time. Pull a few sections from time to time as the mane begins to lengthen. If you wait until it's too long, you'll be in for a big job. If you are starting with a full mane, pull it over several sessions to make it easier on your horse. *Tip*: Pulling the mane is easier on the horse after a ride because his pores will be open and the hairs will come out more easily.

Equipment needed: Pulling comb (smaller than a mane comb), gel and small rubber bands (for braiding stand up manes), sponge (for damping), mane pulling device or clipper blade.

Getting a Shiny Coat

A burnished coat starts with good nutrition, but a little elbow grease can really bring out the shine. The currycomb brings dirt to the surface and loosens

Use a short-bristled body brush to remove surface dust and loose hair.

shedding hair. Starting at the top of the neck, just behind the horse's ear, work your way down and along the body, making small circles as you go. When the comb fills with hair, brush it out with the stiff brush. The currycomb can be used all over the horse except for the legs, face, mane, and tail.

Follow in the currycomb's path with the stiff brush. The stiff brush's job is to remove the debris the currycomb produced. The stiff brush can also be gently used on the legs. Use the currycomb to sweep the hair loose from the stiff brush.

The job of the soft brush or body brush is to remove the last bit of dust and hair. It can be used all over the body, including the face. When using the soft brush on the face, brush in the direction of the hair growth.

For that extra glossy coat, finish with a conditioning spray, leaving the saddle and girth areas free—coat polishes can be slippery!

Equipment needed: Brushes are inexpensive and available in all sizes, shapes, and colors. You will need a currycomb, a stiff brush, a body brush, and coat conditioner.

Clipping and Trimming

There is nothing wrong with the more natural look of long whiskers and fuzzy ears. In fact, some clipping isn't necessary for the average riding horse. Leaving the hair inside the horse's ears will provide a defense against gnats. The long hairs around the eyes can also act as protection of the horse.

However, removing hair in other areas, such as the bridle path and backs of the pasterns, can make grooming, bridling, and judging the health of the legs of your horse easier. Unless your breed dictates a longer bridle path, try to keep it as small as possible—enough to accommodate the crownpiece of the bridle.

Stand on a step so you can see what you are doing, and pull the horse's halter back from the area you want to clip. Point the clippers toward the tail, and starting at the poll, run the clippers through the mane. Depending on the length of the mane (say you're starting a new path) you may have to repeat the clipping to even everything up.

Clip the back of the horse's jaw by combing the clippers with the growth of the long hairs. Go slowly and lightly so you don't cut into the shorter hair on the jaw. You just want to tidy and contour this area. If

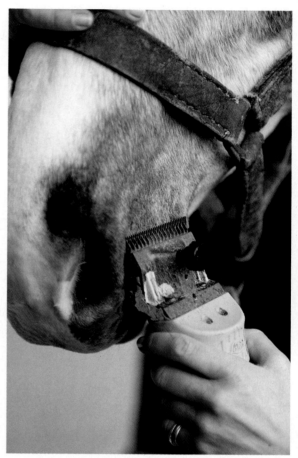

A groomer is using clippers to cut the whiskers around the muzzle. Clipping gives the horse a cleaner look.

you want to remove the whiskers on the muzzle, clip against the growth of the whiskers, gently and with soft pressure. To clip the long hair behind the fetlocks, clip against the hair growth, starting at the bottom and working your way up. The hair can be thick here, so you may have to make several small passes to finish the job. To sculpt the longer hair behind the back of the knee, move the clippers with the growth of the hair in a light, combing motion, working your way down the leg.

Equipment needed: You don't have to spend a fortune on clippers and blades to keep your horse looking tidy. A basic pair of clippers with one blade (a #10 or #15 will do) is all you need. To keep the blade running smoothly, brush off the loose hair with a small brush (an old toothbrush works great), then apply a small strip of clipper oil.

Health

Section 6

If you want to have a healthy horse, learn as much as you can about equine health issues before you bring your first horse home—and continue to build on that learning for as long as you own a horse. Prevention is certainly better than cure, and a lot less expensive. If you know what a healthy horse should look like, as well as the signs and symptoms of ailments, you'll be able to ward off problems before they gain a toehold. To keep your horse healthy, learn basic equine first aid and how and when to administer dewormers and give injections. Find out what plants are poisonous to horses and make sure your pastures don't harbor of them. Explore alternative care as well; you may find other ways to help your horse.

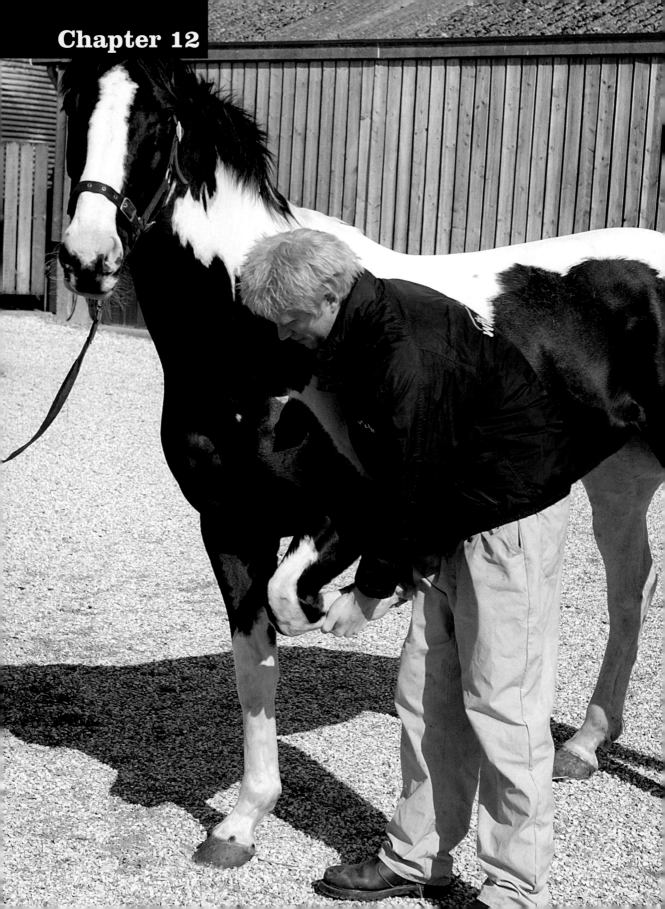

Chapter 12

The Healthy Horse

To ensure that your horse remains as healthy as possible throughout his life, you must learn to recognize the signs of good health, as well as those of injury and illness. You have to know how to evaluate your horse's body weight (is he too thin, too fat, just right?) and take his vital signs. You should be able to offer him basic care and know how that care changes season by season. Of course, you won't be alone in caring for your horse. A host of professionals will be at hand to help you to keep him healthy.

PROFESSIONALS IN YOUR HORSE'S LIFE

A variety of individuals can assist you with keeping your horse healthy and cared for. Some of these individuals are absolutely essential, while others can be called upon as needed.

Veterinarian: This professional is crucial to your horse's well-being. Veterinarians deal with preventive care, such as vaccinations and deworming. They can also diagnose and treat a host of diseases and maladies. Unlike most small-animal veterinarians, equine practitioners must have a good knowledge of soundness and performance issues, as horses are generally required to have a job. Some veterinarians do double duty as dentists and chiropractors. Others may practice some type of holistic healing, as well as traditional western medicine. Veterinarians may be associated with an equine hospital, a university extension service, or a teaching hospital, but most work out of a mobile unit and will come to your home or boarding stable.

Dentist: Essential also to your horse's well-being is the dentist, who provides dental care to an animal whose teeth are constantly growing and changing. An equine dentist will float the teeth (rasp them, even) so that the points and hooks of the teeth are removed. A qualified veterinarian will also do extractions and even restorative procedures. It is especially important that horses have regular dental care because one of the many ways riders communicate with a horse is via his mouth.

Farrier: A blacksmith is an individual who works with iron or steel at a forge but is not necessarily a horseshoer, as is a farrier. Not only do farriers trim the horse's hooves, balance them, and place new shoes on them, but they also keep a horse's feet and legs sound and free of pain. To be successful at their jobs, farriers must have a great deal of knowledge on proper movement of the horse, as well as techniques to help a horse with injured feet or suffering from diseases of the hoof.

Chiropractor: This individual deals not only with a horse's back problems—which are very important to address, as this is where the rider sits—but also with all other joint problems. A chiropractor will have intimate knowledge of the horse's skeletal structure and locomotion and can make adjustments in the legs, neck, or anywhere else the horse requires.

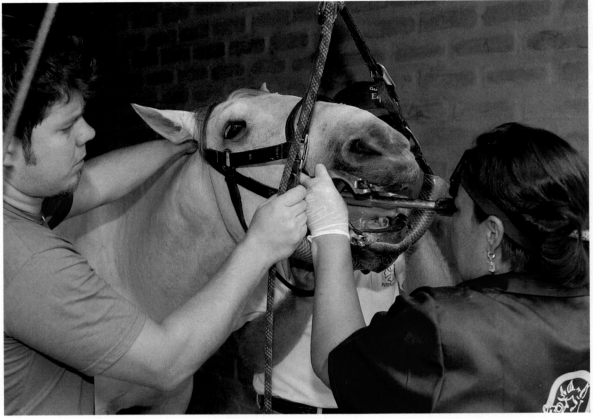

An equine dentist rasps away the rough edges of the horse's teeth while his mouth is held open with a speculum.

SHEATH CLEANING

Geldings and stallions need to have a thorough sheath inspection. You can clean your horse's sheath every few months or so, but only your veterinarian can do a thorough job. Your horse can develop serious infections if this area is not taken care of properly. It's difficult for the layperson to do a thorough sheath cleaning simply because most horses will not voluntarily drop the penis for inspection. Your veterinarian will administer a tranquilizer to entice the horse to relax and drop his penis. After that, he or she will clean the sheath and urethral fossa, which is a small cavity on the end of the penis. Smegma can build up inside this and form a hard bean that, if not removed, can inhibit urination. Your veterinarian will also inspect the penis for abnormalities and tumors. Horses with pink skin are especially prone to tumors and should be inspected more frequently.

Massage therapist: A therapist can provide relief for a horse's soft tissue challenges. Massage increases circulation, relaxes muscle spasms, relieves tension, enhances muscle tone, and increases range of motion.

To find qualified individuals in each of these areas, ask for recommendations from other horse owners, trainers, professional associations, or riding clubs. (*See Searching for a Good Farrier, page 262.*)

SIGNS OF A HEALTHY HORSE

Horse owners take pride in knowing whether their horses are healthy and happy. There is no magic to it; you simply need to know the signs of good health and spend time with your horse.

• A horse's eyes should be bright and have no discharge. The expression should be soft and relaxed. Tight, staring eyes could denote pain or anxiety.

• The nose should be clean, with no copious discharge or thick mucus. A slightly runny nose, if the discharge is clear, is considered normal.

A veterinarian inspects a horse's tendons, checking for swelling and abnormalities.

• Ears should be free of any insect bites or growths. An ear turned outward or flicking constantly may mean the ear has an infection or a foreign object inside.

• Gums should be pink and healthy, not bright red or pale.

• Legs should not be swollen or hot to the touch.

• Hooves should not have cracks; the bottoms should not have a strong odor. Hooves should be cool to the touch.

• Manure should be produced in well-formed balls. It shouldn't be loose or runny.

• A sick horse tends to look depressed or listless. Kicking or nipping at his sides, constant rolling, or sweating for no apparent reason may signal illness.

Body Condition Scoring

You can keep track of how your horse is doing and if your feed amount is on track by body condition scoring your horse often. Body condition scoring is a

How to Pick Out the Hooves

With the hand closest to the leg, squeeze the tendon at the back to get your horse to lift the hoof. If he won't lift the hoof, you can lean into him with your shoulder, pushing the weight off that leg. Hold the hoof with your outside hand and pick with the inside hand. Use the hoof pick to remove debris, working from the heel to the toe. Gently clean the cleft (indentation) of the frog, which is more sensitive than the hoof. For your safety, avoid placing your feet directly under the hoof as you work. If the horse pulls away suddenly, he could drop his hoof directly on your foot. When you're finished picking out the hoof, gently set it back on the ground.

An overweight horse's layer of fat is evident all along the body, especially noticeable in the neck, withers, and over the rib cage.

simple system developed in the early 1980s at Texas A&M University. The system assesses body fat stores and helps analyze gains and losses in body fat by feel and by sight over the ribs; behind the shoulder; over the withers, the back, and tail head; and up the neck.

You can assess the horse by sight, but in the winter, long hair can give a false reading. In fact, horses in low-temperature conditions will actually begin to grow a longer hair coat. It's most important to run your hands over the horse's body, particularly over the ribcage. If you can feel bones easily, you need to increase the horse's feed ration. If you are consistent and follow the body condition scoring below, you'll know if your horse is getting enough nutrients.

1–2 Extremely thin to very thin: This is a dangerous score, usually seen in horses suffering from extreme neglect. The horse will be severely emaciated, and you will feel very little fat. You'll easily be able to see bones, particularly individual vertebrae.

3 Thin: The horse will be thin, but individual vertebrae won't be as prevalent. You'll be able to feel only minor fat over the ribcage.

4 Moderately thin: You'll be able to feel the edges of the spinal column and ribs, but fat will be prevalent over the tailhead.

5 Moderate: You can feel the ribs but not see them. You will be unable to feel the spinal column. Your horse will start to feel the same all over.

6 Moderately fleshy: Fat all along the horse will feel spongy. The withers will also have developed some fat.

7 Fleshy: You'll start to feel fat between each rib.

8 Fat: Fat will start to pile up alongside the spine, creating a crease. Fat also will be felt between the horse's buttocks.

9 Extremely fat: This score means that your horse has become grossly overweight, with bulging fat present all along the body. The horse's flanks will appear filled in completely.

A veternarian reaches inside the jaw of this horse to check his pulse. The pulse should be thirty to forty beats per minute.

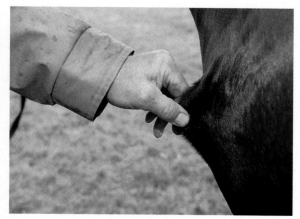

To check dehydration, pinch the neck skin, as shown above. If the skin doesn't spring back, the horse may be dehydrated.

Vital Signs

Vital signs include the horse's temperature, pulse (heart rate), and respiration rate, all taken while the horse is at rest. Vital signs can vary over time and among horses, so it's important to take your horse's signs over a few days to get the average. Record the range in a notebook, and keep the notebook in your first aid kit. (*Note*: results can also vary depending on the weather and whether the horse is excited or agitated.)

To measure the horse's respiration, you will need a watch with a sweep second hand. Watch the horse's flanks, and observe how they move in and out as the horse takes each breath. Find a starting point on your watch, and count how many times the flanks move in

HEALTH

Learning the following terms will help you to talk about horse health issues.

■ **Body condition scoring:** A system that assesses body fat stores and helps you analyze gains and losses in body fat by feel and by sight.

■ **Clinches:** The ends of the nails that are hammered against the front of the hoof.

■ **Floating teeth:** A process in which the veterinarian or dentist rasps the teeth so that the points and hooks of the teeth are removed.

■ **Frog:** The dark, rubbery, V-shaped structure on the sole of the hoof.

■ **Ligamentum nuchae:** The ligament that lies below the crest of the neck.

■ **Rasp:** A file used by a farrier. Also used in dentistry to float teeth.

■ **Smegma:** A thick secretion that can build up in the horse's sheath.

■ **Snow pads:** Rubber or plastic pads applied under shoes to keep snow from building up in the hooves.

■ **Urethral fossa:** The small cavity on the end of the penis.

and out (each in and out counts as one breath) within a minute. The horse should breathe about ten to fifteen times per minute.

To find the pulse, feel for the pulse in the inside of the jaw (just below the large cheek muscle), at the back of the knee, or below the fetlock inside the leg. Count the beats per minute. The pulse should be about thirty to forty beats per minute.

To take the temperature, use a rectal thermometer coated with petroleum jelly, with a tied string attached to the end (*see First Aid Kit, opposite*). Shake the thermometer down to 97 and insert it slowly and gently into the rectum (no need to push it in too far). Leave for two minutes. The equine temperature can range from 99 to 101 degrees Fahrenheit (38°C).

It's also important to know how to check for signs of dehydration. One way this is done is by testing the capillary refill time. Lift the horse's upper lip and press your thumb against the gum and then release. Count how many seconds it takes for the gum to return to its usual pink state. Three seconds is normal; any longer could be the sign of a problem. Another way to test for dehydration is by pinching the skin in the middle of the neck. Hold it for a moment, then let it go. A healthy horse's skin will bounce back immediately; a dehydrated horse's skin will remain standing (tented) or take longer to unfold.

FIRST AID

There are many serious conditions that require a veterinarian's immediate help (*see appendix*). There are other conditions and injuries that you can handle yourself with some practical knowledge, vigilance (so you can catch injuries early), and a well-stocked first aid kit. Wounds are fairly common occurrences that you should be able to deal with on your own. To treat your horse in many cases, you will also need to know how to administer injections.

Swelling and Heat in Hooves and Legs

Swelling in the legs could mean many things: a strain, a torn ligament or tendon, or a simple "stocking up" from lack of exercise. A hot hoof could be a sign of an abscess or laminitis.

Symptoms: Legs are puffed up and can be either very hot or very cold. Hot legs could denote an infection; cold legs could denote endotoxic shock, in which toxins from the intestines enter the bloodstream.

Treatment: Cold hose the hot leg/hoof and phone your veterinarian. If the leg is cold, phone the veterinarian immediately.

Wounds

Sometimes wounds aren't readily apparent. Puncture wounds, in particular, can close very quickly, so make sure to check your horse's body thoroughly each day.

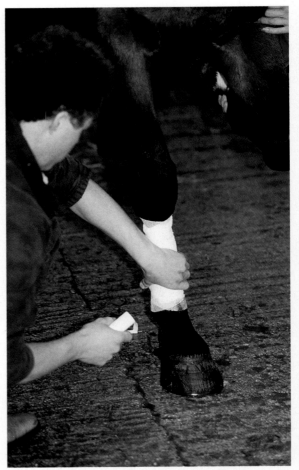

To control bleeding, this man has covered the horse's wound with sterile bandaging and is applying direct pressure.

Symptoms: Wounds can run the gamut from punctures to abrasions and gashes.

Treatment: Cleanse the area to remove foreign objects and other contaminants. A gentle trickle of water is good, but a saline solution is better because the salt helps the tissue resist the absorption of water into the cells, which can lead to edema.

If the wound is superficial, apply an antiseptic spray or ointment. If it's fly season, choose one designed to keep flies away. Wounds located around the hoof or lower leg benefit from a drying antiseptic powder. The drier the wound, the less likely the wound will collect grass and dirt or attract flies.

If the wound is deep or a flap of skin is hanging from it, clean it as best as you can and contact your veterinarian as soon as possible.

Bleeding around the horse's lower leg or hoof could mean a nicked or severed artery. Arterial blood is bright red and spurts in pulses. Have someone phone the veterinarian immediately. In the meantime, place a towel or cloth over the wound and hold pressure on it for several minutes.

Once the bleeding has stopped, replace the towel with a sterile gauze bandage and wrap it with a self-adhesive stretch bandage. If you suspect the wound will need stitches, refrain from using any wound cream or spray, as this will make stitching difficult.

Eye Injuries

Discharge, swelling, and inflammation are the hallmarks of eye injuries. Scratched corneas are common injuries to equine eyes. Conjunctivitis is also a common problem.

Symptom: Conjunctivitis is the inflammation of the conjunctival membrane, which covers the front of the eyeball and inside the eyelid. Many things can cause conjunctivitis, including insects, bacteria, parasites, and pollens. A scratched cornea can be difficult to see, but swelling, a closed eyelid, and discharge are some of the symptoms.

Treatment: Flush the eye with saline solution used for human eyes. Debris may come out in this process. If the problem persists, contact your veterinarian.

To help prevent these types of injuries, use a simple fly mask to keep irritants away from your horse's eyes.

Injections

Although injections are usually given by trained individuals, owners may need to inject their own horses from time to time, particularly when numerous injections are needed within short periods of time. Intramuscular injections should be given in the largest muscle masses.

THE NECK

Inject your horse in the triangular-shaped section made up by the ligamentum nuchae (on the top), the cervical spinal column (below), and the shoulder blade (behind). To aid you when you're first giving injections, use masking tape to outline the area. Inject inside this triangle.

Note: Avoid injecting into the spinal column and the ligamentum nuchae, which lies below the crest of the neck. Nerve damage can occur if you hit the

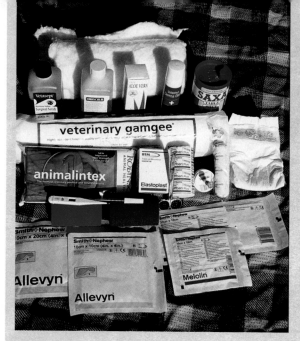

First Aid Kit

A first aid kit is an essential addition to every barn, show trunk, and horse trailer. Items for the kit can be found at the local drugstore and tack shop or through your veterinarian. Keep the items in a waterproof container in an obvious area. It's also helpful to keep emergency contact numbers inside the container, such as your primary and backup veterinarians' phone numbers. Below is a list of suggested items for your first aid kit:

- Absorbent cotton
- Adhesive tape
- Antiseptic scrub
- Disposable razor
- First aid guide
- Gauze dressing pads
- Lubricant for the thermometer, such as petroleum jelly
- Nonsteroidal eye ointment
- Oral syringe
- Pocket knife
- Rectal thermometer (attach a long string tied to an alligator clip on the end)
- Roll of gauze
- Rubbing alcohol
- Safety scissors (for cutting dressings)
- Scissors
- Self-stick elastic bandage
- Stethoscope
- Syringe (without the needle, for rinsing wounds)
- Wound ointment/spray

nerves emanating from the spinal cord. Infection can occur if you inject the ligamentum nuchae, as it has very little blood supply.

THE GLUTEALS (HAUNCHES)

In order to find the best place for your injection, draw a line or use masking tape to mark a line from the top of your horse's croup to the point of his buttocks, and mark another line from his dock to the point of his hip. You'll then have a big cross. Inject where the two lines meet.

Note: The haunches allow for poor drainage, so if there is an infection in this area, it can be difficult to treat. Professionals generally reserve the gluteals for antibiotic injections only.

THE THIGH

Draw your line (or place your masking tape) on the back of your horse's leg (the meatiest part) alongside the tail from about six inches below the top of his tail to the base of his thigh. Inject anywhere along that line in the softest area.

Note: The thigh is the best to inject a horse because it is the most mobile, and movement helps to reduce soreness. The thigh will also drain the easiest if an abscess forms. Unfortunately, the thigh is also the best place for you to get kicked, so proceed with caution.

THE CHEST

The horse's chest is an available site as well, but delivering an injection there is not advisable, as the chest can become quite tender. It's recommended that this site be used only if multiple injections are needed, so the horse isn't continuously jabbed in one or two sites.

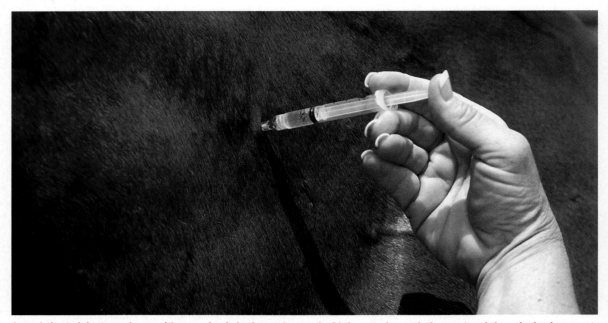

A good site to inject your horse with a vaccine is in the neck area. Avoid the area beneath the crest and the spinal column.

To prepare the injection and deliver it correctly, follow the steps outlined below.

1. Preparing the Injection

Have someone hold your horse or restrain him safely. Then:

- Clean the area well with alcohol.
- Connect the needle to the syringe, and remove the case from the needle.
- Fill the syringe by piercing the top of the medicine bottle with your needle and pulling back the plunger. Hold the syringe vertically with the needle in the air, and tap it to remove any bubbles. Press the syringe until the contents rest at the very top of the syringe.
- Remove the needle from the syringe. (*Note:* Don't insert the needle in the horse with the syringe attached. This way, if your horse jumps, you won't lose both the syringe and medication, as well as the needle.)

2. Making the Injection

If you choose the neck site, you should distract the horse by pinching a fold of skin before inserting the needle. Insert the needle next to the folded skin. Whatever site you choose:

- Insert the needle quickly and straight in—just like a dart. If you push the needle in slowly, it can bend and break off. Don't panic if the needle does bend; start over and use another.
- Make sure that there is no blood coming from the needle. This would indicate that you have hit a blood vessel. If so, find another site.
- If your horse is standing calmly, fit the syringe.
- Pull back on the plunger and check for blood. If you see blood, you have hit a blood vessel. Remove the needle and begin the procedure again (from the alcohol step) in a different site.
- If all is well, press the plunger and release the contents slowly. (*Note:* Your vet will tell you if you need to use more than one site for injections. However, the rule of thumb is no more than 10–15 CCs should be given in one site.)
- Remove the syringe and needle. Replace the cap on the needle and dispose of it safely.

3. Aftercare

Watch your horse for adverse reactions at least ten to twenty minutes after the injection; an hour would be even better. The mildest adverse reactions are swelling and heat at the injection site. The horse may go off his feed if there is swelling in the neck. Rest for a day or two or cold hosing helps resolve this reaction. A more severe adverse reaction, although very rare, can come on quite sudden, within the first minutes or hour after an injection. The horse can be feverish, start shaking, have respiratory difficulty, and can even go down. In this case, a veterinarian must see the horse immediately.

Horseshoe Removal

If your horse's shoe is half on and half off—and this will happen on more than one occasion—it must be removed immediately. If the shoe is left on, your horse can slash a loose toe clip into the other foot or step on the dangling shoe and rip off chunks of the hoof wall. The nails that remain on the shoe can penetrate the sole of the foot and create a serious injury. Nails also pose a risk of tetanus.

To take care of shoeing problems, you should have the following tools on hand, which are available new at most tack and feed stores or secondhand from your

Although a farrier is best qualified to remove a shoe, you should learn to do so in case of an emergency.

The tools for horseshoeing pictured above are (*left to right*) a clinch cutter, nippers, a rasp, a hammer, and a hoof knife.

farrier: a rasp and a farrier's pull-off, which is similar in appearance to nippers (pictured). You'll use a rasp to file down the clinches (the ends of the nail that are hammered against the front of the hoof).

Alternatively, you can purchase a pair of nippers to snip off the clinches, but a rasp is easier for the novice to handle. A farrier's pull-off is used to remove the shoe. Practice is helpful, so ask your farrier if he or she will let you remove a shoe during your horse's routine reset.

To remove the shoe, first assemble your equipment and secure your horse in a safe area. If your horse has lost a shoe on one of his front hooves, pick up the hoof, just as you would to clean it (*see How to Pick Out the Hooves, page 251*), and hold it between your knees. If one of the back hooves has lost its shoe, pick up the hoof and set it on your inside thigh. Then follow the steps below.

1. Thin the clinches: With your rasp, remove or thin the clinches by filing each one individually. This may take a little while to do depending on how many nails are still remaining. If you are working with nippers instead of a rasp, cut off the clinches by working the pincers between the clinch and the hoof. Close the handles and snip the clinches free.

2. Remove the shoe: Once you've thinned or snipped the clinches, pry the shoe away from the hoof wall at the heels with your pull-off. Pry the shoe down and toward the toe, then alternate to the opposite heel and again pry down and toward the toe. Keep moving from one side to the other as you pry off the shoe. Don't try to remove it with one pull, as you may tear off part of the healthy hoof wall. Try to pull the shoe straight instead of at an angle. If you pull to the side, you will widen the nail holes and weaken the hoof wall.

3. Check for signs of damage: Make sure all the nails have come off. If the nails have penetrated the sole, call your veterinarian immediately.

4. Protect the hoof: Protect the hoof with duct tape, a wrap, or a protective boot so the foot will be in good shape when the farrier arrives to put the shoe back on. Place a stretchy, self-adhesive bandage around the circumference of the hoof, and wind duct tape around it, securing the wrap to either the foot or the leg. Put your horse back in his stall to wait for your farrier.

Bent toe clips can cut into the hoof wall or the soft tissue on the opposite hoof. This is a simple fix. You should gently tap the clip back against the hoof wall with a hammer. Be careful not to startle your horse.

A farrier trims an Arabian mare's hooves. Regular farrier care will prevent serious problems.

Whether to ride your horse after he's lost a shoe depends on the environment you will be riding in, the integrity of the hoof wall, and how hard you're going to ride. If you do choose to ride your horse, you'll notice a difference in the gait; if you're riding your horse in wet footing there is an increased risk of slipping. In addition, if your horse is susceptible to hoof wall breakage or bruising of the sole, definitely abstain from riding, as the protection that the shoe provides is gone. The best decision is to wait to ride until your horse's shoe has been replaced; it's not worth risking your horse's health. Wait for the farrier to arrive. However, if your farrier can't make it out to your horse in a timely manner, then try to minimize any potential damage by taping the hoof or using a protective boot. Many horse owners like to keep a hoof boot on hand in their horse's hoof size for emergencies like this.

HOOF CARE AND FARRIERS

There is an old saying in the horse world: "no hoof, no horse." Hoof problems can sideline a horse for a very long time. To prevent hoof problems and leg injuries in your own horse, you as the owner can take two critical steps: pick out your horse's hooves on a daily basis, and provide your horse with a clean environment to promote healthy feet. The bulk of hoof care, however, lies in the knowledgeable hands of your farrier. When it comes to a horse's overall health and well-being, having a good farrier is as crucial as having a good veterinarian.

The Healthy Hoof
Healthy well-balanced feet should have the following:

- Smooth walls without chips, cracks, flared quarters, dished toes, or excessive growth rings.

- A hairline that is straight when viewed from the side, lower at the heel than at the toe.

- Front feet with round-shaped toes; hind feet with slightly pointy toes.

- A hoof-pastern axis (the bone alignment of the coffin joint, pastern joint, and fetlock joint) that is a smooth line (not broken forward or back) and one that is at a natural angle for the horse's conformation.

- Heel bulbs wide enough so there's no deep cleft. If the bulbs are shaped like a valentine heart, they are contracted.

- A wide frog that fills up the space between the heel bulbs. The central sulcus (or cleft) of the frog should be narrow, more like a thumbprint than a deep cleft. If you can stick the tip of a hoof pick deep into the central sulcus, the frog is unhealthy. (*Note*: The frog clefts are the spaces where you run a hoof pick: along the sides and down the center. That central cleft should be shallow enough that the tip of your pick doesn't disappear down a crack.)

- A sole smooth and slightly concave shaped, with short bars, ending near the mid-frog and not too lumpy. The white line (which you can't see in a shod foot) should be tight and narrow. The sole and white line should be free of red spots.

- A well-proportioned foot, one with approximately two-thirds of weight-bearing surface behind the tip of the frog.

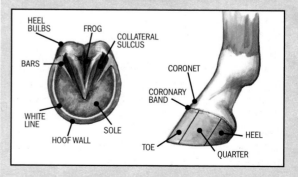

It is important to know that your farrier is doing a good job. This can be difficult to determine because many factors come into play. To help you, here are answers to some frequently asked questions.

Soreness Afterward

Is it normal for a horse to be sore after being shod or having his hoofs trimmed? The horse should always be sound and almost never sore, once the farrier has done his job. However, there is a certain amount of flexion the horse has to do for the farrier, and if your horse has a chronic issue with this, it can make him sore. Go easy and ride him gently at first. In general, though, a horse should not be lame or sore after shoeing or trimming.

Good-Looking Hooves

Should the hooves look good after being shod or trimmed? Preliminary appearances are not always indicators of a proper job. Even a nonskilled person can

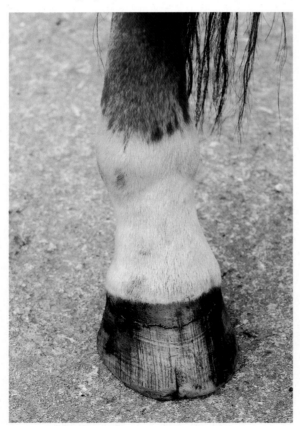

Without proper care, this hoof wall crack will worsen. Always seek a farrier's or veterinarian's advice.

trim a foot, tack on a shoe, rasp the hoof, and make it look good for that day. What's more important is how the hooves look in four to six weeks. Are they still healthy and together? Or are the clinches coming out, the hooves growing over the shoe, the edges of the hooves fraying, and the shoes loose? If a shoe moves on the hoof, the nails damage the wall and increase the incidence of shoe loss, which can cause lameness and wall damage.

As a general appearance, the normal horse should have a 3¼- to 4-inch toe length. The toes shouldn't be dubbed (flattened) off. The walls should be straight, and the farrier should not have rasped into them to shape the feet. Each foot should be trimmed to its own physiological parameters.

Matched Pairs

Should the feet match? Feet should be fairly matched, but horses can naturally have differently shaped feet. Shaping them to look alike can go against the conformation of a horse, which can lead to lameness. Shoeing a horse isn't carpentry, so don't apply carpentry rules. Artificially, you can make everything match, but that's not always a good idea.

People are often concerned, for instance, with the angle of the hoof (as it falls from the fetlock and down the hoof). Although this is something to be aware of, it shouldn't be a high priority. A perfect angle may not be what the horse needs. One hoof may be a club foot, and another may be low slung. Since one is sloping and the other is upright, they will never look the same. Wedging (using pads to lift the hoof) a horse up into the ideal angle can be detrimental, especially if the horse has a weak interior structure.

If your horse doesn't have conformational problems and his feet still look off, talk to your farrier. It could be that you went too long between shoeings and your farrier is working to bring the hooves back into balance. Your farrier may have a treatment plan that will correct the foot over the course of a few appointments. The farrier has to rebalance the hoof in stages and salvage the amount of horn that's there. Your farrier may also be limited in how well he or she can shoe the horse if your horse has pain somewhere in his body that compromises how long he can stand for the farrier.

Imbalance or poor hoof horn quality is not always the farrier's fault. Poor hoof quality, such as cracks, chips, slow hoof growth, or a weak horn that won't hold nails, may stem from nutritional problems. Hoof

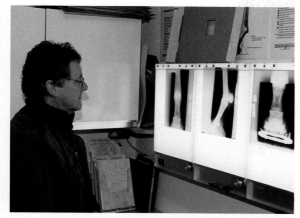

A veterinarian studies a horse's leg and hoof X-rays. Healthy legs and feet are paramount to a horse's soundness.

imbalance may stem from pain that causes the horse to stand or move with uneven weight bearing. These horses will have unbalanced wear no matter what kind of hoof care they get. Discuss these issues with your farrier and veterinarian. They will help you come up with the right solution.

Sole Changes

What should the bottom of the hoof look like? The hoof wall, white line, and the frog should be the only parts of the hoof that touch the ground. The sole is not designed to touch the ground, nor should the inner web of the sole touch the shoe. The shoe should be on the hoof wall and the white line.

Again, this is for a normal foot. If your horse has unmanaged feet that are underslung, there are times when that sole is stretched out past the bone so the farrier can utilize some of that sole as he works on gathering the foot.

One of the most important points to consider for soundness is the distance from the coffin bone to the ground. Studies have shown that the best length is three-quarters of an inch. Taking a horse shorter may not make him sore, but over time doing so can cause vascular damage and constant inflammation. Farriers judge this by looking at the frog. The grooves give a sense of depth. The point of the frog gives the farrier a sense of concavity.

The way to determine the distance from the coffin bone to the ground is with an X-ray. Doing this once a year will give your farrier a clear picture of what's going on, especially if there are lameness issues.

Landing Patterns

How should the hooves land as the horse moves? Equine research shows that feet land flat or heel first, depending on the speed, terrain, and shoeing differences. The human eye cannot resolve that final moment of hoof impact; what we see instead is the approach. What you see on approach is not always correlated with how the foot actually lands. But if while walking on a flat, even surface, the foot approaches the ground in a profoundly toe-first direction, stabs the toe on landing, and tends to stumble, this is abnormal. Toe-stabbing landing is usually not the farrier's fault. It indicates pain somewhere, usually in the back of the foot. Pain elsewhere, in the legs, shoulders, withers, or back, can cause the horse to be short strided, and what we see is the toe stab.

If the Shoe Fits

How do you know if the shoes are the right size? Proper fit is important. If the shoe is inside the hoof wall (for example, the hoof is growing over the shoe), either the shoe is too small or it's too long between resets.

The shoes should be flush to the wall on the front half and be a little bit wider toward the back. As a general rule, the back half of the shoe should extend beyond the wall on the inside and outside of the foot—approximately 1 to 3 millimeters. This is referred to as "fitting a shoe for expansion." The length of the horseshoe should extend beyond the heels the same amount. Shoes fit with expansion, if possible, are desirable to allow for the normal foot expansion when bearing weight and to allow for growth of the foot during the shoeing cycle. This fit gives the hoof wall room as it loads and expands, but it also gives room for the foot to grow out. Hooves should be wider at the base than at the coronary band, so they get naturally wider and migrate forward as they grow down. However, things can change with different conformation. There are no hard rules here.

Going Barefoot

Does my horse have to wear shoes? Not all horses need to be shod, but going barefoot depends upon the environment, the consistency of hoof care, and the horse's job. (*See All-Season Hoof Care, page 265.*)

The sole exfoliates and sloughs away sooner than the wall does. The sole should not be the primary weight- bearing structure; the hoof wall should. As the hoof wall keeps growing, it gets wider and wider. If the wall doesn't have any sole to connect it and keep it

A farrier is an important part of your horse's care, but when he finishes and drives away, hoof care is up to the owner.

SEARCHING FOR A GOOD FARRIER

Horseshoeing is an exacting skill that takes years of education and practice. Farriers have to learn anatomy, biomechanics, metallurgy, horse-handling skills, and conformation. So, ask the farrier about his or her education and experience.

Find out if the farrier is certified. Look for an American Farriers Association (AFA) Certified Farrier (search through their Farrier Finder at www.americanfarriers. org). In America, there is no license requirement for farriers. The AFA is the only organization that requires farriers to take a stringent test to become certified. It's a basic test, but it requires the minimum knowledge a farrier needs.

Your farrier should get along with your horse. If both are having a miserable time and the shoeing session becomes a wrestling match, then the job will be counterproductive. Be upfront with the farrier about your horse's issues. If necessary, your veterinarian may have to tranquilize your horse.

together, it flares out and flattens the sole. The hoof wall then breaks, and the sole touches the ground. Regular trimming and staying ahead of the growth keeps that from happening and also creates a toughened sole.

If you decide to let the horse go barefoot, inform your farrier. He or she will then trim the hoof and round off the edges, staying ahead of the growth of the wall and the building of a thicker and harder sole.

If your footing is abrasive and wet or if your horse is very active, it will be difficult to remain barefoot. If your horse wears away more foot than he's growing or if he's coming up sore, that's a good indication he is not able to grow a thick and calloused sole.

Much like a human's running shoe or work boot, the job of the shoe is to provide protection and support. When a horse moves, there's a lot of movement in the hoof capsule. So, while a shoe may impinge that movement, the advantages may outweigh the drawbacks. Consider what is worse: getting a horse so sore that he has chronic inflammation, or putting on shoes that restrict some movement but give support? When wear exceeds growth, shoes are needed. One option is to only put shoes on in the front, not the back. Most horses put more weight on the front feet, which causes more wear.

Consistent Hoof Care

How often should I have my farrier out? It depends upon your horse. It's important to watch your horse right after shoeing, and then monitor any changes until the next appointment. This information will help the farrier determine the ideal interval between visits. Closer shoeing intervals allow less hoof horn to be taken off each time, which means the horse stays closer to ideal balance throughout his life. Healthy feet will grow enough to be reshod every four to six weeks, which is a fairly typical shoeing schedule.

It is critical to keep on schedule because, as feet grow, the position of the hoof under the bones of the limb changes and migrates forward. The longer you wait, the more out of balance your horse will be. Bear in mind, however, that horse hooves grow in spurts. That's why sometimes there is nothing to trim, and other times there is a lot to trim.

The farrier is an important part of your horse's hoof care, but once his truck leaves the farm, the rest is up to you. Keep an eye on any changes and provide a clean, dry living environment and a healthy diet. And don't be afraid to ask questions if you have them. A good farrier will always be happy to answer them.

HEALTH CARE YEAR-ROUND

Every season brings particular challenges. Your horse has different health care needs as the weather changes —from deworming to vaccination to hoof care. You

This horse is well protected from the chill of winter by a turnout rug.

RELOCATING TO COLDER CLIMATES

Many people are reluctant to move their horses from a mild winter climate to a cold winter climate, thinking that the horses won't be able to deal with the change. In fact, horses adapt well to the cold, and a change to a colder winter climate is easier than moving to a hotter summer climate.

When it comes to winter, if horses get to the cold area early enough in the year and are kept outside, they will grow heavy winter coats to match the climate. Along the same lines, heavy-coated ponies from northern climates will grow a lighter winter coat when moved to hotter southern climates.

also need to think about whether to turn out your horse in the winter and what kind of protection and shelter he may need throughout the year.

Vaccinations

Many owners feel an injection is all that's needed to safeguard their horses, but there is a difference between immunization and vaccination. It's important to choose the optimum time for vaccination to stimulate immunity in the horse, so when a disease-causing organism comes along, the horse's immune system will be able to reject it. Timing and geographic location are also considerations. If insects transmit a certain disease, for example, and you live in an area with a year-round insect population, you may need to consider vaccinating more than once a year. Additionally, it's important that you vaccinate far enough ahead of time, before traveling or attending an event, to provide adequate protection.

VACCINATION GUIDELINES

There are certain vaccinations that every horse should receive regardless of where he lives. This is because of either the uncontrolled risk of the disease or the concern that the outcome of infection would be harmful. For example, the American Association of Equine Practitioners (AAEP) now considers rabies a core vaccine for large animals.

Although not many horses get infected by rabies, the risk of exposure usually being fairly low, once a horse contracts this disease, it's fatal. That's why it is a significant public health concern. Other core vaccines include: tetanus, eastern/western equine encephalo-myelitis, and West Nile virus. Your veterinarian will help you determine when and how often vaccination for these are needed for your horse.

Risk-based vaccinations vary regionally and within populations. A veterinarian will help you evaluate whether vaccines in this group are right for your situation. Resist vaccinating for everything "just to be on the safe side"; this practice can actually be detrimental for a horse's immune system. A better strategy is to evaluate an individual's risk and vaccinate appropriately. Some risk-based vaccinations are anthrax, botulism, equine herpesvirus (rhinopneumonitis), equine influenza, equine viral arteritis, Potomac horse fever (PHF), rotaviral diarrhea, and strangles.

For a complete list of vaccination guidelines, visit www.aaep.org/vaccination_guidelines.htm.

WHO SHOULD ADMINISTER VACCINES

If you're conscientious and careful about the procedure, vaccinating your own horses is certainly appealing from a cost-savings standpoint. There is, however, some risk associated with doing so.

Work with your veterinarian to determine the most appropriate vaccination schedule for your horse.

Vaccinations have strategic components that are more fully understood by veterinarians than by laypeople. As previously mentioned, there are nuances of timing involved. For instance, vaccinating for Potomac horse fever in March is not effective because PHF usually occurs in July. By the time summer rolls around, the horse may not have enough protection left for a reduction in disease severity.

Some vaccine companies have started offering guarantees for certain vaccinations; that is, a company will pay to treat a horse that got sick despite being vaccinated, but only when the vaccination was administered by a veterinarian. If you do it yourself and your horse contracts the disease, you are out of luck.

In addition, the strength of protection and likelihood of vaccine reactions vary between brands and vaccine types. This is something veterinarians are more likely to have knowledge about than are laypeople. It can also be difficult for horse owners to figure out if the vaccines they bought from the feed store or online are still viable by the time they are ready to be used. A vaccine must be kept cold until it is administered. Those that come through mail order are shipped with a cold pack, but sometimes those packs fail. An owner also may not know if the vaccine has frozen during shipment, rendering it nonviable. If there is a bad batch of vaccines or any other issues a horse owner may not be aware of it, the horse may suffer the consequences.

If, with all those caveats in mind, you do decide to vaccinate on your own, make sure you know how to inject a horse and that he is healthy when you do so. If a horse has been ill or has a chronic aliment or is a senior horse, seek your veterinarian's advice. If a horse has previously had a vaccine reaction, discuss this with your veterinarian before choosing to vaccinate on your own.

Follow the directions on the vaccine closely. The label will instruct you about where to put the vaccine. Some vaccines are to be administered only in the gluteal muscle; most horse owners, however, don't like to inject there because they can get kicked. But certain vaccines need to be administered in the horse's gluteal muscle because they are more reactive than others (that is, producing more side effects) and they need to be in a big muscle to lessen those effects. If, for instance, your horse gets a sore neck from an injection administered there, he may stop drinking. Be

This horse has been shod with rim-style snow pads. He is also wearing removable studs for better traction.

aware that, if you have an intranasal vaccine, it can be very challenging to give to your horse because some horses won't like the administration canula.

Do not mix vaccines in the same syringe. Stick to one syringe per vaccine, and never put more than one injection in any spot. Record where you put each vaccine, because your horse may have a site reaction to a particular vaccine. If you don't remember which one you administered in that area, then you won't be able to tell the veterinarian which vaccine produced that reaction. You should also record the vaccine lot number in case your horse has a reaction that should be reported to the vaccine manufacturer and the government.

One bad reaction to vaccination is called *purpura hemorrhagica*, which is an immune-mediated reaction triggered by products in the vaccine. The horse develops inflammation around small blood vessels, which can cause the loss of big patches of skin. It can also result in plugged blood vessels to critical organs. Most of the time, you'll see swelling of the legs, chest, and underside of the belly. There will also be small pinpoint hemorrhages of the gums, eyes, and sclera. This can happen with any vaccination, but it is more common in the strangles vaccinations. This is one of the reasons

why some veterinarians are hesitant to vaccinate for strangles if the horse has had strangles recently.

Be aware that, if you are administering vaccinations yourself, you are responsible for disposing of the needle and syringe safely. You can buy a sharps container at the local pharmacy. The pharmacist can tell you the proper way to dispose of needles in your state.

All-Season Hoof Care

In snowy areas of the country, decide how you are going to use your horse during the winter. If your horse has healthy hooves and you aren't going to be riding on roads or on icy, snowy ground, consider removing the shoes for the winter. Shoes can be problematic because ice and snow balls up inside the shoe, causing the horse to slip.

Horses can get bad tendon and leg problems with ice in the hooves, so many owners do choose to pull the shoes during the winter. But if your horse can't go without them, consider snow pads.

Snow pads come in two styles: a rim pad, which is a hollow tube full of air that sits inside the rim of the shoe and pushes the snow out as the horse moves, or a popper, which is a full rubber or plastic snowball pad with an inverted cup that pops the snow out as the horse walks.

For extra traction on snow and ice, horseshoe nails with a carbide tip or welded-on borium taps will help prevent slipping. If your horse is kept in a herd, however, carefully consider whether borium taps would do more harm than good. If he kicks at his herdmates, he can injure others with the taps on his back hooves.

Horses shed their frogs (the dark, rubbery, v-shaped structure on the sole of the hoof) at least twice a year, usually in spring and fall, so be on the lookout for thrush, a foul-smelling, tarry black substance that can rot the frog and cause lameness. When horses shed their frogs, a pocket can develop and bacteria can get in, so it's important for your farrier to remove the old frog. Thrush can get into the heel bulbs, as well. Thrush can be treated with an over-the-counter medication available at feed and tack stores. Gently clean the crevice between the bulbs with a Popsicle stick and follow with the thrush medication.

Thrush can be prevented with good horsekeeping. The bacterium that causes thrush thrives in wet and unsanitary conditions, so stalls should be clean and dry, and hooves should be picked out daily, year-round.

In the summer when the ground can become sun-baked and hard, keep shoes on if you're planning to do a lot of trail riding or showing. In drier parts of the country, the ground can get really hard, and soles can get bruised, so consider asking your farrier about using pads or a shock-absorbing hoof pack called Equithane.

Shelter from the Elements

To turn out or not to turn out is another winter concern for horse owners, but given a choice, a horse will often choose to remain outside, even in the most

FIBER FOR THE FURNACE

Horsepeople who live in frigid areas know that, when it's very cold, they can warm their hands by pushing them deep inside a horse's coat. The heat felt is produced inside the horse's digestive tract and is stoked by fermentation. It's true: our horses have their own central heating system. You can help keep that warmth flowing by providing the right materials. This means adding fiber to the horse's diet, which in the horse's case means forage such as hay or sugar beet pulp.

MEASURING FOR A BLANKET

Follow these simple measuring tips, and you'll have the right fit the first time out.

STEP 1: Buy yourself a quilter's measuring tape. This long tape is available at most fabric and craft stores. The tape is flexible, so you can get the most accurate measurement.

STEP 2: Measure your horse. Have a friend hold the tape just under the neck in the center of the horse's chest. Run the tape parallel to the ground around the horse's shoulder, barrel, and hip to the point of the butt. Do not measure under the tail, which will make the blanket too big.

STEP 3: Assess your horse. What kind or breed of horse do you have? Certain brands of blankets tend to be made for certain breeds and body type. Find out what type of horse breed or body type a particular blanket brand best fits.

inclement conditions. Horses will naturally find the best weather conditions in their environment. They stand along streambanks where the temperatures are warmer because of the flowing water and the wind is quieter due to the protection of the banks. They huddle together with their backs to the wind.

If you choose to keep your horse outside during the winter, it is still your responsibility to give your animals a choice, so provide a windbreak and a place where they can get out of the snow and wind if they desire. Large round hay bales are a good way to provide a windbreak, as well as a constant source of forage. You can place several in a row or even in a large square or circle, with an opening. Tall trees at the edge of a field won't create an adequate windbreak if the bottom branches are taller than the horse.

In the summer, a horse will need shelter from the heat and sun. If the horse is indoors in the barn, ventilation is very important. Some barns keep all the doors and windows open and use fans to keep air circulating. Stalls can be made airier by the use of a webbed stall guard in place of the stall door. Horses in pasture should have trees for shade or a run-in shelter.

Winter Wear

Whether to blanket is always a question when cold weather draws near. But as shown in our chapter on nutrition (see chapter 11), horses can produce their own heat, and healthy horses left with their natural hair coat will not need blankets. If a horse has a way to get out of the wind and he can stay dry, his own winter coat is adequate. However, do blanket if you're showing and your horse has a short coat or has been clipped, or if your horse is elderly and needs help to stay warm. Elderly horses have a harder time maintaining their body fat, so they can't get warm as easily as they did when they were young. If you choose to body clip your horse, he will need a wardrobe of blankets, from day sheets to hoods.

Horses begin to grow their coats when the days get shorter—not as the temperature drops, as many people believe. This is because a horse must begin to develop his coat long before winter sets in. It works like this: as the days shorten, the horse's retina receives fewer hours of daylight. This triggers the brain to release extra melatonin, a hormone that prompts the hair follicles to produce more hair. Show horse owners who want their horses' coats to stay short for the winter show season often install electric lights to prevent the triggering of the hormone.

If you're going to blanket your horse, get the fit right. Follow the instruction in the *Measuring for a Blanket* sidebar on the opposite page.

PUTTING ON A BLANKET

The belly straps keep the blanket from shifting or twisting. There should be about a hand's width between the straps and your horse. You don't want them so tight that they press into the skin, but you also don't want them so loose that your horse can get a hoof trapped in them. The leg straps should be adjusted so they hang in a slight arc. Not too tight that they will rub, and not too loose that your horse could catch a leg in them.

Buckle the chest straps, leaving enough room for you to slide your flat palm between the blanket and your horse at the chest and withers. After you've finished your adjustments, take a moment to watch your horse walk; then put a carrot on the ground and evaluate the fit as he stretches to eat his treat.

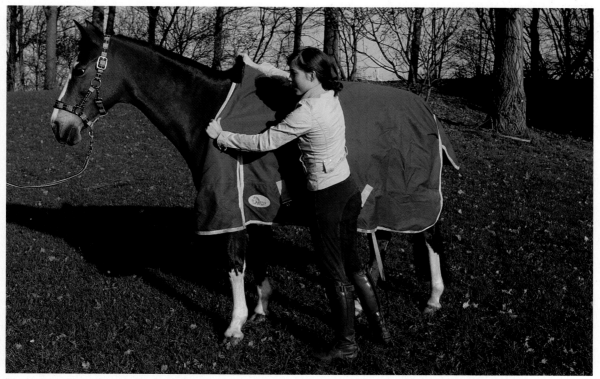

Blanketing will provide additional protection from the elements if your horse does not have a sufficient winter coat.

You can have several problems with blankets:

Blanket slides back from the neck: Either the neckline has stretched, or your blanket is too big in the neck. Have a seamstress or blanket repair person put a dart in the front of the blanket.

Horse can't reach an object on the ground, or blanket constricts his neck as he reaches: The chest straps are too tight, or the blanket is too small. Adjust the straps. If the problem remains, have material added to the blanket and darts put in (room for the horse's shoulders to move)—or buy another blanket.

Rub marks on chest: The blanket is too small or just too tight around the chest. Adjust the chest straps. If the problem still occurs, choose a larger blanket or have it altered by adding material and putting in darts. In addition, consider buying a stretchy knit undergarment that prevents rubbing.

Blanket creases in the shoulder when he moves: It's too tight. Adjust the chest straps. If it still creases, select a larger blanket or have a professional adjust it.

Mane is rubbed out: The neckline is too big, or the blanket is riding up too high on the neck because the blanket is too small. A cut-back-style blanket is often a good solution.

If the blanket still doesn't fit, most tack shops and catalog companies will exchange size mistakes; however, check before you buy, and try the blanket on a clean horse. If you cannot locate a blanket that will fit your oddly shaped horse, then buy a blanket one or two sizes too big and hire someone to customize it for you.

Summer Wear

In the summer, fly sheets, leg wraps, and fly masks are all effective pest barriers during fly and mosquito season. Fly wear is great because it keeps biting insects off the horse. (See chapter 13 for more information.) As is the case with all blankets and masks, don't just put them on your horse and leave them for days. Check daily to make sure that they aren't twisting or chafing the horse.

A farmer gives his horses extra hay in the winter. Digesting hay helps horses generate more body heat in cold weather.

A Warmblood in Florida drinks from a water trough. Offer your horse plenty of water in hot weather to avoid dehydration.

Other Weather Tips

Here are some further tips for making sure your horse stays comfortable and healthy in the deep cold and the sweltering heat.

COLD WEATHER

Be cautious when turning out stabled horses in icy or snowy conditions. In their exuberance to be out, they may get up too much speed and slip and fall.

On very cold days, rather than putting on an extra blanket, toss your horse an extra flake of hay to stoke the internal equine furnace with the slow, steady heat of fermentation.

Since a thin horse will have a harder time keeping warm than a heavier horse, horses should start the winter with a body condition score of at least 5. (*See Body Condition Scoring, page 251–252.*) However, a chubby horse shouldn't be ignored either. During some really cold spells, you will see your horse drop some weight in an effort to stay warm. Be sure to get your hands into his coat and feel if he's losing too much weight.

Provide an adequate, drinkable water source in the winter. Don't count on your horse eating snow to fill his daily needs.

HOT WEATHER

Water should be plentiful in hot weather. Check that your automatic waterers are working daily. Some horses will be reluctant to drink from an automatic waterer located in the sun, as the water delivered will be warm and the device itself may burn them. If this is the case, have a bucket of fresh water available in the shade.

Ride your horse in the coolest part of the day. Normally the early morning hours are best, as residual sun will keep the temperature hot in the evenings in desert climates.

FLUFFY COAT

Why is it that during a cold snap your horse suddenly appears fluffier? Although it may seem that his hair multiplied overnight, it really didn't; that fuller coat is just your horse's clever way of creating insulation. The hairs lift up to trap the warm air against his skin. Even the shorn hairs on a body-clipped horse will lift up if he feels cold, giving his coat a velvety texture. If you blanket your horse, you'll discover that the hair will lie flat. If you leave the blanket off, his hairs will eventually fluff up and look longer. The coat is the same length; it's just raised to try to trap the warm air.

Chapter 13

Pests, Parasites, and Poisonous Plants

Horse owners wage a pitched battle against nature each year, particularly in the summer when insects are active, poisonous plants are growing, and parasites are in their infective stage. Although it's impossible to eradicate them all, it is possible to minimize them so the effects have less impact on a horse's health and well-being.

PESTS

As mentioned in chapter 12, you can protect your horse from pests to a certain degree with fly masks, leg wraps, and fly sheets. Fly spray, traps, and bug zappers are also traditional methods of managing flying pests, but technology has advanced, and now there are many products that can help you control pests during their most vulnerable time: breeding season. Timing is important; you want to have your program in place before pests become active.

Use the strength of nature by releasing parasitic fly wasps wherever flies breed, such as near manure piles and under troughs. The tiny wasps are harmless to humans and animals, but they destroy flies in their pupal stage. The wasps are shipped as pupae mixed in sawdust. To catch the beginning of fly-breeding season, spread the mixture after three days of 40-degree temperatures. You can spread the mixture later, but it will take longer to see a difference.

Feed-through products have an insect growth regulator that passes directly into the horse's manure. The growth regulator prevents housefly and stable fly larvae from developing exoskeletons, and as a result, they can't take wing. Be aware, though, that it can take four to six weeks before the full benefit is seen, and all horses on the property must be treated.

Two flies that make riding in the summer nearly impossible are the horsefly and the deerfly. These creatures have no fear of fly spray or fumigation. The females of both species are vicious and will bite humans and horses hard enough to draw blood—their food of choice. Eliminating breeding areas in your stable is important for managing these flies. They tend to breed in wet areas, so don't let your water tank overflow, and get rid of muddy areas around your stable. They also don't like the shade, so consider bringing your horse inside during the day and turning him out at night.

PARASITES

Parasites have been part of the horse's digestive system, present long before humans ever thought of domesticating horses. Parasites weren't a problem for the wild horse because he never grazed the same patch of land long enough for the parasite population to escalate beyond normal loads. Because a wild horse would graze over miles of land, as the eggs passed through the horse's manure onto the ground, the horse, acting as the host, would leave before the parasite's life cycle could start again. It is also believed that wild horses may have built up immunity to worms.

The domestic horse is not as fortunate. Life with humans means smaller pastures, paddocks, and stables, and a horse is often left grazing on the same length of field day after day. And since the horse can't move away, the parasite load is left to escalate to the point at which every blade of grass the horse eats is saturated with parasite eggs. If left unchecked, parasites can multiply in your horse's digestive system to such amounts that they can rob him of essential nutrients, cause diarrhea, depression, mouth lesions, internal ruptures, and even death. Studies have even shown that 80 to 90 percent of all colic is due to large and small strongyles.

Since domestic horses can no longer handle the parasite buildup, we have to do it for them with the use of anthelmintics—which are more commonly known as dewormers.

Deworming

Fifty years ago, the equine deworming programs consisted of little more than primitive natural remedies of tobacco and extracts of fern, and caustic doses of carbon disulfide and petroleum distillates. They were marginally effective and difficult to administer—basically just shots in the dark. In the 1970s and

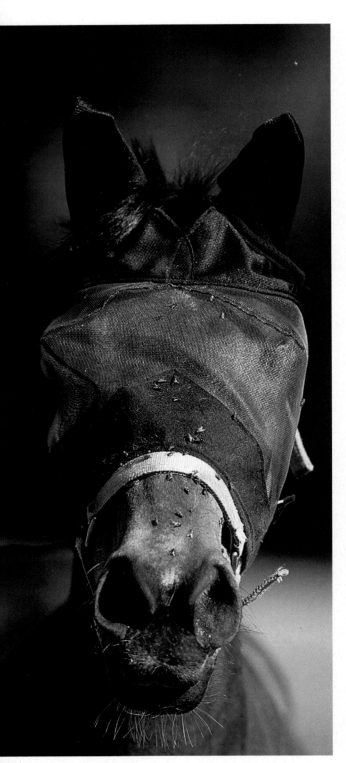

A fly mask helps to protect this horse's eyes from a swarm of flies.

STRATEGIC DEWORMING

Don't rely on chemicals alone; get strategic. Take a many-pronged approach to treating your horse, including removing manure regularly. Excrement is the main vehicle of parasite eggs, so remove manure from your pasture regularly. Since most parasites exit animals through feces, it's extremely important to keep your dry lots and pastures clean, even if they are not overstocked. In the developed world, humans have very little problems with parasites because we have flushable toilets. So the same concept goes with animals. If you can clean the environment, you'll have fewer problems. If it's not possible to pick up all the manure in the pasture, you can drag it. Pull a 12-foot post behind a tractor during a warm, sunny day. This will break down the manure and expose larvae to the sun. This works especially well if the weather is hot and dry. But don't do this if it's wet because you'll only be giving parasites free transportation to other areas of the pasture. Once you drag an area, keep the animals out of it for a few weeks.

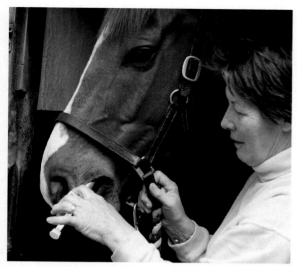

A horse receives deworming paste. Work with your vet to create a deworming program for your horse(s) and facility.

1980s, paste anthelmintics (dewormers) came on the scene and changed the face of parasite control. We now had an arsenal that we could use to prevent a horse from being robbed of nutrition, to stop parasites from reproducing, and to reduce the causes of many types of colic. It was a wonderful development for horses, but all these magic bullets lulled us into a false sense of security. In the last few years, studies have begun to show that parasitic resistance to anthelmintics is on the rise.

Rotation used to be the gold-standard method when deworming horses. The idea was that, if we rotated to another class of chemicals every two months, parasites wouldn't get a chance to develop resistance. Early on, dewormers were not broad spectrum, meaning they didn't knock a wide range of parasites on the head in one shot. We needed a team of drug classes to cover all bases. Some killed strongyles, while others killed ascarids or bots. So we rotated our chemicals to compensate for the deficiencies. But with the advent of macrocyclic lactones (ivermectin and moxidectin) in the early 1980s, we got broad-spectrum dewormers, and rotation took on a new role. Rotation was now used to prevent, or at least delay, the resistance to chemical classes, to keep the selection pressure on the worms from several directions, since the selection mechanism for each class of drugs is different.

Today we have only three chemical classes of drugs for equines: benzimidazoles, pyrimidines, and macro-

cyclic lactones. Ninety-five percent of the small strongyles are resistant to benzimidazoles, and 50 percent of them are resistant to pyrimidines, which leaves only the macrocylic lactones consistently effective.

That means 50 percent of the herds in North America no longer have the option of rotating. Recent studies have found that entire classes of dewormers have stopped working against the serious equine threat of small strongyles.

Indications are strong that not only does rotation no longer works but it may actually be contributing to resistance. Think about it this way: you would never randomly use an antibiotic to treat a suspected infection. That same idea must now be applied to equine dewormers.

HOW RESISTANCE HAPPENS

Individual worms that are resistant to the activity of certain dewormers occur naturally and spontaneously. But they will most likely remain rare in a population unless they gain some advantage. These worms have gained an advantage, but how? Are the drugs themselves the culprits?

No. It's how we use the drugs that is to blame. Drugs and the use patterns we employ are simply selection tools that, when applied to a population of parasites, allow those parasites that can handle the drug to become dominant. In other words, much as the misuse of antibiotics is believed to have led to the development of the so-called superbugs, so the misuse of dewormers may have led to the rise of these parasites.

For instance, if we deworm our horses indiscriminately, worms that are not resistant to that chemical die off. Those that are resistant are left to reproduce. Suddenly, we have a bunch of worms that aren't touched by our once-effective chemical. And it isn't enough to say just use a chemical class that doesn't have known resistance. It's only a matter of time before resistance develops. In Canada, researchers have reported ascarid resistance to ivermectin. And many dewormers are now useless on sheep and goats, which researchers believe to be canaries in the coal mine. If it has happened to these animals, it will eventually happen to horses.

Adding to the issue, a study that was conducted by Drs. Chris Proudman and Sandy Trees of the University of Liverpool in the United Kingdom found that tapeworms cause more problems than we once thought; in fact, they are largely responsible for three types of colic. So now tapeworms are included on the deworming menu.

EFFECTIVE DEWORMING PROGRAMS

There are two ways to go about deworming: treat every horse on the property the same or customize a program for each horse.

INVOLVING YOUR VETERINARIAN

Parasitologists agree that deworming is not an easy concept for the layperson to grasp; worms are ever evolving and becoming resistant to certain chemicals. As all situations are not the same, ask your veterinarian to tailor a parasitic control program for your horse or herd. Exposure is dramatically different throughout the country. For instance, Southern California has a much lower incidence of parasites because horses are kept on "dry lots" rather than pastures. But if you go to the central coast of California where horses are kept on irrigated pastures, all bets are off.

A veterinarian can also help you check the efficiency of each drug class by using fecal egg count reduction testing (FECRT), in which fecal egg counts are taken at the time of treatment and one to two weeks after to determine the percentage of worm eggs killed. The veterinarian can also help you identify the heavy egg shedders in your herd by conducting fecal egg counts per animal. Heavy shedders can be treated more aggressively than low shedders, who should be treated minimally or not at all. Your practitioner can also advise you in pasture management. Your area extension officer can also help with information and advice.

A zoologist makes a detailed drawing of a previously unknown tapeworm. Tapeworms can usually be easily controlled.

Customizing a deworming program means that you have to identify which horses have the most problems with parasites. Your veterinarian can do that for you by conducting a fecal flotation test. You can then discover which horses are heavy carriers and deworm only them.

When customizing a program, you can rank the horses in a herd by their fecal egg count tests. The bottom 50 percent won't need much deworming, the middle 30 percent will need a little bit more, and the top 20 percent will need the most.

Deworming only the horse or horses with the most worms and not the others will also reduce the contamination in pastures. Deworming isn't just about getting parasites out of the horse; it's also about getting them out of the pastures. And by treating only those horses that need it, you reduce the pressure of drug resistance.

Another option is to deworm your horses every six months with a broad spectrum dewormer (ivermectin, moxidectin, and fenbendazole at double the dose for five days); this dewormer should include praziquantel once a year, which is for tapeworm control. In most climates, the optimal time is spring and fall. This will eradicate large strongyles and give adequate control of tapeworms, pinworms and small strongyles for most horses.

However, the top 20 percent of horses who lay 80 percent of the eggs can be treated more often. You would be ahead financially and do a better job of controlling parasites.

There is a geographic component to deworming programs, too. Horses in the northern states become infected during the summer. That's when your treatment should be concentrated, with less treatment in the winter. Horses in the summer states become infected in the autumn and spring, but not much in the summer because it's too hot.

Types of Various Parasites

Parasitologists agree that deworming is not an easy concept for people to grasp; worms are constantly changing and becoming more resistant to certain chemicals. However, understanding how parasites affect our horses can help us to develop a good deworming program and save money. The following worms are the ones that are most commonly found in horses.

DETERMINING DOSAGE

Administer the correct dose of anthelmintic for your horse's weight. As underdosing is a problem, manufacturers urge horse owners to first determine the weight of their horses. You can do that by using a weight tape, which can give you a close approximation. Do not measure by height because the weight of a Thoroughbred that is 16 hands high will be different from the weight of a Friesian of the same height. Once your horse's weight is established, examine the worming syringe. You'll notice that there are number marks running down the side of the tube. Slide the latch or spin the dial until it rests on the number denoting the weight of your horse. Put a halter on your horse, and either tie him or have a friend hold him. Make sure that your horse's mouth is empty of all grain and hay. Place the syringe into the interdental space (where the bit normally sits) and point it toward the back of the tongue. Press the plunger and administer the dosage on top of the tongue. Then hold your horse's head slightly upward until he swallows.

LARGE STRONGYLE, OR BLOODWORM

The large strongyle (*Strongylus vulgaris*, *S. equines*, and *S. endentatus*), also known as a bloodworm, can grow up to an inch and a half long. It lives in the horse's colon and cecum. The large strongyle does the most damage to the horse's insides during its migration through the body, making lesions in the walls of the blood vessels, causing blood clots, intestinal colic, and gangrene of the bowel. *S. equinus* and *S. endentatus* can cause liver damage and peritonitis. Because of this extensive damage, large strongyles used to be the biggest challenge for horse owners, but thankfully they have been largely eradicated throughout North America. It remains important, however, to continue to deworm for large strongyles.

Treatment: Broad-spectrum dewormers for the parasite's adult stages; moxidectin, ivermectin, and fenbendazol (administered at double dose for five days in a row) for the migrating phase.

SMALL STRONGYLE

The small strongyle (cyathostome) is a threadlike worm, less than an inch long, which lives in the cecum and large colon. Small strongyles differ from the large strongyles;

they form tunnels into the lower intestine and turn into cysts, where they can either develop or remain encysted this way for many years.

Small strongyles are now the parasite of most concern. Up to a million can infect a horse at one time, and 2,000 eggs per gram can be found in the manure of a severely infected horse.

There are estimated about twenty-five species of small strongyles, which vary in size. The different species live in different parts of the intestines, and some of them are less susceptible to drugs than others. Most of the damage occurs when the parasites emerge from the cyst, releasing all the waste product that has been built up inside. The tissue fluid has protein in it, so if the cysts are in the large intestine,

The result of excessive scratching, this horse's rubbed tail is a telltale sign of pinworm or other anal discomfort.

the horse can lose protein when they burst. When the horse loses protein that far back in the intestine, he cannot recover it, and the horse will be protein-malnourished. Without protein, the horse can't build bone and muscle and won't have a healthy skin condition. The body will also mount an inflammatory reaction against the small strongyles and will use the available protein for the fight instead of for building bone and muscle.

Since all encysted larvae emerge from their cocoon at once, if there are many, diarrhea, dehydration, and colic can occur.

Treatment: Broad-spectrum dewormer for the parasite's adult stages and moxidectin and fenbendazole (administered at double dose for five days in a row) for the adult and encysted stage. Some populations have an inherited resistance to benzimidazole (95 percent do) and pyrantel (50 percent do), so check with your veterinarian before you use a dewormer from this group.

PINWORM

The female pinworm grows larger than the half-inch male, to nearly two and a half inches long, with a pin-like tail. They form in the colon and are passed on into the anus, causing the horse to scratch his tail and create bald patches. Although pinworms are irritating to the horse and dismaying for the owner, internal damage is rare. However, some sensitive horses can be restless, stop eating, and lose condition.

Pinworms (*Oxyuris equi*, common pinworm; *Probstmayria vivipara*, minute pinworm) affect all ages but used to be most prevalent in young horses. However, pinworms are becoming more common in older horses as well.

Treatment: Broad-spectrum dewormers usually kill pinworms; however, if a given treatment doesn't remove all the pinworms, then change to a different group to kill the rest.

TAPEWORM

The tapeworm (*Anoplocephala perfoliata* and *A. magna*) can grow up to two inches long inside the body, depending on the species. It develops in the cecum, colon, and small intestine. Pasture mites are the intermediate host. The horse eats the infected mite, and then the larva develops into an adult inside the intestinal lining. If there is a light infestation, you won't see any signs. During a heavy infestation, the horse will be depressed, will lie down frequently, or will become colicky.

Treatment: Problems caused by tapeworms are due to a heavy infestation, but they can be controlled with a single treatment of praziquantel once a year. Many broad-spectrum dewormers have versions that includes praziquantel in the dose.

BOT

Adult bots (*Gasterophilus intestinalis*) are flies with yellow and black–striped bodies. The female bot attaches her eggs to the horse's body. There are two varieties: the common botfly and the throat botfly. The common botfly lays eggs on the horse's shoulder and forelegs. The horse then licks the eggs off of his skin and ingests them.

The throat botfly lays eggs on the horse's chin and throat. They hatch on their own, burrowing under the skin into the mouth. Both species remain in the lining of the tongue and cheek for a month before moving down to the stomach. A bad infestation can cause ulcers in the mouth.

Treatment: Ivermectin and moxidectin. Because bots overwinter inside the horse, deworm in early winter after the first frost. In the summer, remove the eggs from the coat with a special bot knife.

POISONOUS PLANTS

There is nothing more satisfying to horse owners than watching their horses graze peacefully on a lush pasture. But lurking alongside all that healthy vegetation could be toxic plants that bring serious health problems to any horse unlucky enough to take a bite. There are hundreds of toxic plants in North America; fortunately, only a few are harmful to horses. For a more extensive list of toxic plants near you, contact your local US Department of Agriculture Cooperative Extension office.

Bracken Fern

This tall (two to three feet high) perennial fern (*Pteridum aquilinum*) has triangular leaves. Look for it growing in bunches in wooded areas and on wet open lands. It's seen throughout the country except in dry desert climates. Bracken fern can cause neural dysfunction, including depression, loss of coordination, and blindness. If the poisoning is discovered early

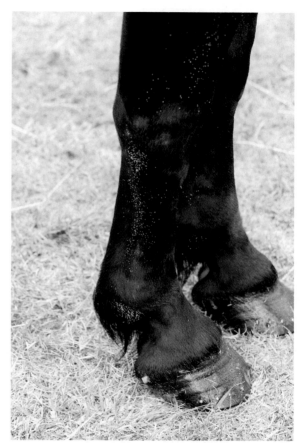

Tiny bot eggs can be seen on a horse's legs. Botflies cause oral and intestinal problems once they've burrowed into the body.

enough, large doses of thiamin administered over one to two weeks can help in recovery.

Hemlock

This perennial, *Conium maculatum*, can grow up to six feet tall. It is multistemmed with serrated parsley-like leaves and bunches of small white flowers. Hemlock is sometimes confused with Queen Anne's lace, so look for purple spots on the hemlock stems, usually near the plant's base. Hemlock grows alongside roads and in open areas throughout North America.

If a horse eats hemlock, signs of distress will appear within an hour of ingestion, including: nervousness, trembling, loss of coordination, colic, reduced heart rate, and heavy respiration. There is no cure, and death usually occurs from respiratory failure. If the horse has consumed only a little, he may get better with supportive care from a veterinarian.

If a horse ingests hemlock, he will suffer loss of coordination and colic; death occurs from respiratory failure.

Johnsongrass/Sudan Grass

Johnsongrass and Sudan grass (*Sorghum spp.*) are very tall, coarse-stemmed grasses with large, veined leaves and multibranched seed heads. The grasses are usually seen in the South alongside roads and in open areas. Symptoms are similar to those of cyanide poisoning: fast breathing, tremors, constant urination or defecation, and convulsions. Drugs can help slow the effects of cyanide poisoning.

Locoweed

This short, leafy perennial grows in tufts from a single root. Flowers are either white or purple without a leafy stalk. Several species of locoweed grow throughout dry, sandy areas of the West and Southwest. Locoweed (*Astragalus spp.* or *Oxytropis spp.*) causes a horse to act oddly, lifting his legs high, staggering, or nodding his head. Severity depends upon the amount of locoweed the horse has consumed. There is no treatment; symptoms are permanent.

Milkweed

The Monarch butterfly could not exist without milkweed (*Asclepias spp.*), the butterfly's primary source of food. But milkweed is deadly to horses. This perennial plant is found in dry and swampy areas throughout the United States. It exudes a milky sap when cut or broken. Narrow- and broad-leaved species exist, but

This beautiful oleander is toxic to humans and equines. If ingested, it causes irregular heartbeat, respiratory problems, and death.

the narrow-leaved variety is most toxic. The fruit is a capsule filled with seeds born on a silky tuft. All parts are toxic, even dried. Signs of milkweed poisoning include salivation, loss of coordination, seizures, and colic; death often occurs from one to two days. Treatment includes gastrointestinal detoxification and treatment for heart arrhythmias.

Oleander

A favorite ornamental in hot arid climates, oleander (*Nerium oleander*) is a tall-growing evergreen shrub. Flowers grow in large clusters at each branch and can be pink, white, or red. Symptoms occur shortly after ingestion and can last for several hours. These include colic, respiratory distress, slow or fast pulse, and an irregular heartbeat. Early treatment is key, and supportive veterinary care is important, including administration of activated charcoal to slow toxins and the use of drugs to stabilize the heart.

Red Maple Tree

The red maple (*Acer rubrum*) is a quite common medium-size tree with green leaves that have bright red stems. In fall, the leaves turn scarlet. These trees are found throughout North America, growing in the wild and as ornamentals. If a horse eats the leaves, especially if they are dead or wilted, effects will begin to appear anywhere from a few hours to several days after.

A horse suffering from red maple toxicity will be lethargic and refuse food; his urine will be dark red or even black, and his mucus membranes will advance from pale yellow to dark brown. Other signs include dehydration and a rapid heartbeat. Treatment involves supporting the system with large amounts of IV fluids, as well as a blood transfusion if necessary. Sometimes, the horse may recover, but recovery depends on how many leaves were eaten and the promptness of care.

Other toxic trees include: cherry (*Prunus spp.*), black walnut (*Juglans nigra*), black locust (*Robinia pseudoacacia*), the buckeyes of horse chestnut (*Aesculus hippocastanum*), the acorns of oak (*Quercus spp.*), and Russian olive (*Eleagnus angustifolia*).

Russian Knapweed/Yellow Star Thistle

Yellow star thistle (*Centauria spp.*) is a tall annual weed that has round yellow flowers set inside stiff spines. Russian knapweed has no spines, and its thistlelike

Seven of the 112 species of ragwort in America are a danger to horses. Once liver failure presents itself, death is inevitable.

flowers can be purple or white. Both grow throughout the western United States from Missouri to California and both are found along roadsides and in fields and pastures. Signs include clenched facial muscles and the inability to chew properly. There is no treatment, and damage is permanent.

Tansy Ragwort

Tansy ragwort (*Senecio spp.*) is a biennial weed, which spends its first year in a rosette stage then progresses to a multistemmed flowering plant in the second year before it sets seed and dies. The flower heads are topped with clusters of small, flat daisylike flowers. One-hundred-and-twelve species of this genus are in the United States, but only seven are toxic to livestock. Tansy ragwort is found in pastures and along roadsides. It's difficult to tell if a horse has consumed tansy ragwort until signs of liver failure appear. There is no treatment.

Yew

This evergreen shrub, *Taxus spp.*, has flat needlelike leaves and red or yellow berries dotted with a dark seed on each end. Yew is found throughout the United States and is a common ornamental. Yew trimmings are often thrown into pastures mistakenly. Signs of ingestion include trembling and rapid heart rate. Sudden death is common. There is no treatment.

Chapter 14

Alternative Care

We all know the importance of taking good care of our horses. Today, more and more studies about sports therapy, nutrition, and health care are coming out with the idea of extending the working lives of our horses. But horses are more than just performers, more than just a pair of hocks, and more than just healthy feet. They are emotional, living beings with instincts and desires to live a lifestyle other than the one we impose upon them. We can change our ideas about the way we work with horses, not only to keep them healthy and performing well, but also to help make their lives happy and content. We can do all this by adopting a more holistic approach in our day-to-day horsemanship and care.

Holism means that you see the horse as one inter-related being—what is happening in his hocks may be affecting the rest of him as well. Is his lackluster performance a direct result of changing his trainer, or is it because he's unhappy with his management? Holistic horsemanship addresses all these issues and helps a horse fulfill his potential—mentally, physically, and emotionally. You can treat holistic horsemanship as a philosophical approach to total health care. Whether you choose a holistic approach to help enhance your horse's performance, to help ease your old friend's pain naturally, or to adopt a more organic and chemical-free lifestyle, incorporating these techniques into your lifestyle with your horse will help create a better partnership and develop a greater understanding between the two of you.

FUNDAMENTALS OF A HOLISTIC LIFESTYLE

It used to be thought that holism was just quackery or a philosophy that only health nuts adopted. But attitudes are changing. Today, research and trends in human holistic health care are motivating veterinarians to include natural adjuncts to veterinary care, and they are encouraging their clients to consult such professionals as chiropractors, physical therapists, and massage therapists. There are a number of forms of alternative care that can enhance a horse's health and well-being and augment traditional veterinary care.

Many practitioners choose to study alternative healing arts, such as acupuncture and homeopathy, because that knowledge allows them to consider both a medical approach and an alternative approach for the benefit of the animal. Alternative care is also known as *integrative* or *holistic* medicine. Practitioners who examine all aspects of an animal's life while using both traditional and alternative care are called holistic veterinarians.

Top-level riders are also realizing that a holistic approach helps them in their sports, particularly at the upper levels, where a zero-medication policy is in effect. These riders have alternative care practitioners help alleviate the muscle soreness that can make or

A physical therapist uses stretching exercises to examine a patient and increase the horse's mobility.

ALTERNATIVE CARE

Learning these terms will help you talk about alternative horse care.

■ **Acupressure:** This stimulates acupuncture points without the use of needles.

■ **Acupuncture:** The use of needles to release blocked energy. It is used to strengthen the immune system, relieve pain, and improve organ function by stimulating acupuncture points throughout the body.

■ **Chakras:** Centers on the body where energy is concentrated.

■ **Endorphins:** Hormones secreted by the brain that have painkilling and tranquilizing effects.

■ **Herbal medicine:** A type of Chinese medicine in which the entire system is taken into account.

■ **Homeopathic remedies:** Matches the patterns in the disease and treats the causes of the illness.

break a competition. Most importantly, you can tackle a problem at its basic level, perhaps before one even starts, and prevent it from returning by implementing holistic techniques in your day-to-day routine.

To treat your horse holistically, you need to become a detective and begin to ask questions. Are you hanging his hay in a position that makes it difficult for him to eat comfortably? Is his noseband too tight? Just as a person works at an ergonomic workstation to avoid repetitive injuries, consider how your horse's environment affects his mental, physical, and emotional health.

THE NATURE OF YOUR HORSE

The best place to start is by understanding the nature of your horse. So many owners are unaware of how different the stable environment is from the horse's natural habitat. Horses are meant to move and graze twenty-four hours a day, not be shut up in a stall. Yet we've changed that by feeding them only two to three times a day and letting them out of their stalls only a few hours a day. Most behavioral problems arise out of management and physical problems; however, since we have to be practical as well, we can incorporate the horse's natural instincts into our world through compromise. Studies have shown that horses are less likely to weave if they are in a stable with more activity. In addition, as horses are trickle feeders—meaning they eat small meals frequently—weaving can be allayed by keeping hay in front of the horse constantly or at least by offering it throughout the day.

Feeding

Consider the ergonomics of feeding your horse. If you look at your horse's front teeth, you'll notice that the horse has a slight overbite (not to be confused with parrot mouth, a conformational flaw). This is for a functional reason. When a horse grazes, his bottom jaw comes forward to meet the top so he can take a bite.

Many people, however, feed their horses from a manger or from a hay net hanging overhead. Horses are not giraffes; it is unnatural for them to eat above their heads. In fact, you are damaging your horse by feeding him this way because as he lifts his head to eat, his jaw goes out of alignment; as a result, problems such as headaches and decreased performance will develop. He'll also have a greater risk of getting hay, dust, or chaff in his eyes. It's best for the horse's health to feed him on the floor in a rubber feed tub.

Herd Dynamics

Horses have evolved to graze on open land in herds; they do this for protection. Since horses gather in a herd for safety, a horse on his own may consider himself in danger. If you turn your horse out, consider what emotions he goes through while out on his own. A horse alone is unnatural. Even if he doesn't express his anxiety outwardly by walking the fence line or calling, he may suffer a low-level anxiety that can manifest itself in other ways. This translates to his stable environment as well.

A study found that having at least three windows in a horse's stall stopped him from weaving because he could see other horses nearby. Not everyone can afford to chop windows in their stalls, but the study found that attaching a mirror to the inside of the stall stopped the horse's weaving as well. Experts don't believe the horse sees the mirrored image as himself. Rather, the horse sees a moving object, and that comforts him.

Since a horse is designed to be on the move, it's best holistically for your horse's health to have him outside as much as possible. However, make sure he is happy with the situation. Give him a buddy for company, a well-drained paddock, ample shelter, and plenty of forage. Don't force him to stay out if he shows signs of stress, such as pacing or running along the fence line.

Stabling

How you bed your horse's stall is important from a physical-therapy standpoint. Many keep bedding away from the door for cleanliness, but the horse spends most of his time standing at the front looking out. If you don't want to put bedding there, consider placing a mat to cushion his feet and to keep out the cold. In addition, if the horse stands with his front legs on this empty space by the door and his hind legs on the bedding behind him, he's spending a lot of time standing on an incline—not the sort of posture you want to encourage.

The best stall should have a firm, level base of compact dirt or gravel, preferably topped with a rubber mat and enough bedding to prevent sores, but not too much so that the horse can't walk through it easily. Ventilation of the stable is important to remove airborne diseases and dust and to maintain a healthy respiratory system.

When horses eat from a ground-level manger, their heads are in a grazing position, the best eating position for jaw alignment.

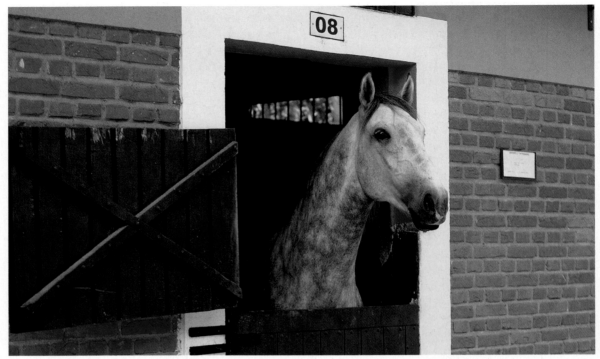

Like most horses, this Lusitano stands looking out his door for long periods. A well-placed mat will give his feet needed cushioning.

NATURAL HEALTH CARE CHOICES

Rather than using Western medicines or practices when you are treating your horse, you can choose from a number of different systems. These include homeopathic remedies and Chinese herbal medicine, as well as massage therapy, acupuncture, and chiropractic.

Homeopathic Remedies

Homeopathic medicine was developed in the mid-1800s by Samuel Christian Hahnemann, a German medical doctor. Homeopathy works on the principle of "like cures like." Large doses of toxic substances can cause death, but if a diluted dose of the same substance is ingested, it can prompt the body to heal itself. Homeopathic remedies are made from plants, minerals, drugs, viruses, bacteria, and animal substances. Homeopathic remedies match the patterns present in the disease and treat the causes of the illness.

Homeopathic remedies are easily purchased over the counter, but be sure to consult a qualified homeopathic veterinarian before using them. Overuse can result in your horse actually beginning to show symptoms of the disease. You can also muddy up a true diagnosis by giving conflicting remedies. Your horse

Chinese herbalists seek to restore emotional-physical balance by a precise combination of herbs, like those being weighed above.

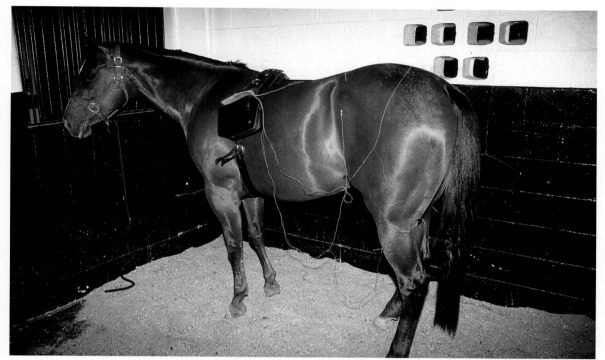

Acupuncture needles stimulate the horse's energy-flow meridians, strengthening his immune system and relieving pain.

may be showing symptoms from the incorrect remedy rather than from his true ailment. Feeding mints or nutritional supplements such as garlic can also act as an antidote for the remedies, so overall it's much wiser to start a homeopathic health regime under the direction of a professional.

You can implement natural remedies into your day-to-day work by keeping a homeopathic first aid kit at hand. A wonderful healing remedy for superficial cuts and scrapes is ten drops of mother tincture of *Calendula* (which is the pure, nondiluted form) added to a bucket of water, and sponged over affected areas. (*Note*: *Calendula* should not be used in deep wounds, as the remedy causes swift healing. Deep wounds need time to heal from the inside out. Please consult your veterinarian with suspect wounds or ones that need suturing.)

You can purchase *Arnica*, *Rhus toxicodendron*, and *Ruta* in a liniment form for your after-work cooldown. Dr. Bach's Flower Essence Rescue Remedy is wonderful for times of stress (for you and your horse). Readily available at most pharmacies, a few drops of Rescue Remedy on the tongue or in a bucket of water can help calm a horse.

Chiropractic, Massage, and Acupuncture

Chiropractic care can help horses' back pain. Vertebrae can get stuck or become limited in their range of motion. If this happens, muscles around the vertebrae work much harder than they need to, becoming sore, inflamed, and limited in motion. When a vertebra gets stuck, the condition is known as a subluxation. This can be caused by daily stress, a traumatic injury, a fall, becoming cast (stuck against the wall) in the stall, or even ill-fitting tack. Chiropractors use their fingers to assess the muscle and explore any pain reactions. They then perform the bone adjustments, using their hands to thrust or push the bone into place. Make sure your doctor is accredited with an organization such as the American Veterinary Chiropractic Association.

Massage therapy is also helpful to horses, just as it is to humans. It's a noninvasive treatment that can help increase the blood supply, nourishing muscles and flushing out built-up toxins. It can relax muscles and heal soft tissue. It can also help create a better range of motion.

Acupuncture has been used in China for 3,500 years, and it is the main health care treatment for a quarter of the world's population. The primary goal in veterinary acupuncture is to strengthen the animal's immune system. Acupuncture can also relieve pain and improve organ function by stimulating acupuncture points throughout the body. The concept of traditional Chinese medicine says that *chi*, the life force in the body, moves along channels of energy-flow called meridians. Disease blocks the normal flow of chi in these meridians; acupuncture needles help release the blocks. In acupressure, practitioners stimulate these points with their fingers, instead of with needles.

Chinese Herbal Medicine

Chinese herbal medicine is a very complicated medical system and only a well-trained practitioner should practice it. A certified practitioner can examine the horse and give you the best advice on which supplements, herbs, and foods are right. With proper assessment, most horses can benefit from herbal medicine in the treatment and prevention of diseases. Herbal medicines address health issues in a gentler way, from a Chinese perspective, a way that may not be addressed in traditional western medicine. Emotions, for instance, mean something in Chinese medicine. If all blood work and x-rays for the horse are fine, but he is still unserviceable, Chinese herbals go a step further to take the horse's emotional state into account. For example, anxious or spooky horses may be reacting to liver or heart disharmony and could benefit from calming herbs. The herbs are intended to balance the horse, rather than drug him. But they can be just as potent as medication, so follow the practitioner's directions wisely; natural doesn't always equal harmless.

HOW HEALING WORKS

Everyone understands the comfort associated with a hug after a stressful experience. A healing touch, such as a hug, releases endorphins, hormones secreted by the brain that have painkilling and tranquilizing effects, helping relieve negative emotions. We all have the natural ability to comfort and heal one another.

Eastern philosophy states we all, including horses, are composed of a vibrating energy field through which runs our life force. This life force runs through all the cells of the body connecting everything together, so when anything good or bad happens on a physical, emotional, or mental level, the reaction flows throughout the body, and all its parts are affected. It is through this energy field that we affect others. This is one of the reasons why stress is both mentally and physically debilitating. When you touch skin, which is so sensitive, you make direct contact with the whole body and bring about healing changes in the body and in the mind.

The energy field within the body radiates an electromagnetic field called an aura. A type of photography called Kirlian records auras, revealing that all living organisms radiate this aura. A healthy body will display a strong, bright aura, while an unhealthy body, in turn, has a weak aura. Although only some highly sensitive people acknowledge seeing auras, many people say they can feel it during healing, a sensation a bit like pins and needles. Practically everyone reacts to auras, whether they realize it or not. Consider when you like or don't like someone immediately, because you get a feeling about them. You may not like even being in the same room with some people because their energy is so bad, while others with good energy make you feel happy, safe, and good about yourself. That energy you are responding to is the person's aura. Horses, being the extremely sensitive creatures that they are, are like sponges. They soak up these vibrations from humans and add them to their own. Have you ever had a bad day and tried to ride while you are upset? Sometimes your horse will react to your energy and respond with the same emotions.

When we touch a horse, we connect physically with our fingertips, creating an energy-field to energy-field connection. With this connection, we tap into a third, universal energy field. The horse will work with that energy field for healing benefits.

Where to Heal

Hindu philosophy teaches that there are centers of energy on the horse called chakras, where the body's energy field is concentrated. You can touch the horse anywhere during healing to be in contact with it physically and emotionally, but the chakras directly relate to particular organs and glands in the body. If one or more chakras are blocked, it indicates which organ or gland is having trouble. The spine is the major area for chakras (there are four along the back) in the horse. Therefore, saddle-fitting problems, bad riding, and ill-fitting girths greatly affect the horse's life-force energy.

Reactions to Healing

Horses react to the endorphins released from healing touch by getting drowsy, closing their eyes, dropping

Owners and holistic practitioners can enhance a horse's well being by moving their hands gently on the chakras along his body.

their heads, or relaxing their faces. Their mouths lose tension—signs of this include licking, yawning, or chewing. Some horses' legs have even buckled underneath them during more intense sessions. The horse's respiration also changes, with breaths lengthening and deepening.

Horses will also respond emotionally to alternative therapies. Since they can't cry or talk to release their stress, horses hold their emotions within, which results in stable vices, behavioral problems, changes in performance, and other negative actions. After treatment, horses often show signs of being happier and more at peace.

WHO CAN HEAL?

Eastern philosophy teaches that the opportunity to heal another being is there for everyone. All that is needed is an open mind and a desire to help. That desire starts as a natural instinct, but it is a basic skill that holistic practitioners must develop. If you are interested in learning these techniques, you'll need training and practical experience. You can find more information online. Here are some tips to help you start channeling your horse healing energy for your horse's benefit:

• Practice regularly—for at least half an hour every week.

• Make sure you have some privacy and quiet.

• Pick a time when you are not rushed or stressed.

• Keep any items containing magnets away from you and your horse during sessions, as they can interfere with the natural healing energy. You can replace them afterward.

• If you are comfortable with your horse loose, feel free to treat him in his stall. If not, have a friend hold him.

• Start at the horse's side by his neck and head.

• Slowly move your hands over the chakras; however, don't feel you have to touch these areas only. Healing energy can travel through the entire system, no matter where you touch. A chakra is simply the strongest point you can access.

Healing energy doesn't promise a cure, but it is usually beneficial on some level. Often its role is to help the horse emotionally and relieve his suffering.

Rider Instruction

Section 7

Every rider, novice or experienced, wants to improve on his or her riding skills. Whether riders are trying to learn to post the trot, to jump a cross-country course, or to tackle a difficult reining pattern, they will find that there is no better way to improve riding skills than through formal lessons. To get the most out of your lessons, begin by finding the right instructor; who that instructor will be depends on how experienced you are and which discipline you wish to learn. You will also want to make sure the facility at which the instructor teaches is well run.

Chapter 15

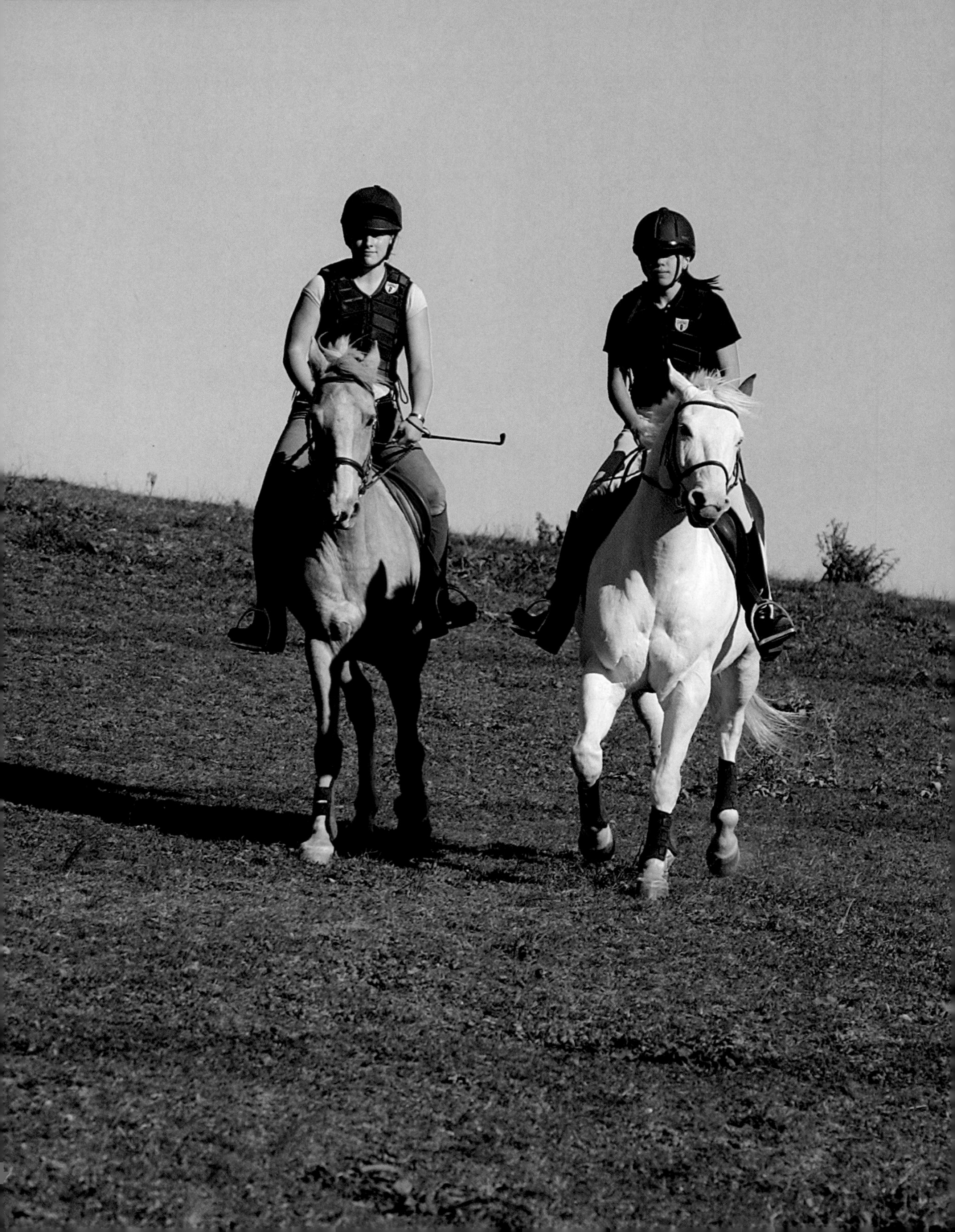

Riding Instructors, Safety, and Fitness

Most novice riders go about finding an instructor the wrong way, often choosing one who is conveniently located, is famous, or is the first one available at the time. None of these is the best way to hire a professional who is right for your experience level or riding goals. It's a waste of time and money to ride with a trainer who isn't furthering your education. And having to end a relationship with a trainer—even the wrong one—can be awkward, especially if he or she is in your boarding barn or horse community. So choose your trainer wisely. Consider your riding level and goals carefully, decide what kind of trainer you need, then go looking.

INSTRUCTOR CERTIFICATION

Is it important for a US instructor to be certified? Not necessarily. Certification, which is new to the United States, is not universally required in this country nor is it as well established here as it is in countries such as Great Britain and Germany, which have the credentialing system down to a science. Many US instructors choose not to be certified because they do not like the system of teaching required to achieve a particular certification. There are a number of systems for teaching dressage, for instance, and an instructor may not want to change the one he or she is using.

The purpose of instituting certification in the United States was to make it more uniform, as it is in Germany. There are some US organizations that offer instructor credentials, both multidiscipline and discipline specific. Two of the multidiscipline organizations include the American Riding Instructors Association and the Certified Horsemanship Association. Discipline-specific programs include the United States Dressage Federation Instructors' Certification Program and the United States Eventing Association.

The American Quarter Horse Professional Horsemen program bypasses the problem of standardized teaching with its referral program. The organization provides a list of prescreened, reputable trainers/instructors of American Quarter Horses. Trainers/instructors must have a certain number of referrals to make it into the program. They must also be well established, successful, and professional.

The bottom line is that there are talented instructors who are uncertified as well as talented ones who are.

THE RIGHT INSTRUCTOR FOR YOU

If you're a beginning rider or one who has never taken formal lessons, don't worry about finding an instructor for a specific discipline. Instead, try to ride with the best beginner's instructor you can find. Once you have the basics down, you can easily switch over to a different riding style if the one your instructor is teaching isn't for you. Many aspiring show riders start out in 4-H and schooling shows and then work their way to a show barn. So even if showing is your

INSTRUCTOR EVALUATION

We've all heard of the stereotypical drill sergeant instructor who pushes students to the emotional brink. It's important to enjoy yourself in lessons. If you're frustrated or upset, it's not worth it. An instructor must build confidence in a rider. If there is a problem, the trainer should try to work through it and discuss what the student is doing right and wrong.

You must trust your instructor enough to be open about any trepidation or fear you felt during a lesson. An instructor may miss the fact that you are riding with white knuckles. Many trainers have never experienced fear on a horse and often forget that other people do.

Once you have talked about your fears, if an instructor tries to push you to do something that scares you, then you may have to consider getting another instructor. It's true there are people who do better with a certain motivation or pushing, but that can have the opposite effect for others who don't.

It's also counterproductive if your instructor comes down hard on your horse and blames him for riding problems and insists you fight with him constantly. Occasionally, horses do need correcting, but the attitude toward the horse must always be positive.

If you feel like you aren't progressing, try talking to your instructor first. Sometimes that's all you need to

goal, you don't have to start out with a noted showing trainer and a slick facility.

If you are ready to specialize, then definitely find an appropriate trainer. Don't expect a jumping instructor to teach you dressage if that's not his or her specialty.

When interviewing instructors, be honest about your experience level. Students who want to jump ahead right away or take lessons from a specific instructor who only teaches advanced riders will sometimes inflate their skill levels. Don't do it. Once you're in the saddle, your level of experience will be readily apparent.

Top-level instructors focused on high-level students may not be good choices for inexperienced students. Even if such instructors are willing to teach beginners, they might not be available to give novices the attention needed because their pursuits lie elsewhere. Find the instructor who fits your needs.

You must be able to trust your instructor enough to communicate openly about flaws or frustrations in your riding.

The horses at this stable look well cared for. A conscientious trainer often influences just how safe and tidy a barn is kept.

do to make a situation better. After all, how will your teacher know you're unhappy if you don't tell him or her? Find some time to discuss your issues outside of the lesson. If you feel as though you can't talk to your instructor, then you have a problem anyway. A good instructor should always be approachable. If the promised changes don't happen, then consider moving on.

It's never easy ending a relationship, but do so with integrity—let your instructor know as soon as you decide. That person deserves to know what his or her schedule and finances are going to be like for the month. Don't just disappear. If your original instructor finds out that you went down the road to someone else, the situation will create bad blood between everyone.

No instructor wants to be blamed for poaching another's student. It's true that your instructor doesn't own you, but the class act is to let him or her know you're leaving. The horse world is very small.

FACILITY EVALUATION

Where you learn can be as important as from whom. Visit the stable where your potential instructor works before you book a lesson. Phone the instructor and ask if you can watch a lesson. First impressions mean a lot, so make

sure you're comfortable there. The lesson horses should be in good condition and well cared for in clean stalls. A neat and tidy barn speaks volumes about the instructor and shows that he or she is a caring horseperson. The instructor's dress also tells you a great deals. Instructors shouldn't teach lessons while wearing shorts and flip-flops. Appearing neatly and appropriately dressed sets a good example for students.

Overall, the facility should feel safe, careful, and kind. Places that tie their horses correctly because it's safer, or only let riders up to a certain weight ride their horses, or require boots and helmets when riding will be a place that takes care of you.

Every barn has its own culture, which can be casual or highly competitive; that culture can affect your comfort level and ability to learn. If people look down on you because you're not showing, the joy of riding can soon fade. If gossip and backbiting abound or if people are impatient and harsh with their horses, then walk on by. No one can learn in such a dismal atmosphere.

Consider taking a few private lessons before you decide whether to join a group. One-on-one attention from the instructor can help to allay any fears or concerns that you may have.

INSTRUCTION AND CERTIFICATION

Learning these terms will help you talk equine instruction and certification.

■ **American Riding Instructors Association (ARIA):** National association promoting safe riding through the certification of knowledgeable instructors. Certifies instructors in three levels and fifteen disciplines.

■ **American Quarter Horse Association (AQHA) Professional Horseman Program:** Referral program that includes a list of prescreened, reputable trainers and instructors who belong to the AQHA.

■ **Certification:** A system of testing to prove a riding instructor's competence, safety, and knowledge. Certification is relatively new in America.

■ **Certified Horsemanship Association:** North American association certifying instructors, accrediting equine facilities, and publishing educational material.

■ **Instructor/trainer:** An instructor usually gives riding lessons only; a trainer instructs riders and trains client horses. Terms used interchangeably in America.

■ **US Dressage Federation Instructors' Certification Program:** Designed to educate dressage instructors and inspiring instructors in the classical concepts of dressage through four different levels.

One of the warning signs of a poor instructor is the promise of quick advancement. Learning to ride is a slow process. The faster the better isn't true with horses. Learning will take as long as it takes, and riders have to school themselves to be patient, calm, and willing to listen.

SAFETY AROUND HORSES

Safety should be a top priority for you, as well as for your instructor. Be certain, before doing any riding, that you and your horse have the right tack and safety equipment (*see chapter 8*). You should also arm yourself with as much knowledge as possible. To keep you and your horse safe, learn about the nature of horses and how to avoid accidents. Knowing what frightens a horse is essential.

What frightens a horse first and foremost is the possibility of predators. Horses have evolved behaviors that help them see and avoid predators. Basically, they are always looking out for danger. Because of their wide-spaced eyes and long necks, horses can see things from many angles. Their very mobile ears mean they can hear sounds from faraway. Your horse might overreact to things that seem like nothing to you, but millions of years of evolution tell him to be wary.

When horses are startled, they react in three different ways: freeze, flee, or fight. Recognizing these signs will help you identify the problem. A horse freezes to be less noticeable to a predator and to give himself time to figure out the situation. You will notice the horse staring at something with his head up and ears forward.

If a horse decides to flee, he can do so in any direction, and he can rear, spin, or trample anything that gets in his way. If a horse feels cornered, he will fight, using his teeth and hooves to defend himself. Never corner a panicked horse or shout or run around him. Stand quietly well out of the way, leaving him alone until he calms down.

To avoid triggering these reactions in your horse, don't make any sudden movements or loud noises, especially when your horse can't see you. Doing so is a sure way to trigger that flight or fight response, and you may get trampled or kicked. Always approach a horse from the front, so he can see you coming. Talk calmly to him and touch his neck and shoulder. Always try to make your movements predictable. Horses are calm when their handlers are calm.

Bucking and rearing are never fun. Rearing, in particular, is very dangerous. A rearing horse can flip over and crush his rider. If your horse has developed the tendency to rear while under saddle or while leading, seek professional guidance rather than trying to fix the problem yourself. If your horse has only given you a small, half-hearted rear, denoting his reluctance to move forward, distract him from your intended direction by pulling him to the side and then making him go in a small circle. What you do not want is for your horse to think his rearing will end the work.

Bits of Safety Advice

It may seem odd, but the closer you are to a horse the less damage he can do if he kicks. Horses kick straight out and back, so stay close to a horse's side as you work around him. Do not walk directly in front of a horse when leading him. If he shies and runs forward, he could bowl you over. Never tie a horse to a temporary, loose object. A frightened horse can pull the object down and drag it behind him, becoming seriously injured. Never wind a lead rope or longe line around your hand. If your horse should bolt, you won't be able to let go and you will be in danger of being dragged or losing a finger or even a limb.

A show jumper rides out a rear, leaning forward and releasing the reins so the horse doesn't fall over backward.

Make sure that you never pull back on the reins if the horse starts to rear while you're riding. If you do so, you will make your problem much worse; in fact, you will cause the horse to fall over backward.

Some horses buck because they are excited or feel overly energetic. Others buck because they are in pain. A horse has to have his head and neck down to buck. To stop a buck, pull up with the reins and lean back if you feel your horse begin to pitch his head down. It's important to seek your veterinarian's advice if your horse has begun bucking or rearing for no apparent reason. He may have an underlying health problem.

If your horse hasn't been ridden or turned out for a while, warm up slowly, longe in an arena, or work your horse in a round pen prior to your ride to burn off any excess energy.

If you're planning on going on a trail ride, make sure you know who you are riding with and what their plans are for the day. Some riders like to go fast while others prefer a leisurely pace. Mixing the two doesn't always make for the safest ride. Your horse may fret or want to rush to catch up with riders leaving the group. Always carry a mobile phone on your person. Do not keep it with your horse, in case you fall off and are separated.

RIDING AND FALLING

All riders fall, no matter their age or experience. It's a fact of life for horsemen. So take every precaution to ensure a safe ride by making sure your tack is in good repair. It's particularly important that the stitching and buckles on your stirrup leathers are not worn or in danger of coming loose.

Improperly fitted saddles and chafing girths can make the horse uncomfortable and invite a misbehavior, as can ill-fitting bits and bridles.

A fall can lead to a loss of confidence in your horse and in your riding ability. The way to regain that confidence is in a step-by-step fashion. Stay within your comfort zone until your fear subsides. For some riders this may mean keeping away from jumps for a certain amount of time; for other riders this may mean staying at the walk.

Once you feel confident, you can begin to add elements of your old routine back into your riding. Positive visualization can also help. Think about the day you fell and go over what happened. Ask yourself what you could have done differently that might have stopped the fall.

This may mean understanding how your horse spooks and readying yourself for it. Perhaps he shies at traffic cones. If so, practice with cones at home until he accepts them. Maybe your horse tends to buck during a canter depart. Then ask a trainer to help you work through this problem. Fear creates a tense seat, which transmits tension to the horse, creating a self-fulfilling problem. Going back to basics for a short period can make you feel secure again.

Longeing is a great way to get your confidence back quickly, with you in the saddle and a knowledgeable instructor or friend at the end of the longeline. This offers you the safety of having someone else in control of the horse. Have your friend longe you in an enclosed arena.

The saddle can be equipped with a safety strap to hang on to, or you may loop a leather strap around the neck of the horse, which prevents you from clinging to the horse's mouth when you are uneasy. Start with your feet in the stirrups. You can drop them and practice without them as you gain greater confidence. Ride at the walk and trot as usual while holding the reins. The only thing that will be different is that another person will be helping you to control the horse.

Next, tie the reins in a knot and lay them on your horse's neck. Practice holding your hands up over your head and out to the sides as you walk and trot. Moving your arms around will engage your core muscles and give you the knowledge that you can move around without losing your seat.

Next sit the trot. If you feel yourself getting tight, lean back in the saddle so that your shoulders are just a fraction behind your hips. This mini sit-up will engage your core muscles and relax your seat. Now sit with your shoulders over your hips while holding your core muscles. These powerhouse muscles will help hold you to the saddle.

It can be intimidating to canter on the longeline; work in a small arena or in a round pen. Do not attempt to canter out in the open until you are comfortable cantering in an enclosed area.

Just like novices, experienced riders, such as the competitor here, occasionally fall. Don't let falling shake your confidence.

THE FIT EQUESTRIAN

Certainly, you've heard how exercise and getting fit are important for improving your time in the saddle. Unfortunately, most exercise programs don't address the specific needs of the equestrian. Riders require a combination of strength, balance, feel, and endurance. Strict weight training isn't going to do it, nor is pure cardio. Riding is such a specialized sport that it requires attention to specific muscle groups to really see improvement. Your ultimate goal is to become a total rider-athlete.

Exercises

Exercising three days a week, for bout half an hour each day, is all it takes to target what you need to help you improve and get you to the barn. You don't need to invest in expensive equipment. You can train on air-filled balance disks and balance balls, speed rope, and a set of 10-pound dumbbells.

When doing your workout, your posture should mimic your position in the saddle: your eyes should be up, your back straight, your core activated, and your body relaxed. This will help you translate these moves effortlessly to your riding.

FORM AND BALANCE

Putting your weight deep in your stirrups is Riding 101, but it takes time and patience to automatically and consistently keep your heels down. You can use a stair or a curb to help stretch and lengthen your Achilles tendons, but unlike these platforms, the stirrup is not stationary. Doing heel stretches on a balance disk not only helps you put your weight in your heels but also improves your balance because the disk gives you a moving surface.

Heel stretches on a balance disk exercise: Stand on one leg with the arch of your foot over the center of the disk. Press your foot forward until your toe touches the ground; then flex your foot back until your heel touches the ground. Repeat back and forth as many times as you can in one minute, then switch legs. Hold dumbbells to increase difficulty.

CORE AND GLUTE STRENGTH

The elusive half-halt is hard for most riders to master because the muscle groups needed to execute it are usually underdeveloped. You need to work your core and glutes for better half-halts. Once you have

Heel stretches, as shown above by horsewoman and author Moira Reeve, are an excellent exercise to promote balance.

BALANCE, CORE, AND LEG STRENGTH

Balance, an activated core, and a strong leg position are the hallmarks of a good jumping position. Many timid jumping riders gain confidence over fences after they've put in the time to strengthen their two-point (the position used in jumping). Galloping to that first fence is less of a nail-biter when you know you're secure.

Squats on balance disks exercise: Place one foot on each balance disk, toes pointed out slightly. As you go deep into the squat, keep your knees apart and pointed directly over your toes, even when things get wobbly. As your strength and stability improve, add weights by holding a dumbbell in each hand at shoulder level. Do as many squats as you can in two minutes.

UPPER BODY STRENGTH

While grooming and lifting feed bags can help you to improve upper-body strength, that's not the whole picture. Strong arm muscles, as well as upper- and lower-back muscles and core muscles will improve your posture and position.

Raised leg crunch exercise: This modified sit-up works legs and abs without stressing the neck. Keep your arms in front of your body and bring your legs up, bending them as you perform the sit-up. Then extend your legs and body until both almost touch the floor. Remain balanced throughout the entire exercise. Complete as many crunches as you can in two minutes.

Lateral dumbbell press exercise: Hold a weight in each hand. Bring one weight up into a hammer curl, then continue the motion into an overhead press, keeping your elbow in and driving the weight straight up. The weight should be in line with your shoulder as your arm is extended overhead. Alternate sides. Complete as many presses as you can in two minutes.

LEG STRENGTH

Who doesn't need to build stronger legs? It's important at any time to have the security of a strong leg, but it is even more crucial when you need to press a dull horse into the bit.

Reverse leg raise with ball exercise: Begin in a push-up position with your feet on the ball. Raise one leg off the ball as high as you can toward the ceiling. Next, lower your leg so it's parallel to your other leg before moving it out to the side by 90 degrees. Return you foot to the ball and repeat with your other leg. Complete as many raises as you can in two minutes.

Moira demonstrates repetitive squats on balance discs, an exercise that will increase core, balance, and leg strength.

the ability to utilize the right muscle groups, your half-halts will become clearer and more precise.

Leg curl with balance ball exercise: Start with your back on the floor and your ankles and calves resting on top of the balance ball. Raise your pelvis so it's in line with your feet and shoulders. Draw your knees up and your heels in as you roll the ball toward your rear end, then extend your legs straight out as you roll the ball out. Do as many of these curls as you can in two minutes. As your strength and stability increase, do your sets one leg at a time.

ENDURANCE

Some riders hit the wall when they are competing. Part of this may be due to nerves, but part of it can be attributed to conditioning. You can improve lung function, endurance, concentration and coordination with five minutes of speed (jump) rope exercises.

Rope skipping exercise: Rope skipping is a boxer's conditioning staple. It's not as easy as it looks, but don't get discouraged. First, work on the skill in short but frequent intervals. Try five rope rotations at a time, then ten. Add footwork once you get the hang of rope skipping. Begin by jumping for five minutes, then work up to ten minutes.

The Total Rider Athlete

For many riders, the goal of an exercise program is not to lose weight, but to increase athletic ability. Here are a few points to keep in mind if you're serious about being an equestrian athlete:

• Riders need to practice to perform well. Train efficiently for your discipline, and vary your program in and out of the saddle. Doing so can help you avoid burnout and boredom, as well as reach new levels of performance.

• Strength training is a critical part of athletic performance. Increasing muscle strength and endurance improves your ability to exercise more efficiently, to avoid sustaining an injury, and to recover from an injury faster.

• Many riders rely on sports psychology techniques to get a mental advantage over the competition and learn how to improve focus.

• The right nutrition can fuel your body to help you to recover faster, to reduce injury risk, and even to decrease muscle soreness. Proper sports nutrition—including drinking plenty of water—can also help avoid dehydration, hitting the wall at shows, and general exercise fatigue.

• Athletes often look for nutritional supplements to improve performance. Double-check with your doctor first to make sure your supplement or vitamin program is sufficient.

Here she performs the reverse leg raise exercise builds muscle that comes in handy when riding a lazy, unresponsive horse.

Chapter 16

English Riding Basics

English riding offers many variations (*see chapter 5*). It is all about sport and movement. English riders can explore the balanced precision of dressage, in which the relationship between rider and horse is raised to an art form. Show jumpers can take on the sheer exhilaration that comes with navigating jumps at speed against the clock. Hunt seat riders can display their elegant equitation in the show ring on the flat or over fences. Eventers can embrace the athleticism of jumping solid fences over hill and dale. Saddle seat riders get to experience the stylishness and the high-stepping action of the park or five-gaited horse.

Each discipline calls for a certain type of horse with movement that suits that sport's demands. All English disciplines require the rider to post (or rise) to the trot, but that is about the only thing they have in common. The position the rider utilizes and the philosophy of training and riding can be very different.

THE HORSE'S GAITS

There is a place in English disciplines for many types of horses. Dressage calls for a horse that is flexible and strong, yet still elegant. The hunter's long sweeping and comfortable gaits mean a rider can follow the hounds through the countryside all day. The saddle seat horse, with its flashy action, would rarely be seen galloping cross-country, but the athleticism in his movement is every bit as evident.

The Gaits

English horses move in the walk, trot, canter, hand gallop, and gallop. There are variations within these gaits depending upon the discipline. For instance, collected, extended, and medium paces are required in some dressage tests.

Some saddle seat horses, depending upon the breed, are gaited, and so their movement may include the rack, the slow gait, or the running walk.

Walk: An English horse's walk must cover the ground and have a clear four-beat rhythm that carries forward with energy and enthusiasm, neither racing ahead nor lagging behind. Jigging or dancing in the walk is a fault and is a sign of tension and anxiety.

Trot: The English trot is very forward and free in movement. The type of trot varies by breed and discipline. The saddle seat horse must pull his knees up in a high-stepping trot, which is the hallmark of these show horses. The hunter's low-moving "daisy cutting" trot is a nod to his worth in the hunting field, where an easy-to-ride, ground-covering trot was a must for an all-day ride. The dressage horse's trot must be flexible, with the ability to extend and collect in gaits.

Canter: The canter, a three-beat gait, is also ground covering and forward moving. The jumper's and eventer's canter is strong, building from powerful hindquarters needed for jumping. The dressage horse has an elastic canter, able to eat up the ground in an extension and nearly canter on the spot as he turns in a pirouette.

GAITS

Understanding the following terms will help you talk horse gaits.

■ **Collection:** The horse brings his hind legs farther underneath his body and carries more weight in his haunches.

■ **Covering the ground:** When the horse moves with long strides, he is covering the ground.

■ **Daisy-cutter trot:** A long, low-moving trot. The hooves stay close to the ground.

■ **Free rein:** The English rider gives the reins completely to the horse, allowing him to stretch his head and neck out fully.

■ **Jigging:** A vice where the horse "dances" in place instead of moving forward as the rider asks.

■ **Lengthening:** A slight extension of the walk, trot, or canter.

■ **Paces:** A variation within a gait, such as collected, medium, extended.

■ **Passage:** A highly collected trot with a extended moment of suspension. The weight is borne on the hindquarters.

■ **Piaffe:** The highest demonstration of collection. The horse's legs move up and down in diagonal pairs, while remaining on the spot. The hindquarters bear the weight; the forehand is light and free.

■ **Rack:** A fast, lateral four-beat gait with lots of knee and hock action.

■ **Slow gait:** A slow, four-beat gait, slightly faster than the walk with knee action.

Hand gallop: The hand gallop is a three-beat gait, slightly longer in stride and a little faster than the canter. It is used in lower levels of eventing; it can be used as a schooling exercise for dressage horses to create a more forward and expressive canter. The gait is used often in show jumping competition. Hand gallop is also shown in hunt-seat classes, such as the hunter hack, and in saddle seat classes, depending upon the breed.

Gallop: The gallop is a four-beat gait, and it's the fastest of all the gaits. Event riders do most of their cross-country work at the gallop. Very rarely is it called for anywhere else outside of racing.

Paces within the Gaits

There are varying paces within these gaits: collected, lengthening, extended, medium, and free (in the walk only). In the first stages, riding is all about the working (meaning the everyday, usual gait) walk, the trot, and the canter. But as horse and rider advance, paces are added. The dressage rider will begin to add extended, medium, and collected walk, trot, and canter, then the passage and piaffe (which comes at the pinnacle of

Anky van Grunsven and Hanoverian Keltec Salinero, Olympic dressage gold medalists, perform at the Aachen festival.

training). The jumper rider will begin to learn to adjust the horse's stride within the speed of the canter, hand gallop, or gallop. The equitation rider will also learn collection and extension of the gaits.

Collected: Riders often mistake collection with riding very slowly, but in truth, the tempo (speed) of the particular gait alters little. When a horse is collected, he brings his hind legs farther beneath his body and carries more weight in his haunches. His neck must not get shorter; it shouldn't change in length. Instead as the horse's strength improves bit by bit, his outline should be more uphill because the horse's front end elevates as he takes more weight back onto his haunches. The hallmarks of collection are the ability to ride movements with ease and smooth and steady transitions between the gaits. The working gait does not require the horse to collect. The working gait also lacks a certain amount of impulsion from and engagement of the hind legs.

Lengthening: This is developed from a working gait. Lengthenings are the precursors to extensions, but lengthenings are often confused with changes of speed rather than length of stride. When you lengthen a gait, the horse's frame needs to lengthen as well.

Medium: Medium pace builds from a collected gait. When you ask for the medium gait, you'll get more impulsion from the hind legs and a bigger stride. Extended trot is a bigger medium with a larger reach of the front end and stretch of the topline.

Extension: The more collection and strength a horse has, the bigger your extension will be.

Free: In dressage, the walk includes the free walk, medium walk, and extended walk. In the free walk, the horse is allowed to lower and stretch the head and neck on a loose or free rein. The medium walk should have an overstride, and the horse should stretch to and stay on the bit. In the extended walk, the horse covers more ground and stretches the head and neck out while maintaining contact. All walk extensions should march forward with energy and have a purity of rhythm.

RIDER POSITION IN GENERAL

English riders strive to attain an independent position, which means that you're in total control of your own seat at all times, no matter what your horse does, even if your horse spooks or bucks. If you have an "independent seat," you will be able to move with any horse's stride and with your legs held lightly against your horse's barrel, instead of gripping for balance. You will

This hunter/jumper rider leans slightly forward in the rising trot.

also have the ability to maintain a steady contact with your reins, instead of using them as a way to stay on. An independent seat means you can time your aids correctly and use them effectively.

Dressage requires you to sit in a vertical line, over your feet. Your knees will be slightly bent, and your hips will be over your feet so your foot, hip, and shoulder all line up. Hunter, jumper, and event riders sit in a similar fashion but with more of a closed hip angle. Your foot (in the stirrup) should be underneath your hip, with your knee bent to whatever degree your kind of riding dictates—dressage riders have a longer stirrup compared with hunter and jumper riders' shorter stirrup. In dressage, the middle of the foot should be underneath the hip; hunter, jumper, and event riders will be slightly forward. All English riders must be balanced laterally side to side and have equal weight in the sitting bones and in the stirrups. Shoulders should parallel the horse's shoulders, and hips should parallel the horse's hips.

Saddle seat riders sit slightly behind the horse's center of balance. This allows the rider to encourage the horse's high front leg action, which is required in competition.

When jumping, the rider adopts a "forward" seat, which is also called the "two-point" position. Captain Federico Caprilli (1868–1907) is considered the father of the modern forward seat, adapted in the nineteenth century. In this position, the rider rides with a shortened stirrup, sitting with the body forward so the weight can be carried over the horse's center of balance. In this way, the horse is not encumbered by the rider and is able to have free movement while jumping. The rider's hand also follows the mouth, and the leg position helps in stability. In this position, the rider conforms to the outline and forward movement of the horse.

Hands

Good hands are important because hands are communicating the rein aids to the horse's mouth. Unyielding, stiff hands will be like static on a telephone line—not many messages will get through. Plus heavy hands will cause discomfort to the horse and lead to training problems. *Feel* means that you can feel the horse's mouth through your reins and in your hands; it should feel soft and supple and ready to yield to slight pressure.

To achieve good hands, you must learn to feel the position of your arms, hands, and fingers. To begin, your shoulders should be soft enough to allow your arms to

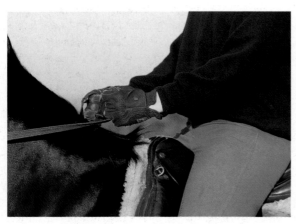

Good hand positioning should be relaxed yet firm in order to more easily communicate with your horse.

hang at your sides, and there should be a slight bend to your elbow. Dressage and saddle seat riders should have thumbs turned up. Hunter and jumper riders should hold hands slightly inward but never rolled flat. Saddle seat riders must hold hands high and out in front of them, relative to their horses' head carriage. There should be no tension in the biceps in these examples; instead the triceps should hold the arms in position.

Your fingers should hold the reins toward the first knuckles of each finger and not at the back of your finger toward your palm. This will give you a more sensitive grip. Your thumb should press down on top of the rein against the knuckle of your index finger. How your arms follow the horse's head is important.

At the walk, your arms should move forward as your hips move back, and then back as your hips move forward. Try not to exaggerate this, as it may look like you're rowing a boat. Instead, let the horse's natural motion guide you. At the trot, arms should stay still. At the canter, arms should follow the motion of the hips.

Legs

The use of the leg is a hard concept to grasp. There are many muscles you can use to aid your horse; the wrong ones can cause your horse to stiffen his back. Beginning English riders often think they must grip with the whole leg to stay on, but in actuality, thigh muscles should stay soft and relaxed. The lower legs are the key players and should be held against the horse and remain there at all times. This may vary between disciplines, particularly with saddle seat, where legs are held in a more forward position away from the horse's body. To use driving aids, squeeze your calves with enough pressure to press into the horse's sides slightly. If he doesn't respond, don't increase the pressure, as you'll teach him only to respond to strong aids. Ask again, paired with a gentle tap from your whip behind your lower leg.

Riders often want to know how much leg pressure they should use. Remember, a horse can feel a fly on his coat, so your aids don't have to be huge to get a response; however, they must be definite. Therefore, whenever you use your legs, make sure that what you are asking is clear and that you receive the correct response right away. Don't wait; repeat the question. Your horse may require a tap from the whip, paired with the leg, until he understands that you mean what you say. Spurs are a good aid, but used incorrectly they can make your horse dull or—worse—frightened of your leg.

In English riding, keep the upper thigh soft and relaxed and the calf firmly against the horse.

CORRECT CONTACT

The rider must also understand the correct contact and acceptance of the bit for her or his discipline. And no matter what that is, the horse should never resist the contact, come up against the bit, or drop behind the vertical. Trying to get the horse on the bit by seesawing the reins is a grave error. The horse must take the contact, stretching into the rider's hand. This contact develops from the active use of haunches over a swinging back, free of tension, and the stretching of the entire topline toward the rider's hand, which in turn lightly balances and directs. Many riders think their horses are resistant to the contact, when really the riders aren't giving enough room within the reins to make that stretch.

A horse properly on the bit will stretch into your hand, and his topline muscles will begin to work—the back lifts and becomes a comfortable place for you to sit and a shock absorber for him. You'll achieve connection, which means the haunches are "connected" to the front

end so you can influence your horse more easily and with lighter aids. You're going to feel a different horse underneath you. The poll should be at the highest point of the neck, rather than at the third or fourth vertebrae, a common sign when the horse's neck is incorrect. The neck muscles will bulge along the top while the bottom muscles hang slack. The horse should move freely forward. If a horse is forced onto the bit, his back will be hollow. He'll be rushing because you won't have influence over his haunches. You may have to pull the reins to slow him down.

RIDING EACH GAIT

It's not just the quality of the gaits that matters, but also the way they're ridden. Move with the horse and allow him to move his whole body. The hands must move with the horse's head. One thing all English riders have in common is that each discipline requires the rider to post the trot and hold the reins in both hands.

Walk

You can practice the various paces of the walk at home by teaching the horse to march around the arena or within the pattern required in your test. Keep the horse moving forward, in front of your leg. In the free walk, the rider must allow the horse to stretch as far as he wants. Never neglect the walk or allow your horse to become lazy, even when warming up or finishing your schooling session.

Sitting Trot

In the sitting trot, some horses are very smooth and their riders won't have to do too much. Other horses are very bouncy, and many riders try to sit as still as possible. But this never works. The horse is moving, so you must move with him. As a horse is in his phase of suspension, the sitting trot feels like you are catapulted up, and when he comes down from the suspension, you fall back into the saddle. If you are sitting the trot correctly, you'll learn to absorb that up/down part of the movement in your knees and hips and slightly in your ankles. If you brace your leg out in front of you, no shock will be absorbed. While sitting the trot, keep your upper body upright, rather than leaning forward.

Try to sit so your upper body is disturbed as little as possible. Since your upper body is high above the horse, you have an influence on his balance. If your shoulders are flopping from side to side, you can cause

Sitting the trot, this dressage rider uses her core muscles to balance.

him to stagger sideways, in which case he may throw his head up into the air for balance. This is where your core muscles come in. If you flex your back and stomach muscles, you will hold your seat to the saddle. Never hold onto the reins to stay in the saddle.

Posting Trot

The posting trot (also known as the rising trot) is an important skill to perfect, as English riders often use it. Hunter and jumper riders should lean forward and post forward and backward. The angle of the rider's shoulders in the posting trot should be about 30 percent in front of the vertical, so the rider can move with the horse's motion, which will also allow the horse to trot out better. Hunter and jumper riders also use this trot to get off the horse's back and allow him to stretch his neck out and forward.

A dressage rider should sit over the vertical with shoulders and hips in one line. Thighs should hang as straight as possible with the knees slightly bent. The angle of the shoulders should never come forward but remain straight. The hips should rise out of the saddle and forward over the pommel, then land back in the saddle in the same place. In this position, you are able to keep your lower leg quietly against the horse's barrel throughout the phase of the posting trot so that you can use it when needed. This position also helps the horse arch his frame and encourages his haunches under.

The rider is said to be on the "correct diagonal" when he or she rises as the horse's outside leg moves forward. You can see this by glancing down at the shoulders. If you're on the wrong diagonal, simply sit one stride or two beats and then come up on the correct diagonal. This takes practice, and you may need to have someone on the ground who can tell you when you are correct.

The saddle seat rider posts the trot somewhat differently from other English riders. He or she sits further back in the saddle with legs in front. The rider's legs will "pump" as he or she posts.

Canter

Riding the canter is basically the same in all English disciplines. Your seat must move in balance with your horse. Some instructors may urge you to ride the canter in a forward and back motion, similar to riding a child's rocking horse. This is a good visual for the beginning stages of your riding career, but you will need to add another piece later. The horse's leading leg will cause the twist of the saddle to be canted more to one side than the other. Therefore, your inside hip will twist further forward than your outside hip. Hold your shoulders still, and allow your lower back to be soft, and move with your horse.

Hand Gallop and Gallop

The hand gallop and gallop positions are similar to the forward seat or two-point position. This means you'll no longer sit in the saddle. You'll take the seat, your third point of contact, away and ride from the heel up to the knee. Most importantly, your hands should be low; the foundation of the position should be in your lower leg. Don't use the horse's mouth for balance.

The reins are held in the usual way during the hand gallop. In the gallop, you can hold your reins in either the half bridge of or the full bridge. In the half bridge, you stretch one rein across the neck so that you're holding two pieces of leather in one hand. For the full bridge, you stretch both reins across the horse's neck so that both of your hands are holding two pieces of leather. The reins will be pulled across the horse's crest instead of hanging in a loop alongside the neck. Bridged reins are also useful in terms of safety. If your horse stumbles, the bridge can catch you from falling because your arms won't collapse on either side of the horse's neck.

Like this rider, allow your lower back to remain soft so that you can move with the horse at the canter.

The term *hand gallop* means the horse is still "in hand," or in control. Lengthening the stride and slightly increasing the speed is what you want.

The best way to begin the gallop is slowly. Some horses can get high on the speed and a fast start can undo the hard work and training you've put in to teach your horse obedience and adherence to your aids. When you're galloping, begin at a trot, then a canter, then a hand gallop, and then the gallop to ensure your horse is still listening and rideable. To practice galloping while jumping, gallop toward the fence, then slow down or balance up a few strides in front of the fence to allow your horse to get his legs underneath him to jump.

Galloping asks a lot from your horse and can cause serious injury if his body isn't used to foot concussion at top speed. Think about whether galloping is right for you and your horse. Galloping is for those who ride their horses at least five times a week and for those horses that are in top condition. The once-a-week rider should not gallop. Work up to galloping by conditioning your horse with ten to twenty minutes of trotting, then begin with three minutes at the hand gallop, two at the walk, then another three minutes at the hand gallop. After a few weeks or months (consult a trainer if you're unsure how to test your horse's fitness), progress each gallop for three

With rider on board, a Thoroughbred gallops down the field. Galloping should only be done with horses in top condition.

minutes, with your walk in between, then a faster set for three minutes, walking in between. This interval training builds the cardiovascular system and soft tissue. Refrain from galloping in wooded areas; it's hard to gauge speed and see what's coming in the other direction. Before you gallop, walk the area to check for holes and debris, and make sure it is not too hard, deep, or slippery. You also have to recognize that the speed you can go will depend on what the land allows. You can only turn so sharply or go downhill at so fast. As you go faster, the balance of the horse should always stay the same.

Chapter 17

Western Riding Basics

There's a sense of complete freedom when riding western. Everything in the discipline is about comfort, liberty, and ease. Expansive, scenic vistas become all the more accessible, and relaxation on the trail beckons to the leisure rider. Like English riders, western riders have a wide variety of activities they can engage in, both competitive and noncompetitive.

Competitive riders can envision the exhilaration of putting their skills to the test against their peers in shows such as cutting and reining. Rodeo riders get the biggest shot of adrenaline, pitting themselves against rough stock or against the timers. The horses, too, are very select. At one end of the spectrum sits the performance horse—agile, athletic, and responsive, a true all-terrain vehicle. At the other end, there is the western show horse, with perfect manners and refined gaits. These horses would rarely do any ranch work, but they are as integral to today's western riding as the cow horse.

Western riding is truly a liberating activity. It's not grounded in centuries of tradition and ridged methodology. It's fluid, relaxed, and deceptively easy looking. But despite appearances, western riding does require a great deal of fine-tuning, for both horse and rider. The tack is designed to keep the rider in comfort for long days in the saddle. Even the horse's way of going has been bred over the decades to be pleasing and easy to sit over long hours and rough terrain.

THE HORSE'S GAITS

A horse naturally has four basic gaits: walk, trot, canter, and gallop. A western horse has these gaits, but they are modified slightly for the type of riding done and the original conditions of the West; therefore, some have names that are slightly different from those used in English riding: walk, jog, lope, and gallop.

Walk: The western horse's walk is smooth, purposeful, and easygoing. The horse may not cover a great deal of ground, but he gets safely to his destination by picking his way carefully through craggy canyons and over rocky bluffs. What he may have sacrificed in ground-covering stride, he more than makes up in sure-footedness.

Jog: Western riders developed the jog because a free-flowing, active trot is simply too difficult to sit for long, too quick for a leisurely trail ride, and too fast paced to work within a herd of cattle. The English trot and the western jog are essentially the same, but the jog is slower, covering less ground yet maintaining a steady rhythm. Because the competitive show horse that exhibits excessive speed at the jog is penalized, the show horse is trained to travel slowly, but still with impulsion coming from the hindquarters, his engine. The show horse should exhibit a smooth, flowing, two-beat gait with little knee action.

Lope: The leisurely lope, cousin to the canter, is a three-beat gait. This gait should be a pleasure to sit, with its trademark easy rhythm and relaxed, smooth

This young rider's loose rein and deep seat are the hallmarks of a good western position.

stride. The lope is slower than the English canter, but it still exhibits a good deal of activity as the western horse steps under himself to propel his body forward. A good rider helps the horse create his lope with energy from the hind end; the rider should not merely try to slow the horse down by the reins. This can cause the horse to move in a disjointed manner. A good lope is an energy-conserving gait, made for long-distance riding. Instead of using all his power up in fast bursts of speed, he can cover more ground by keeping a slow, steady pace.

Gallop: A gallop is a gallop, no matter where you are in the world or what kind of saddle you're sitting in. Due to the conformation of the stock horse, the western gallop has a stride that is shorter than that of the English gallop. Yet the western horse is still capable of lightning-fast (western) speeds. And as discussed in chapter four, the world's most popular horse, the American Quarter Horse, is the fastest horse in a sprint. At a quarter-mile, this breed is unmatched by any distance horse, even the Thoroughbred and the Arabian.

RIDER POSITION IN GENERAL

Through culture, purpose, and style, the western horse and tack have developed for the utility of the rider. Still, there is more to riding a stock horse than just swinging into the saddle. Every horse needs to be able to perform his tasks efficiently and promptly, but this is especially true in western riding. Riding in harmony with a finished (well-trained) western horse can only be achieved if the rider understands how he or she influences the horse with skill and position.

It's easy to see the physical differences between English and western, namely the tack and the stocky, compact build of the horse, but there are other distinctions as well. Most English horses are ridden with steady, consistent contact on the bit; western horses are ridden on a fairly loose rein.

The rider trains the horse to become tuned in to the position and the balance of the bit in his mouth and to respond to the lightest of pressure—even just picking up the reins means something to the western horse.

This western rider's weight is placed in the heels, but not so much that the foot is shoved forward.

Combining that sensitivity with the rider's natural aids (seat, leg, and hand) keep the horse "on the bit" and ready to react.

The rider's weight is in the saddle at all gaits, sitting the walk, jog, and lope. It's not unusual to see a western rider extending the trot and posting to it, but mostly, the horse's gaits are slow and smooth enough for the rider to easily absorb the motion.

Legs and Feet

The position of the western rider in the saddle is similar to, but not the same as, that of his or her English counterpart. The leg position is longer, resulting in a more secure seat position. The rider remains naturally upright and poised at the walk, jog, and lope, instead of leaning forward and closing the hip angle during the faster gaits. The seat of the saddle slopes gently back, which puts the rider in a deep sitting position, opening up the pelvic bone so that his or her weight rests on the seat bones.

WESTERN RIDING

Understanding the following terms will help you talk western horses and riding.

■ **Bosal:** A type of bitless bridle.

■ **Falling in (or out):** The horse loses balance and drifts in or out of the figure, line, or pattern.

■ **Jog:** A slow and easy to sit trot.

■ **Lope:** The slower, western version of the canter.

■ **Quirt:** An ornate extension on romal reins.

■ **Romal reins:** Braided, stiff reins that are closed or held together in the middle.

■ **Split reins:** These open reins are unconnected, which is a benefit if the rider parts company with his or her horse because there is no chance of entanglement.

■ **Stock horse:** A western breed, such as the Quarter Horse or the Paint. The name refers to the horses' traditional use of working livestock.

There is balance in the western position, as in the English, with alignment from the ear, shoulder, elbow, hip, and heel in a straight line. The rider's back is flat and straight, with no arch or slouch. The ball of the rider's foot rests in each stirrup, and weight is placed into the heels, but not so much that the heel is forced down dramatically or artificially. Stirrups should be adjusted so that the rider does not have to reach down with the toe to keep the foot secure, but not so short that the leg is thrust forward into a chair seat.

The leg communicates lightly with the horse. Thighs and knees are pressed along the saddle. Calves only come into play when supporting the hand while giving an aid.

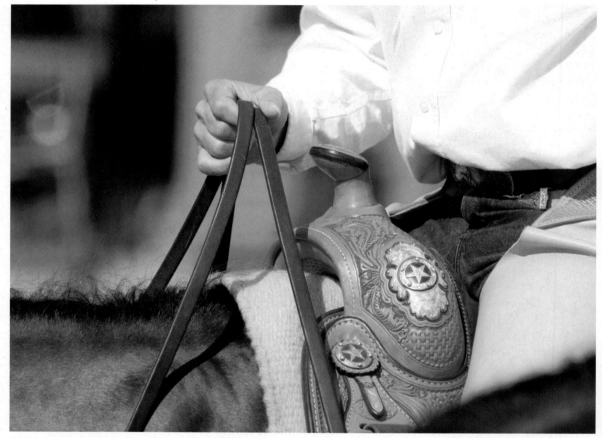

As this horse is ridden in a curb bit, the rider holds his split or open reins in one hand; his arm in a correctly relaxed position.

Arms and Hands

The arms of the rider are held in a relaxed position, with an imaginary line going from the rider's elbow to the bit of the bridle. A leisure or trail rider has plenty of options on how to hold the reins or how to position the hands, depending on the type of bit he or she is riding the horse with or on the horse's level of schooling. Western horses can be ridden in a curb bit, a snaffle bit, or even bitless in a bosal, a type of hackamore.

Generally, when the horse is ridden in a curb bit, the reins are held in one hand, usually the left, and the other hand rests naturally on the leg or at the rider's side. Reins are of two general types, open (split) and closed (romal). Most cowboys prefer to use open reins. One advantage of split reins is that because they are not joined together, the rider is not in danger of becoming entangled in the reins if he or she is thrown. The reins are held two different

The Western Horse's Frame

A typical western horse is built compactly and carries himself in a small, powerful package. Working horses, performance horses, and show horses all have fairly level toplines, but the refined show horse has it to the extreme. The game, little, speed-event horse would be on the other end of the spectrum, nearly upright in his carriage. Conformation and breeding have a great deal to do with it, but there is also much in the training that influences the horse's carriage. What is not accepted by judges is a horse with an inverted frame, where the back is hollow, the neck is concave, and the horse's gaits are strung out. Similarly, the exaggerated show horse, with a too-low neck and head and shuffling gaits, is becoming a thing of the past, as show organizations and passionate breed enthusiasts continue to work diligently to change what was once the industry standard.

ways. The first method has the reins held with the hand knuckles up. The index finger is placed in between the two reins and the ends of the reins thread out of the bottom of the hand. In the other method, the reins are held in a loose fist. The rider keeps the thumb on top, which stops the reins from slipping through the fingers. The hand is held in front, just above the horn of the saddle.

With the romal, reins are braided and stiff and joined together where the rider's hands meet with an ornate extension called a quirt. The rider's left hand holds the reins just below the romal knot and the right hand holds the quirt against the right thigh.

Reins can also be held two-handed. Horses wearing a snaffle bridle are ridden two-handed, and it has even become the fashion to ride some horses in a curb bridle two-handed. Reins can be crossed or bridged according to the preference of the rider. The hands are held about four inches apart.

In crossed reins, each rein is held like an English snaffle rein, threaded through the fourth and pinky fingers and out through the top of the fist, with the extra length of each rein crossed over the opposite side of the neck. That means that the end of the left rein hangs on the right side of the horse's withers, and vice versa.

For bridged reins, the hand position is the same, but the ends are held in each opposite hand, bridged with the rein in a shallow loop.

Regardless of how the reins are held, a rider with educated hands is never cruel with the bit. This is particularly true when using curb bits, which can have a high port and a lot of leverage. A rider must be skilled to use only enough pressure in the cues to get a response from the horse. It's a mistaken impression that the western horse is controlled mostly by the mouth, because in reality, the rider strives to control through the shoulder and the hindquarters.

Being able to move or elevate the horse's front end gives the rider maximum responsiveness and influence. If the rider has the horse's haunches in gear, he or she can ride with energy and be able to collect and extend any gait.

RIDING THE GAITS

Cueing a stock horse to transition in and out of his gaits is fairly simple. To move from the halt to the walk, the rider nudges the horse lightly with the calf

Only fit and well-trained horses and experienced riders, such as this pair, should gallop. Always gallop over safe terrain.

This Paint pivots around a traffic cone. Western horses are trained to respond to very slight aids.

and adds a bump of the heel if needed. The seat of the rider should easily follow with the motion of the moving horse.

In order to put the horse into a jog, the rider squeezes equally with the legs and the upper thighs, adding more pressure with the heel if that proves necessary. The one-two beat of the jog can be absorbed by a rider with a relaxed, supple back and a deep seat.

The jog should never get so fast that it becomes uncomfortable for the rider to sit. If the rider desires greater speed, then the horse should be put into a regular trot, and the rider can post, just as an English rider does.

For the lope, the rider picks up the reins with a very subtle contact and sits with a deep seat to push the horse into the gait. The horse's head is turned slightly toward the inside of the ring. The outside leg of the rider is placed well behind the girth, asking for the correct lead of the lope, while the inside leg remains at the girth to support the horse and keep him from falling in.

The lope is an effortless gait to sit because it is relaxed and rhythmical, like a rocking horse. The rider usually finds it easier to sit the lope by simply relaxing the thigh and seat muscles and moving with the motion. A little bit of even pressure with both legs will keep the horse loping easily.

For the fastest gait, the gallop, a rider asks in the same manner as with a lope, but in contrast uses a driving seat by tightening the abdominal muscles and squeezing with the buttocks and upper thighs.

A good series of bumps with the heels can also help increase speed, as can using the seat in a driving motion. The rider shifts the position forward while still remaining fairly upright. The horse is encouraged to go forward by the rider's hand, which gives free rein and is held halfway up the horse's neck instead of just in front of the saddle horn.

STEERING

For some novice riders, the thought of steering a horse with only one hand seems very odd. Additionally, the fact that the majority of horses are ridden with a very loose, even flopping rein can be something to adjust to for many riders. A good western horse, however, listens intently to his rider. The horse will respond to very light cues if he had been trained properly.

To steer a horse and make it turn, the rider uses neck-reining. He or she holds the rein hand slightly above and just in front of the saddle horn, and then lightly pulls the hand in the direction he wants to go. The outside rein comes into play, making contact with the horse's neck. This tells the horse to move away from the pressure and turn. Most riders will support the hand cue with outside leg pressure as well, so that the haunches of the horse travel straight and do not swing out.

If a rider wants to make a left turn, for example, the rein hand moves from the center position over the saddle horn a few inches to the left. The right rein places slight pressure against the horse's neck. The rider supports the turn with the right leg, yet keeps the left leg slightly on as well, just to support the bend.

Rider eye contact is also important. When the rider looks into the turn, his or her balance shifts and a horse will know where the rider's attention is focused and where he should be going.

As a rider cuts a cow away from the herd, he holds on to the saddle horn for stability and puts his rein hand down on the neck to cue the horse to keep the cow separated.

Riding a Cutting Horse

Riding a cutting horse is quite different from riding a western pleasure horse. With a cutting horse, the rider does little to direct the horse; instead the rider allows the horse's training to take over as he cuts a cow from the herd. To order to accomplish this task, the rider's body must move with the horse's body at all times to help him do his job.

Relaxation is key. If the rider sits in a rigid position, the horse won't be able to move and stay with the cow. A stiff rider will fall forward and lose balance, causing the horse to lose balance, too.

The rider also must keep an eye trained on the cow otherwise he or she won't be able to anticipate when or where the horse will stop and turn to stay with the cow. Taking their eyes off the cow is one of the biggest reasons that riders fall off.

Riders must also sit the stop correctly. When the cow stops and goes the other way, the rider must sit down into the saddle to tell the horse to stop. Sitting the stop also keeps the rider's balance during the many abrupt stops and turns. To sit the stop, the rider rounds the lower back, but keeps the upper body straight; neither leaning forward nor leaning backward.

Horse Training

Section 8

Training methods used in the Old West were harsh ones. Fortunately, horsewomen and horsemen have moved on to more humane and methodical approaches to training. There are many ways to train a horse, from western training to classical dressage principles to natural horsemanship. Although their techniques may differ, all horse trainers have the same goal: to instill good, solid basics in their equine students. Basic training, better known as the basics, is the cornerstone of every horse's education. No matter which discipline the horse will specialize in, he must first learn how to submit to his rider; balance himself with the rider on his back through the walk, trot, and canter/lope; move on straight lines and bending lines; and understand a rider's cues.

Chapter 18

Training Philosophy and Early Handling

When a horse misbehaves, it's important that you correct his behavior rather than punish him. In the rider-horse relationship, one is a smart partner and the other is a strong partner. The rider should always be the smart partner. You don't want to switch roles by trying to force a horse to do your bidding while he is figuring out ways to get out of work.

Horses learn best through consistency, so ask the horse for the desired behavior the same way each time—but make sure he responds to your request every time. Never let him ignore you, and never move on to the next lesson until he grasps the current exercise. Skipping over crucial steps will only leave holes in your horse's education and lead to problems down the line. For instance, if your horse will not respond to your leg aid for a trot, then he certainly won't listen to your aids for a canter. If he won't let you ride him straight ahead, then chances are he won't let you ride him in a circle either.

Horses respond well to rewards. And you have to reward the slightest positive move your horse makes, even if it's the most tentative step forward in response to your leg. Remember, in the early stages of training, you should be looking not for perfection, but for understanding. You must be quick with your reward so that the horse links the praise with the behavior. To reward the horse, you can pat him on the neck, speak soothingly, or slow to a walk to allow the horse to stretch his neck and back. Or you can end the lesson. Above all,

always end the lesson on a good note. If your horse is having problems grasping a new concept, go back to something he knows, get the correct response, and then end the lesson. Horses have amazing memory, and a good or bad ride will affect his next lesson.

Never punish a horse out of anger or vengeance, and never punish him long after the misbehavior has occurred. Your horse will not understand why he is being punished. Tying him up for an hour or smacking him with a whip long after the behavior has occurred will confuse and frighten him. Your job is to teach, not to punish. If your horse is acting up, think of ways to correct the behavior. A few are discussed below.

Work on the longe: If your horse is leaping around and won't settle to his work, consider working him on the longe or in a round pen to expend some energy.

Change locations: If he is shying an object at one end of the arena, work him on the other end and deal with desensitizing him to the object on another day.

Reiterate your position: If the horse is flat-out challenging your position as herd leader, you can reiterate that position in different ways. Sometimes this can be as simple as a sharp "No!" accompanied by a short pop on the halter (if you are leading). You can back up the horse a few steps and make him stand (while leading or under saddle). Or you can apply your crop or whip at the same time you use your leg aid. (*Note*: whips and spurs should only be used as adjuncts to your aids; never use them as punishment.)

PHILOSOPHY AND IMPRINTING

Learning the following terms will help you to talk about horse training philosophy and imprinting.

■ **Break a horse:** To introduce him to the saddle, bridle, and rider's weight.

■ **Butt rope:** A looped rope placed around the foal's buttocks, over the tail, and above the hocks, to give the handler control.

■ **Desensitize:** To teach a horse to ignore certain stimuli.

■ **Imprint training:** A way for humans to bond with newborn foals and to sensitize and desensitize the foals to certain stimuli.

■ **Leg aid:** A touch or pressure of the rider's leg to cue the horse into an action.

■ **Paddock:** A small enclosed space used for turning out or riding.

■ **Quick-snap release halter:** A safety halter that breaks away under pressure.

■ **Turnout time:** Leisure time for the horse in a paddock, an arena, or a field.

Your body language speaks volumes to your horse. If you are timid and hesitant in your actions, your horse will read that you are the submissive one in the relationship. Holding yourself tall and moving confidently shows the horse that you are the dominant leader.

IMPRINTING A FOAL

A brand new foal is a blank slate and an opportunity for you to instill good skills for life. It's important to understand that, when a foal is born, he is able to get up and run shortly after birth. He can even defend himself with his tiny baby hooves if he needs to. A young horse is very close to his instincts (his natural wild state) so he will be quicker to react to perceived dangers and threats. Understand that a foal can move fast, wheel around, and kick out. Therefore, always move carefully, calmly, and slowly around a foal.

Robert M. Miller, DVM, formalized imprint training more than thirty years ago as a way for humans to bond with newborn foals and to sensitize and desensitize the foals to certain stimuli. The minds of newborn foals are open to training right from the start, and imprint training can help when handling the horses later in life. Training a foal immediately after birth will connect him with humans—imprinted foals recognize people as herd members and not as predators.

Imprint training puts the power in the hands of the trainer because the foal will recognize that person as the dominant herd leader. This means the foal will naturally relax and willingly submit to the trainer's cues, an approach so much better than forcing the foal to give in out of fear of punishment. Psychologically, this is a very important type of relationship to build between horse and human. A horse must be submissive to a human, or else he is untrainable. Submission should always be gained through the horse's dependence on a kind leader.

Imprint training desensitizes the foal to most sensory stimuli (sight, hearing, touch, and smell). The foal can also be introduced to the types of touch that will be critical in training, such as moving away from pressure. It is important to train the foal to respond to the pressure of touch at least a day or so after birth. This sensitivity to pressure will allow the trainer to lead the foal, rather than the foal blindly following his mother.

The only downside to imprint training is that it has to be done right after birth, which can be difficult for some large breeding facilities. Of course, the mare also has to be happy to let humans touch her foal. When you first approach a mare and foal, especially if the foal is her first, do so slowly. Once the mare understands you're not a threat, she should calm down quickly, especially if she's used to being handled herself.

To begin the process, towel-dry the foal. Kneel beside him and gently turn his muzzle back toward his withers, which will keep him from standing. Do not restrain the mare; let her near her foal. Rub the towel all over the foal. Next, using your hand, gently rub the head and ears. Touch the mouth and nose until the foal calms

A foal, such as this Quarter Horse, should become accustomed to being handled at a young age.

Like this Welsh Cob mare, new mothers want their foals at their sides. When you begin foal training, have the mare within sight.

TURNING OUT

You can turn out the mare and foal in a paddock alone or accompanied by a familiar mare and her foal. Be sure to supervise this turnout time closely. A foal can get caught under a fence or escape the paddock. Foal-safe fencing, such as diamond-weave mesh or four-bond wood is important. Never leave a halter on the foal. It can get caught up on an object easily and possibly hurt the foal. You may find that it takes longer to catch a foal without a halter, but he will be much safer without one.

down. Now move down the body, touching and rubbing as you go. Only move on to a new area once the foal relaxes, then go back to the areas you've already rubbed to ensure the foal is comfortable with your touch. Do not rub the sides of the horse where the rider's legs will go. This will desensitize the horse to the cues the rider needs to give the horse under the saddle. Now work your way down the foal's legs, bending, straightening, and touching the sides and bottoms of the hooves. The whole procedure may take an hour or longer.

In the next day or two, begin sensitizing the foal to pressure. Place a leather halter on the foal. (*Note:* Nylon halters can be used as well, but never leave one on an unattended foal. Foals are inquisitive by nature, and halters can easily get caught up in objects. A leather halter will stretch or break; a nylon halter will not.) The halter should have a quick release snap, just in case you need to remove it quickly. To put the halter on, approach the foal quietly. You may find the foal will

Wearing his very first halter and lead rope, this Quarter Horse foal is ready to learn to lead.

move away from you. Keep changing directions and moving toward the foal until he lets you put your arm around his neck. Loop the lead rope under his neck to help control him as you put on the halter. Unbuckle the halter and slip it over the foal's nose, then behind his ears. Take your time; give the foal a chance to get used to the feeling of the halter. Always have a great deal of patience when you're working with a foal. Make sure each experience he has with a human is a good one.

TEACHING A FOAL TO LEAD

It's a good idea to allow the mare to be close to her foal during training. Have an experienced handler lead the mare. A mare can get upset when her foal is not by her side, particularly a new mother, so it's important for everything to go smoothly.

It's crucial to teach a foal to lead while he is still small enough to handle. He will not walk alongside you, as a mature horse will. Never pull against him or yank him forward. Instead, teach the foal to give in to pressure by leading him with a butt rope. This rope will give you some control and leverage over the foal. Never pull the foal forward with the halter: this can cause damage to the neck.

Start in an enclosed space at first, just in case the foal or mare gets loose. Take a long lead rope (6 to 8 feet), and clip it to itself so that you make a large loop; hold the loop in your right hand. Snap a lead rope on to the foal's halter, and hold it in your left hand. Stand on the left side of the foal, and put your left hand (with the lead rope held slackly) over the foal's chest and opposite shoulder. With your right hand, gently place the loop around the foal's butt, over the tail, and above the hocks. Have your handler lead the mare forward while you follow. Try to keep the foal moving in a straight line. Your left hand guides the foal, while your right hand provides a little bit of pressure to encourage him to walk forward. Make sure to ease up on the pressure when the foal moves forward. Don't drag or force him forward with either of the ropes. Your goal is to get him to understand that he must give in to pressure.

Keep using the butt rope until the foal is weaned, even if you feel he doesn't need it. You don't want to put yourself in a situation in which the foal gets his way and learns he can go where he wants. And always lead the mother a few feet ahead, keeping her in the foal's sight. Never move the mother so far ahead that the mare or the foal panics.

Chapter 19

The Training Lessons

There are many lessons a horse must learn, among them: how to tie, how to longe, how to carry a rider. This chapter covers the basic dos and don'ts of teaching your horse these lessons.

TEACHING YOUR HORSE TO TIE

One of the most problematic vices a horse can have is the inability to tie. A horse that refuses to stand tied can't be trailered safely, will be a liability at shows, and will force you to take time off work to assume the role of human hitching rail for the veterinarian and the farrier. Plus, you'll only be able to groom and saddle the horse when he is in his stall.

A horse that won't tie is dangerous—to himself and to the humans around him. He can flip backward on you (or whatever happens to be behind him) or rear up and lunge forward. Horses that will not stand tied will eventually learn to fight so hard that they develop a habit of leaning back on their haunches, or "sitting back," to break free from their halters.

Fighting constraint is a natural instinct for a horse, while standing tied and restrained is not. This behavior must be learned. Remember that the fight-or-flight response is what has kept the equine species alive for so long. Don't consider it a character flaw if your horse doesn't know how to tie, and never assume that your horse won't someday have a problem standing tied just because he never has. His habits may change abruptly if he steps on a trailing lead rope, if something spooks him, or if he gets his head tangled in the lead rope while he's tied.

Bad Training Concepts

Don't let anyone convince you that any of the following training concepts is the way to introduce your horse to tying. You are opening yourself up to a world of trouble if you go down any of these routes.

Never tie a horse to a big tree, and let him fight it out. Unfortunately, this is a frequent practice. You are risking your horse's life and your own safety by doing this. A horse will not learn by exhausting himself, and he may be seriously injured in the process.

Never teach your horse to tie by using an inner tube. Many people teach foals to tie in this manner. They assume that because the inner tube has some stretch to it, they can tie a foal to the tube. The reasoning is that if the foal sits back, the inner tube will give some, and the consequences won't be as severe or as tiring for the horse. This training concept is still extremely risky.

Your goal with a horse that sits back when tied should be to teach him to yield to pressure. It doesn't make sense to teach him not to sit back by allowing him to sit back against the pull of an inner tube. You are just reinforcing the behavior. Many horses have broken their necks by using this concept; the risk is simply not worth it.

Do not use an elastic, or bungee, trailer tie. If your horse sits back, the tie can break. If it does, it will act like a rubber band and spring back, possibly hitting the horse or a bystander with great velocity.

Do not use training halters or chains. These devices will twist and torque painfully when a horse sits back, and they're likely to cause significant injury. Usually, the reason a horse sits back when tied is because he is panicking. Because these devices create even more pain, the horse will only continue to panic and pull, thereby escalating the problem.

Do not rely on breakaway halters. You cannot substitute safety equipment for proper training. A good rule of thumb: don't tie a horse unless you are positive he won't sit back.

Training Steps

Your horse must learn to understand and yield to the pressure of a rope to be tied safely. Be sure that your horse is comfortable with each step before you progress in your training.

Dropping his head to the ground: Outfit your horse in a halter and a lead rope. With your hand, pull down on the lead rope and put about 3 pounds of pressure on the halter. Slide your hand down the rope and wait for the horse to drop his head. As soon as he does, release the pressure and reward him with a pat. Reinforce this cue by repeating the exercise over and over again. When you're certain the horse understands yielding to this type of pressure, remove all the slack from the rope by stamping on it really hard—as if the horse were stepping on it. You want him to react by dropping his head instead of by throwing his head up, training him to be unafraid of possibly stepping on or tangling his lead.

Putting the rope over the horse's poll: When a horse's lead is too long, the horse may get the rope hooked over the top of his head. You may have seen the nightmare that occurs when a horse in this situation panics and tries to sit back. Simulate this scenario by putting the rope up and over your horse's poll, teaching him to yield to this type of pressure.

Turning and facing: For this exercise, attach a longeline to the halter, and put the horse in an enclosed area, such as a round pen. Send him away from you at a walk, then add pressure to the rope, asking your horse to turn and face you. He should immediately yield to pressure and turn toward you. Many horses,

Because of the fight-or-flight instinct, restraint is unnatural for a horse, so take the time to teach him the concepts of tying.

when they feel the pressure on their halters, will react by dragging their trainers off. If that happens, it's a sure sign that you shouldn't tie your horse yet. Master this exercise at a walk, going both directions. Repeat the lesson from both sides at a trot and a canter. It is important to do this exercise at all three gaits because the horse's excitement builds when you add speed. This exercise will work on controlling the emotional side of the horse. When a horse sits back, he is feeling fear or anxiety.

Turning to pressure: Guide your horse to the middle of the round pen. If you are standing on the horse's left side, put the longeline along the right side of his body and back around the hip. If you're on the horse's right side, put the longeline along the left side of his body. Turn him slowly and teach him to follow the pull of the rope. When you first do this, you may find that you have to step around the horse as he turns from the same side that you started—especially if he panics about having the line around his hips. Do this exercise at a standstill, and then at a walk, a trot, and a lope, with the horse circling around you, turning to the pressure of the longeline.

Tying test: Make sure you take plenty of time to train your horse on all of the other steps, and ensure that they are solid before you attempt this final exercise. Attach your long lariat rope or longeline to the horse's halter. Find a safe fence, such as a pipe fence or a round-pen panel, loop the rope through the top rail, and hold the loose end. This simulates the horse's standing tied without actually tying the horse. Stand in the middle of the pen (behind the horse—but out of kicking range!) holding the end of the line slack. Flip the rope back and forth, up and over your horse's body from one side to the other, slowly taking up the slack of the rope.

If you've done your work effectively with the other four exercises, then your horse will give in to the pressure and move closer to the fence. However, expect your horse to be nervous about this process and want to move away. If he sits back, he will take up the slack of the rope, and you can give him room to move away. If he panics, you can release the rope, wait for him to relax, and then repeat the exercise. If he is still nervous and uncooperative, backtrack to one of the other exercises until he feels more comfortable.

(*Caution:* Make sure you only simulate tying initially—by simply looping the rope through the rail and holding the end—before you actually attempt to tie your horse. Each horse works on his own timetable. It may take several hours or several weeks to move your horse through all the exercises. Give your horse sufficient practice, with every opportunity to regress, before you actually tie him. If all is well, and you feel confident that your horse now yields to pressure as second nature, go ahead and tie him. If you've been diligent with your exercises, tying him should be no trouble. Now you'll have a safe, happy horse on the end of the rope.)

Even if you believe your horse will never pull back, you're going to have to use some kind of knot to tie

Quick-Release Knot

For a quick release, it's best to use a cotton rope knot because it will release freely. Some nylon style ropes can catch and tighten rather than release. To tie this knot, follow these steps.

1. Loop the rope around a sturdy hitching post or through a tie ring.
2. Turn the free end of the rope underneath the length of rope attached to the horse's halter.
3. Bring the free end of the rope back over the attached length, and pull a loop up through the circle.
4. Pull the rope taut, removing any slack.
5. To release the knot, pull on the rope's free end.

A lead rope is tied to baling twine with a quick-release knot; if the horse panics, the twine will break or the rope can be untied.

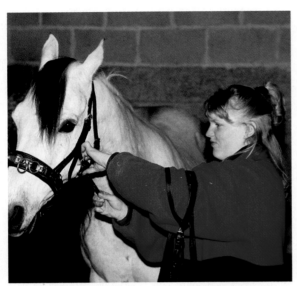
A rider adjusts the longeing cavesson, which is a piece of equipment used for longeing that is fitted over the bridle.

Tying Safety Tips

When tying your horse, always keep safety in mind. Here are a few tips for keeping your horse safe while you're tying:

- Don't use a rope halter to teach your horse to tie. It's too severe. Choose a solid nylon or leather halter.
- If your horse sits back, don't try to run up and save him. Stay calm and wait for him to calm down. If a horse is panicking and you try to jerk him loose, you may get seriously hurt. A panicked horse throws his body around everywhere. Wait it out.
- Don't tie your horse to a horse trailer that is not hooked up to a vehicle.
- Never tie your horse to a movable object such as a jump gate.
- Don't tie your horse to a flimsy fence or anything that he can uproot and take with him.

If you are still concerned about your horse's standing tied, you can tie him to a loop of baling twine or string that is then attached to the tie ring. If the horse sits back, the twine will break, which minimizes the chance of injury. There are also innovative slow-release tie rings on the market, which will pay out the lead gently and slowly as the horse pulls back.

him. Always use a quick-release knot, which is a type of slipknot. As with a helmet, you never know when you may need it, so it's best to be safe than sorry.

LONGEING A HORSE

Longeing is an exercise in which the handler stands inside a circle and controls the horse with a long line (called a longeline) and a long longe whip. The horse moves around the handler in a circle, responding to verbal cues, whip cues (light touches on the hock or flicks on the ground), or squeezes on the line. Although longeing will be used throughout the career of the horse, in the early days it establishes a novice horse's understanding of what is required of him and develops the muscles that are needed to support and carry the rider. Longeing instills a degree of self-carriage (balance) in the horse and gives you skills needed to school your horse better. Because longeing is so beneficial, it should be part of a horse's training program for life.

Longeing introduces the young horse to the saddle and the weight of the rider; it also helps you to evaluate how training is progressing and how the horse's muscles are developing. You can improve transitions and observe whether the horse is falling onto his forehand (shifting his balance toward his front legs), if his hocks are stiff, or if he's flattening his back. Longeing also helps teach the horse to lengthen and shorten his stride. This is because the arc of the circle helps the horse understand how to use his hind legs properly.

Most of all, you can fully school your horse in twenty-five minutes. You can even school a horse or pony that is too small for you to ride.

Tack for Longeing

Longeing the horse in a saddle (or surcingle), bridle, and side reins is important in order to be able to longe the horse into a contact. The term *contact* denotes when the horse actively holds the bit in his mouth (versus the bit passively resting in his mouth).

Without contact, the horse will be free to carry his head and neck as he pleases. You will have little control over the horse, and it will be easy for him to fall onto the forehand.

The side reins mimic the rider's hand in the saddle and also keep the horse on track around the circle. Without side reins to restrain him, the horse is able to overbend his neck to the right or left and to fall in or out of the longe circle.

This English rider is attaching the longline to the bit rings, which is a good alternative to the longeing cavesson.

How Often to Longe

For a green horse, you should aim to longe twice a week, but if the horse's rider is also a novice and isn't able to help build the correct musculature of the horse when under saddle, he or she should longe the horse before riding. Horses that are weak through the back also benefit from longeing before riding. Older horses should longe once a week or once every ten days. It all depends upon your goal and your horse's muscular frame. To loosen the muscles properly, longeing sessions should run from fifteen to twenty minutes long.

Some trainers are anti-longeing, but this is often because they don't fully understand the skill. It's better not to longe at all than to longe poorly because horse and handler can get injured if longeing is done incorrectly. If you're unsure how to longe, you should ask for help from someone who can longe competently and has done so for a number of years.

Bridle: Use the horse's normal riding bridle and bit.

Cavesson: It can be difficult to find a cavesson that fits well, so a good option is to run the longeline through the back of the bit, making a simple knot on the side you are going to longe, and attach the spare piece of the longeline to the opposite side of the bit. This secures the longeline without pulling on the outside and causing the horse to tilt his head.

Some trainers run the line from one bit ring over the poll and clip it on the opposite bit ring, but this puts too much pressure on the horse's head, making him reluctant to move forward into the harsher contact.

Gloves: Always wear gloves; your hands won't get burned if your horse spooks or pulls the line away.

Longeline: Many trainers prefer the safety and feel of a cotton longeline or rope to a nylon one.

Longe whip: A longe whip should feel balanced and comfortable in your hand.

Saddle or surcingle: You will need to tack your horse up in either his usual saddle or a surcingle in order to attach the side reins Secure stirrups so they won't slide down and knock against the horse's sides by knotting them with the leathers.

Side reins: Reins that have a rubber doughnut for length adjustment in the middle of the rein are preferable because they offer a little extra give. Adjust the rein when the horse is standing naturally. The horse's nose should be slightly in front of the vertical, meaning that the nose should be inclined further ahead of the forehead. There should be contact, but the reins should not restrict the topline in any way. Attach the side reins to the saddle, going under the girth straps and threading the side rein back under the first. This ensures that the side reins do not slip down. The height of the side rein will depend on how advanced the horse is, but as a general guide, attaching the side reins so that they are level with the horse's shoulder is a good starting point.

Training Steps

The most difficult (and potentially dangerous) part of longeing is getting the horse out on the circle, but it doesn't have to be that way if your horse can lead in hand correctly. If he doesn't do so, tap him with the longe whip so he walks along with you. If your horse understands how to be led, it's easy to begin.

You want to be in a position to drive the horse forward: starting at a walk with no side reins attached, take a step behind the horse's girth, and ask the horse to move forward with a slight movement from the longe whip toward the hindquarters. Because you are driving the horse forward, your body needs to be in this position. People often stand at the shoulder, and as a result, they are unable to get the horse to move on the circle.

Think of the longe whip as an extension of your hand. Don't crack it to startle the horse into moving. The horse should move from the whip to the hand in the same way he moves from the leg to the hand when you are riding. If he's resistant to moving forward, you can tap him on the heavy muscle between the hip and the hock to encourage him to move. The horse should stay out on the circle without dropping the longeline. If the horse falls in, point the whip at his shoulder and encourage him to move out.

If you are working with a very green horse, now is the time to teach him voice cues. As you ask for a walk, swing the whip slowly and say in a low and slow tone "Waaaalk." As you ask for trot, again swing the whip slightly faster and say in a bright voice "TEE-rot!" When you want him to slow down or stop, say slowly and calmly "Whoaaaa." When the horse is ready to canter, flick the whip up and say in a more urgent tone "Can-TER!" Teaching voice cues will stand you in good stead when you begin riding; you will simply pair your voice cues with your leg aids.

If your horse responds by bombing around the circle, don't worry as long as you feel safe and the horse isn't going to injure himself. Let him canter around a few times. Don't let him get going too fast,

or you may get pulled around the arena. If he gets out of control, drop the whip and gradually take up the longeline, guiding him in a slightly smaller and smaller circle until he slows. The smaller arc of the circle will prevent him from going too fast. Whatever happens, don't run the horse into a fence or a wall to slow him, because he may jump it or stop too fast and injure a tendon.

Slow the horse by using body language and a light vibration on the line. Turn your body away or drop your eyes. This works quite well if the horse is in tune with your motions. You can tell this if the horse flicks his inside ear, licks, and chews, or looks at you while he is longeing. You can also use your hand as a turning aid, giving little squeezes to encourage him to turn in on the circle as a way to slow down. (*Note*: for the best contact, feel, and control, hold the longeline as you would hold a rein.)

There are many choices when you longe. You can walk with the horse as you longe, varying the size of the circle and where it is positioned in the arena, rather than working the same circle all day. After all, if you were riding the horse, you wouldn't ride him on the same size circle over and over. By changing the degree of bend, the horse becomes suppler. You can also longe him on a fixed circle, which

A rider longes her horse at the trot in well-adjusted side reins and a longeing cavesson.

helps the horse learn how to balance and bend, or you can longe your horse in a round pen on or off the longeline. The wall of the pen helps keep the horse on a constant circle.

However you choose to longe, it's best to keep the side reins equal on both sides, rather than shortening the inside rein. The outside side rein may be tighter, but this is all right because it is preferable to forcing the horse's head to one side.

After a couple of minutes at the walk and the trot on both sides, you're ready to attach the side reins. (*Note:* warming up without side reins gives the horse's muscles a chance to stretch a bit so the horse is ready for the work in the side reins.) Attach the side reins to the bit rings, and move off again on the circle. Don't worry if your horse chooses to trot right away, even if you haven't asked him. Let him trot. He's moving forward, and that's what you want.

When you turn to work the other side, lead the horse straight for a few steps before you make the turn. The side reins will restrict him if you turn him too tightly.

The more efficiently a horse moves, the more likely he is to remain healthy and hardworking for many years. Training should make the horse sounder and fitter and not break him down. This is the goal every rider should have no matter the discipline. Proper longeing will put you on the right track.

TEACHING A HORSE TO CARRY A RIDER

Backing a horse (that is, training a horse to carry a rider for the first time) should be done by an experienced horseperson or under the guidance of a trainer. Even the calmest of horses can become frightened when he feels the weight of a person on his back for the first time.

A horse has to use his neck, his back, and his stomach muscles to support the weight of the rider. A young horse will have to figure out just how to do this, and he will be prone to balance problems and even tripping. You must bear in mind that any mistakes caused by the rider can set back the horse's education greatly.

Training Steps

To mount a horse for the first time, put on a snaffle bridle, which he should be accustomed to from his longeing exercises, and ask someone to hold the horse (preferably someone the horse is familiar with). Stroke

Gait Rundown

Here is how the horse's gaits should look on the longe:

Walk: Make sure the walk is purposeful and has a march to it. If the horse is dragging, encourage him to be more active by swinging the whip toward his hocks. If he's too quick, make the size of the circle smaller to work off energy.

Trot: Encourage the horse to take long, relaxed and bigger steps, which will move and strengthen the long back muscles and help him to carry his rider. The hocks should be active during the trot. If not, try asking for transitions from walk to trot. The horse's first step in the transition from the trot should be a long one, and not a short, stuttering step. The hocks should remain active during the walk as he makes the transition.

Canter/lope: Unless the horse is well-balanced, skip the canter on the longe. Every horse leans in at the canter, but the more schooled he is before he begins this gait, the less he will lean. If you ask the unbalanced horse to canter, he is likely to slip and could severely damage his tendons. The canter is of more value to the advanced horse, as the sequence of the gait helps loosen his back, which creates a better trot.

the horse all over for a few moments, including his back, until the horse appears comfortable. Have your helper lift you up with his or her right hand just below your knee. Lie over the horse's back on your stomach for a few moments, staying as still and as quiet as you can. Next, slowly swing your right leg over and sit down, leaning slightly forward over the horse's neck and taking as much weight off his back as possible. After a few minutes, dismount by bringing your right leg back over the horse and slowly sliding your body to the ground. Repeat a few times, giving your horse some sugar or another treat in between.

Now you're ready to mount using a saddle. Your horse should be used to the saddle if you have been using it during longeing. Work the horse on the longe for a few moments to burn off some energy. Then ask your helper to boost you up into the saddle. As soon as your seat touches the saddle, have your handler move the horse forward on the longeing circle. Sit very still; let the handler give the aids and control the longeline. If the horse trots, let him; try to follow his movement to the best of your ability. If the horse settles down and stops, get off and repeat the whole exercise a few more times. Keep the session rather short and end on a good note with lots and lots of praise.

Lead horses are always good role models for young horses, especially if the horse is a quiet pasture companion. In the arena, a lead horse should walk in front of the young horse, which follows in the lead horse's tracks in a large circle or in long straight lines. Do this at the walk and the trot. Since your horse will have little knowledge of aids, this is a good way to teach the horse how to balance himself on straight lines and circles without a rider's input.

Continue this mounting exercise every other day for two weeks before you move on to something new. Try not to overtire the horse; keep in mind that the horse's muscles will take a while to adjust to and strengthen under the rider's weight. Riding the horse longer and harder will only strain the muscles. Keep to about fifteen minutes the first week, and lengthen to thirty minutes the second week. You can break up your mounting exercise with longeing or round-pen work.

Keep to the walk and the trot for these two weeks. The canter and lope take a lot of balance on the horse's part, so it's best to let him get used to the weight of the rider at the slower gaits. Try to keep your weight off the horse's back as much as possible by leaning forward slightly. Keep your hands low with a slight, steady contact. Stick to voice cues—the same ones you use in round-pen work and when longeing.

Don't use spurs on a young horse as this can make him afraid of leg aids. If the horse should buck and you fall off, make sure to get back on right away. If you're too nervous, have a trainer take over. You never want a horse to learn that he can get out of work by bucking.

Basic Concept Aids

Your horse should understand voice cues from the work on the longe or in the round pen. In the saddle, add a slight pressure with your legs and give your voice cue. If your horse ignores you, do not increase the pressure—this will only desensitize him to your leg aids. Instead, give him a little thump with your calves

An instructor gives a student a leg up so he can lie on his stomach on the horse, accustoming the horse to a rider's weight.

This rider and horse have progressed from trotting over ground rails to jumping cross-rails. Training to jump is done in slow stages.

and repeat your voice cue. This may startle him into scooting forward into the trot or a very fast walk. That's fine; you want forward motion, after all. It doesn't matter how rough it is. Praise him for moving forward.

Now repeat the exercise, this time with the light leg aid and the voice command. The horse should now move forward in response to the lighter aid. If not, repeat with the thump of your calves. As your horse progresses, you can move on to teaching him your trot and canter aids.

If your horse ignores your aids, give him a more difficult task; this a good way to get him to listen. For instance, if your horse is lazy and reluctant to move forward at the trot, make him canter around the arena with good, forward energy. Bring him back to the trot, and use the energy from the canter to push him to the speed you want. Come back to the walk, praise him, and then push him back to the trot. If he doesn't move into the trot at the same pace, pick up the canter again. Soon, he will learn that being lazy brings him more work, not less. As soon as he trots forward willingly, end the lesson.

If your horse won't listen to "whoa" or won't slow down, put him on a circle. It's harder for the horse to rush on a circle because the bend requires him to use his hindquarters more. Circles also drain energy naturally. Keep moving on the circle, making it smaller and smaller until he comes to a walk and then a halt. Above all, never resort to pulling hard on the reins or turning the horse toward a solid object, such as a wall or a fence; he may try to jump it. Soon the horse will realize it's easier to listen to "whoa" than it is to run around on a circle.

Once your horse understands the basic concept of aids, you'll want to finesse each cue according to your discipline, such as teaching the half halt if you're an English rider or the pivot if you're a western rider.

TEACHING A HORSE TO JUMP

As with any other type of horse training, it's very important to take the time to slowly introduce your horse to the concept of jumping. You don't want to make the mistake of being too precise in the beginning; let your horse jump on his own, using his own natural balance.

Four cross-rails make up this grid. Grid work builds confidence in a horse; the set striding means he will meet each fence correctly.

If you are there to set him in front of every fence in training, you will create a horse that depends constantly on his rider. No rider can be perfect in front of every fence, so you want a horse to be a confident partner, willing to make the decision when the chips are down.

The breeding of your horse and how well he has grasped his basic training will determine when he can begin jumping. Thoroughbreds mature at about age four, earlier than do warmbloods, who mature at age five or six. A horse's body must be mature enough to withstand the stress of the jumping action. If you are unsure at all about whether your horse is ready to jump, check with your veterinarian.

A horse should fully understand all the basics of riding before jumping training begins. If a horse won't move forward into a trot from your leg aid, then chances are he won't move forward to a fence. He also must understand how to balance himself under a rider's weight. Jumping a fence takes balanced precision, and if he can't balance his rider at the walk or the trot, he doesn't stand a chance in balancing a rider over a fence.

Training Steps

Trotting over ground rails (poles set on the ground) is a good place to start. Ground rails teach your horse to adjust his balance and stride, which are important skills for him to have for jumping. Begin with one pole. Aiming straight for the middle, walk your horse over it.

Next, add a second pole, about four to five feet from the first. Walk your horse calmly over both. When your horse steps cleanly over both poles without hesitation, add a third one. Continue until you have four to six poles placed at the equal intervals.

Now, pick up a trot and ride through the poles. If your horse tries to rush through the poles or jump over them, slow down by circling in front of the poles at the trot, come back to the walk, and then walk through the obstacles. Only let your horse go through the poles in a calm manner. As soon as you feel him tense up (by speeding up, bringing up his head and neck, or losing the trot rhythm), circle again and come back to the walk. Include this ground-rail exercise in your training sessions for several weeks before you begin actually jumping.

Grid work teaches your horse basic jumping skills. The set striding (space between each jump or ground pole) of the grid means that there are no surprises; your horse will meet each fence correctly, which will build his confidence. Start the grid with one low cross-rail (two jump poles crossed in an X fashion), then begin to add low jumps, slowly increasing the height and width.

Approach the grid in a steady trot, and stay in your two-point position throughout, making sure to have your hands forward as the horse jumps. Let him work out how to navigate the grid without your cues. Make sure to exit the grid in a straight line, and only then ask for a turn or a downward transition. This ensures that the horse stays obedient and never learns to run off after jumping.

Begin with three ground rails with a cross-rail nine feet after the last rail. Ride through calmly, adding a leg aid if you need it. Next, add the second cross-rail eighteen or nineteen feet away from the first cross-rail, and ride through several times. You may find your horse wants to drift one direction or another. Use a wide-opening rein to keep him straight.

Now, add a third cross-rail twenty-one feet from the second cross-rail. Move through the grid. Once your horse understands his job, you can change the second and third cross-rails to one small vertical and then add a back oxer rail to the third. After mastering the grid, you can even add a fourth jump set twenty-four feet after the third.

Important Dos and Don'ts

Make sure your horse is comfortable with the exercise before you add more elements or more difficult jumps. Never overface him (ask too much of him). Jumping is about building confidence, not seeing how high a horse can jump. If a horse is having trouble with a higher jump, then lower it immediately.

Using an experienced lead horse as a visual aid for your horse is a good choice if your horse is timid. Seeing and then following an experienced horse over jumps can boost your horse's confidence. Never let your horse refuse to jump; keep the elements small enough that he can get over them even from a standstill. Don't turn away—try again. Keep the horse at the obstacle with your leg on until he hops over it.

It is important to be as soft and as steady as you can in your riding position because the horse will be learning to adjust his balance. Above all, make sure you never catch him in the mouth (jerk on the reins) over a fence, as this could hurt him and frighten him away from jumping for good.

Horses that are learning to jump often jump too big, or they jump at the last second, making it difficult for their riders to stay with them. You can anticipate this by grabbing your horse's mane just before you reach the jump, as insurance against accidentally jerking your horse's mouth. Some riders also buckle a jumping strap, which is basically a long stirrup-leather, around the horse's neck to grab onto when the horse makes a big or unexpected jump.

TRAINING

Learning the terms below will help you to talk horse training.

■ **Backing:** Riding the horse for the first time.

■ **Contact:** The horse actively holds the bit in his mouth (versus the bit passively resting in his mouth), stretching the reins evenly.

■ **Ground rails:** Poles on the ground, generally used when teaching a horse to jump.

■ **Longeing cavesson:** A piece of equipment used in longeing that is fitted over the bridle. The longe-line is attached to rings on the noseband.

■ **On the forehand:** More of the horse's balance is centered over the forehand instead of the hindquarters. The horse pulls himself along instead of pushing himself forward with his haunches.

■ **Overface:** Asking too much of a horse.

■ **Side reins:** Reins used during longeing that are fixed on the girth or surcingle and attached to the rings of the bit.

Chapter 20

Training Problems and Solutions

A pushy horse is no fun. He bullies you, barges ahead as you lead him back to his stall, yanks your arm out of its socket as he makes a grab for grass, and even jigs along, shying and bolting at every opportunity. Understanding the basics of good manners is important for your horse's safety and for your own.

Horses develop rude manners because timid owners don't correct the behaviors quickly enough. This may not be your doing; the horse may have had a previous owner who let him get away with things. Whoever is at fault, you *can* get your horse back on track. It takes a firm hand and a willingness to regain the control. Horses function in a hierarchical social setting, and they need to know who is the leader. If your horse sense that it's not you, he will assume the role. Impolite horses have gotten the pass to be the leader for too long.

LEADING

If you're having problems with your horse, go back to the basics of leading. Horses often get the upper hand while leading, and people don't realize that this is when problems begin. Unless a horse is bolting out of the rider's hand as horse and rider walk together, bad behavior while leading is ignored all too frequently. Be aware that little problems mount up quickly, and a horse that doesn't respect his rider can be dangerous.

Check your leading stance. Your left hand holds the extra rope while your right hand grasps the lead about four inches below the snap. Hold your bent, right arm slightly away from your body, up under your horse's chin. This allows maximum control.

Keep your horse looking forward as you lead him. If he looks away, chances are he's paying more attention to his surroundings than to you. If he looks at you, he most likely isn't moving forward, and he's probably stepping into you or on your toes. Keeping his head pointed straight ahead keeps him focused on the task at hand—getting from point A to point B. If he's looking at you, push his head straight with your right hand (still holding the lead). If he's looking away, tug his head back. If you may find him reluctant to comply, you may need to use another aid, such as a chain.

Keep your eyes looking forward as well. Don't stare at your horse while you lead him. Some horses stop when people look at them.

To correct a leading problem, you will need a long stud chain (about fourteen inches) and a long bat or dressage whip. Place the stud chain on your horse's halter by threading it through the left-side cheek ring, up over the nose, and through the far-right ring (with the nub of the snap pointing away from his face). Do not put the chain under his chin; you'll create more problems if you do. Trainers use chains by popping or jerking the lead to correct the horse. Popping a chain under a horse's chin can cause him to toss and flip his head up in the air. Only experienced trainers should employ the chain-under-the-chin tactic.

This handler shows the correct way to lead a horse. She walks by his side and keeps his head straight and his attention on her.

By snapping the lead with your right hand, you are putting a bit of pressure over your horse's muzzle and getting his attention. Remember, do not use the chain to hurt your horse. Chains level the playing field by putting a bit of "oomph" on your end.

Correcting the crowder and the balker. The following technique works well with the horse that crowds you when you are leading. Gently use your dressage whip or bat to keep your horse from pushing into you. Hold it in your left hand, point the lash toward his haunches, and tap him if he swings them into you. This is also an effective "come along" cue if your horse is a balker. You can also rattle the stud chain on the halter if he pushes his head and shoulders into you.

Correcting the jigger. If your horse jigs (dances in place) while you are leading him, you should stop, wait until he stands calmly, then continue. Stop each time he jigs. You may have to rattle the chain on his nose or pop the lead with your right hand to get him to stop. You may not get very far the first time, but when he realizes you're in control, the behavior should eventually stop.

Under saddle, horses tend to jig from nerves, so it can be helpful to have a steady equine companion walk alongside him until he becomes comfortable. Many riders find that their horses jig on the trail or when they are away from home. Try to ride with a group of calm horses, and stay at a walk until all signs of jigging are gone. Galloping on the trail to tire your horse so he won't jig will only create more anxiety in a horse and more reasons for jigging.

In addition, make sure that your hands are quiet and you aren't doing anything to cause the problem, such as riding with your spurs turned in, with a braced seat or back, or with reins that are too tight. It's also helpful to teach your horse to walk on the long or free rein in a forward, comfortable walk in the arena. Practice taking up the reins and dropping them as you walk. Any time the horse jigs, halt and wait for him to calm down.

An experienced rider may be able to push the horse out of the jig by encouraging him with leg aids to take a longer stride. At first, the horse may break into the trot, but calmly bring him back to the walk and push him on with your legs again.

BOLTING

Bolting under saddle is extremely dangerous. Usually, horses that bolt are reacting to something that has frightened them (remember the fight or flight instinct).

This bucking horse is also on the verge of bolting. The best way to gain control is by circling because circles use up energy.

The best thing to do when riding a bolting horse is to take the horse in a large circle, then gradually reduce the circle's circumference. Circles help reduce excess energy and give the horse something to focus on other than his fear.

A horse that bolts ahead when leading needs to be put back in his place. If your horse is bolting forward while you're leading him, back him up by tapping him lightly on the chest or on top of his legs with your whip, while applying gentle pressure to the halter with the chain. Once he's backed up obediently, continue on your way. Repeat this each time he gets ahead of you. This is also a good technique for constant jiggers. It keeps them focused on a task that is different from their own.

If your horse has learned to bolt under saddle, bring him back to basics by working in the arena, practicing transitions from walk to trot and trot to canter. On the trail, stay at the walk and only move forward into the trot or canter if you're with a group of calm horses. Never canter back toward home; the horse will realize he's nearly home and will pick up speed.

Never aim a bolting horse at a solid object, such as trees, a fence, a gate, or a wall. The horse may not be able to stop, and you could get thrown into the object. Or, he could be so frightened that he might try to jump it and hurt himself.

Make sure you aren't contributing to the problem by hanging on to the horse's mouth or leaning forward

Leading Don'ts

Leading a horse can be very dangerous if you're not on your guard. Even the safest horses can whirl around, bolt ahead, or jump if frightened. Here are some safety tips for leading:

• Don't wrap a lead rope around your hand. If your horse bolts forward, the loops can tighten quicker than you can drop the lead. People have lost fingers and hands this way. Layer the lead rope, folding one piece on top of the other, in your hand instead of looping it.

• Don't toss the end of the lead rope over your left shoulder. If your horse whirls, the rope can get wrapped around your neck.

• Don't walk with your horse trailing behind you. If something startles him and he bolts, he can run over you.

• Don't let your horse circle around you. Keep him in control at your side at all times. If you let him drop his head to nibble on some grass, he should still stay at your right side as he grazes.

• Don't let your horse walk in front of you. This teaches him to barge ahead whenever he pleases. And again, if he bolts, you could be jerked off your feet.

• Don't let your lead rope get too long. A too-long lead rope gives your horse way too much leeway. How long is too long? The rope should be no more than three feet in length, measured from the halter ring under the chin.

in the saddle and gripping with your legs. This type of defensive position mimics a predator's behavior and can trigger the flight instinct in the horse.

BITING AND NIPPING

If your horse has a tendency to bite or nip, make sure you never feed him out of your hand. If the horse does bite or nip you, tell him "no" in a firm voice as you slap him with of the palm of your hand on his neck. This will remind the horse that you are the dominant partner. Never hit a horse in the face. It will only make him headshy.

This horse is threatening to nip. Make sure to correct this behavior immediately with a reprimand.

REARING AND BUCKING

If your horse rears or bucks, under saddle or in hand, keep him moving forward. This is the cure for most equine nastiness because it's difficult for a horse to walk and buck or walk and rear at the same time.

If your horse bucks or rears as you lead him, use the whip-tapping technique discussed earlier. You may have to give your horse a stronger tap or smack if he digs in and won't walk. If your horse manages to rear or buck anyway, use your chain technique. Make sure you stand to the side of your horse when he rears. Don't turn to face him, because you could get struck and injured.

If your horse bucks under saddle, lift his head with your reins and urge him to move forward by using leg aids and taps with your whip. This can be hard for novice riders because they haven't developed the strong core muscles needed for steady seat and position.

A horse must stop to rear, so push him forward with your legs and whip as soon as you feel him try to stop. If your horse will not go forward and begins to rear in earnest, pull his neck to one side with one rein and force him to walk in a very tight circle. Let him

It's natural for a horse to act up while he's turned out, as this one is. But rearing isn't acceptable when he is handled or ridden.

stand for a moment and then ask him to walk forward again. If he tries to rear, repeat the circle, then follow that with a moment of standing still, and then walk off. He will soon learn that rearing has no payoff.

This training method can be difficult for a novice because timing is crucial. Horses that rear or buck constantly can be very dangerous and should be dealt with only by professionals or experienced horsepeople.

SHYING

Forcing a horse to deal with his fear of an object may pay off in the short term, but the horse will still be afraid of the object in the long term. Soon you'll find yourself in a wrestling match you can't win. Remember, horses are prey animals with a strong fight or flight instinct. Something that scares your horse, such as differently colored dirt, may seem ridiculous to you, but it's a life-and-death matter to him. Kicking or whipping him past a scary object will only ingrain the fear.

Begin by identifying the object causing the shying, such as a garbage can or flapping material. Ask a friend to ride a calm horse that has no fear of the object, and then begin to walk your shy horse near it, as close as he will comfortably move, keeping him as straight as you can. Keep him looking away from the object slightly. If he looks directly at the object, he may swing his hindquarters out, causing you to lose control of his shoulders. If you walk toward the object at a slight bend, you can keep his quarters under control so he goes past the object keeping his body straight. Walk the horse forward in a steady pace. This tells him that you have

TRAINING PROBLEMS
Learning these terms will help you to talk about horse training issues.

■ **Balker:** A horse that refuses to move forward as a rider requests.

■ **Bolting:** Running away at speed and usually with little warning to the rider.

■ **Bucking:** A vice in which the horse jerks his hind legs into the air while the head and neck plunge downward.

■ **Crowder:** A horse that pushes into his handler's space.

■ **Headshy:** A horse that is reluctant to have the face and head handled.

■ **Jigging:** A nervous habit in which a horse "dances" in place rather than walks.

■ **Rearing:** A dangerous vice in which the horse stands on his hind legs.

■ **Shying:** A natural response in which a horse moves quickly away from an unfamiliar object or situation.

A horse shies at a plastic bag. Take your time when desensitizing a horse to an object that frightens him; don't overface him.

no fear of the object and it's no big deal. Since you are technically the leader in your partnership, this should pay dividends.

Your good leadership decisions always give your horse confidence. As soon as the horse walks past the object comfortably, you can start reducing the distance between the target object and your horse. By using repetition and avoiding overfacing the horse, you'll be able to get closer to the object. Soon the horse will lose all fear of the object. Give your horse lots of praise when he does well so he understands that he's done a good job.

Chapter 21

Natural Horsemanship

Although the art of natural horse training has grown greatly in the last two decades, it is really nothing new. Xenophon, an Athenian soldier, wrote *Peri Hippikes*, or *On Horsemanship* (translated also as *On the Art of Horsemanship*) circa 400 BC. Because of its concern for the horse's welfare throughout the teaching process, *Peri Hippikes* could be considered the first book on natural horse training.

Natural horse training, sometimes referred to as resistance-free training, got its start in the United States in the early 1900s with a handful of western horsemen who rejected the severe training method practiced by cowboys of the Old West. This new breed of cowboys wanted to find a different way of training, one based on establishing communication with the horse rather than breaking his wild spirit.

THE ROOTS OF NATURAL TRAINING

Traditionally, cowboys had trained their horses by breaking them—that is, breaking their spirits by dominating them. Cowboys roughly caught and saddled feral horses, then mounted and rode the horses while they bucked and reared. This went on, sometimes for hours, until the horses became either too tired or too weak-willed to fight back.

This harsh method of training was rejected by Tom and Bill Dorrance, who were born around the turn of the twentieth century in the Pacific Northwest and saw their fair share of breaking as young horsemen. Each brother went on to develop his own philosophy and method of training. Bill called his method True Horsemanship, while Tom spoke of his as True Unity. The common thread was the idea that there should be a genuine trust between rider and horse. Both men based their training on that practiced by the late-eighteenth-century vaqueros in the area that would become California and the mid-nineteenth-century buckaroos of America's western and northern regions. The vaqueros and buckaroos trained their horses in gradual steps in order to produce the highly tuned, quality horses needed for working and roping cattle effectively.

Bill, who worked mainly with ranch horses, developed his methods in the 1940s and 1950s, in a time when animals were still considered utilitarian, a means to an end. He had a gentle, solid way of establishing deep contact with horses that he called *feel*. His book, *True Horsemanship Through Feel* (published in 1999, the year that he died), is considered one of the great books for those interested in developing a harmonious relationship with their horses. In his book, Bill wrote, "I've always liked horses, and I've always wanted to help them do their job better, whatever it was that they had to do."

Tom Dorrance's *True Unity: Willing Communication Between Horse and Human* (published in 1987) has become a bible among natural training enthusiasts. "The thing you are trying to help the horse do is

A turn-of-the-century cowboy works to break this feral horse's spirit, riding the bucks and rears until the horse is exhausted.

to use his own mind," Tom wrote. "You are trying to present something and then let him figure out how to get there." Tom passed away in 2003.

Ray Hunt, a student of Tom Dorrance's, took a practical approach to sharing this natural training philosophy with others in the thousands of clinics he gave over the years. On his website, Hunt is quoted as saying: "The human is so busy working on the horse, that he doesn't allow the horse to learn. [People] need to quit working on the horse and start working on themselves. They might get it done, but they don't get it done with the horse in the right frame of mind. The horse usually gets the job done in spite of us, not because of us." Hunt passed away in 2009.

Natural training has many different philosophies and several current practitioners. There is much confusion about what constitutes natural training, because some practitioners believe in taking particular training steps in a linear fashion, while others follow a less structured method, in which learning and understanding are considered to be more organic.

CONTROVERSY

Proponents of natural training are often devout followers of both the training philosophy and the trainers themselves. There is a belief among some fans of natural horsemanship that any method not labeled natural is inhumane; conversely, any method with "natural" tacked on to it is considered humane. Many followers also take natural training to the extreme, which can include riding the horse without a bit in his mouth or a saddle on his back. Yet this is not the real meaning of natural training or natural riding, as there is little that is natural for the horse when it comes to carrying a human.

Today's practitioners of natural methods come under fire from various camps within the industry. Some critics believe this type of training has been done for centuries in other riding disciplines, such as dressage, where the term *starting* a horse is used instead of *breaking*. Proponents of traditional training believe that their way is every bit as effective and kind as that of natural horse training, and sticking the *natural* label on a training method doesn't make it better or more effective.

Natural training methods have suffered most, however, from the fact that this style of training became so popular through the late 1980s into the 1990s that many trainers jumped on the natural horsemanship

Monty Roberts, creator of the Join-Up nonverbal communication training method, works with instructor Kelly Marks.

bandwagon. In the 1990s, horse expositions (giant fairs dedicated to the world of the horse) brought many of these trainers great exposure. Profits soared as consumers saw these methods demonstrated live, often by trainers who acted more like performers. Some trainers later held their own clinics, charging hefty sums for the experience of learning their techniques. Many clinicians made their techniques available on videos (and then on DVDs), often included in large packages that covered entire training systems. And some clinicians were even sponsored by national equine companies. Critics argued that these systems were being marketed to individuals with little actual horse knowledge and that the trainers didn't address the fact that, in addition to learning the training techniques, riders still had to know how to ride.

For opponents of it, this type of marketing existed solely to benefit the trainers' status—turning them into stars within the industry—while the horses' welfare was only a byproduct of their success. In addition, there were, and still are, some so-called natural trainers who have no talent for working with horses but a great flair for putting on shows; they have also damaged the credibility of the truly gifted trainers and the natural training movement itself.

THE BASICS OF NATURAL TRAINING

Natural horsepeople study horses' nature and body language, using equine psychology to teach horses. Although training style differs among natural horsemanship trainers, most of them use a philosophy of communicating with the horse to make the "right answer" to the task easy to do and the "wrong answer" difficult.

A horse is taught through the application of pressure in the form of a squeeze of the rider's leg, a pull of a rein, or a tap with a whip. When the horse moves away from the pressure, he is released. The release is the reward. The horse looks for the release of pressure to know that he has done the task correctly.

Pressure and Feel

Pressure is seen in the natural order of the horse's world. If you watch horses interact, you'll easily see how they "speak" to each other with subtle and not-so-subtle body language. In the herd, horses will posture, become aggressive, and even fight to establish a dominant position—in other words, horses will apply a type of pressure to get what they want. Horses do not want to hurt themselves in a fight, however, so they go in with the intention of merely making their adversaries back down. If a horse tries to get another horse to yield and that horse doesn't do so, the first horse will come up against his own pressure and back down.

If one translates this scenario into horse and trainer, the trainer is the second horse. By not backing down, the trainer forces the horse to put pressure on himself. The horse must figure out how to release it (by yielding). Once the horse moves away from the trainer, the situation goes back to neutral, and the horse regains the space around himself. More importantly, he learns that his "answer"—to move away—was correct.

The horse isn't being punished or disciplined with the pressure method, but sometimes the amount of pressure that is applied may appear harsh to the layperson. Many trainers believe that the pressure should be as gentle as possible yet as firm as necessary to get the horse to respect and cooperate with the trainer. By following this philosophy, proponents of this technique say, the trainer is communicating with the horse in terms he understands.

For instance, because horses use a dominance display to get a subordinate horse to back down, if a handler assumes a dominant posture, standing tall and moving swiftly into the horse's space near his

Icelandic stallions compete for dominance in their herd or territory.

shoulders, the horse should immediately become submissive. If he doesn't, the handler follows up as a dominant horse would, with a physical reprimand that is given swiftly but without anger, such as a rattle on the lead rope. The handler then returns to his or her previous position without dwelling on the incident.

To be effective in applying these teachings, the rider or trainer needs certain requisites. He or she must know not only what to ask of the horse but also how to ask it, and with how much pressure. Feel is crucial for the success of this type of training. A trainer must be able to know precisely when the horse has yielded, to feel that sometimes-imperceptible "give." The rider must also have precise timing—knowing exactly when to release the pressure so the horse understands the communication. Feel is developed from many years of experience and a deep understanding of horses.

Groundwork and the Round Pen

Many of today's horsepeople emphasize that groundwork will translate into a good riding horse. Trainers claim that one should gain the horse's respect on the ground, and then use the lessons taught there while mounted. A horse that pushes or bullies a handler on the ground will not be respectful of a rider in the saddle.

Round pen size is very important when teaching groundwork. It's difficult for the trainer to effectively establish control in a large round pen; the horse can easily get to the far end and be away from the trainer's influence. But cut a 100-foot pen down to about 60 feet in diameter, and the dynamics change dramatically. At

Flooding

Handlers must take care not to create a situation called flooding when they want to desensitize a horse to something that spooks him. In humans, *flooding* utilizes the idea of pushing oneself to endure high anxiety and not retreat from the phobic situation until anxiety is lessened. By contrast, *systematic desensitization* is based on keeping anxiety low in the phobic situation and retreating repeatedly until one becomes accustomed to the stimulus.

For a horse, flooding creates a situation so overwhelming that, any time he feels trapped, he will succumb to the fear. Flooding is actually more about breaking the horse's spirit, teaching him learned helplessness. The horse doesn't become desensitized to the object; he just learns to give in. A horse that is broken in spirit will often, if given some time off, get over the shell shock of learned helplessness with humans and regain enough spirit to try to escape from the same situation; he will have learned nothing and may even become defiant. The key is to have the horse instill his trust in the handler. The horse should be taught that his reactions are indeed normal, but that he will be protected and never put in a sink-or-swim situation.

A round training pen at a clinic eliminates the horse's ability to back into a corner and resist training.

this size, the area is large enough for a horse to lope around the rail and have his own safety zone, but not so large that trainer loses control. And the circle size doesn't put undue stress on the horse's tendons and lower legs. Cut the pen down further, however, and it will be too small for safety; the trainer and horse can be injured if the horse panics and strikes out or if he feels trapped and tries to leap out of the pen.

Round pen training is effective because it works with, rather than against, the horse's flight instinct. It contains a horse but doesn't restrain him. Restraining a horse, by a rope or by confinement, may make him think he cannot escape and trigger his fight instinct. At the worst, the handler or horse can get injured; at best, the horse is no longer in a frame of mind to learn anything. When the horse is confident that the pen's size doesn't confine him, he can focus his attention and intelligence on the handler.

This technique is effective with horses that are truly fearful of humans. In the round pen, frightened or mistrusting horses have room to react to the actions of the trainer, and they can learn how to respond correctly and calmly before the trainer ever places a hand on them. This type of training is also beneficial for horses that have been neglected or abused or were once feral.

A common fallacy about round-penning a horse is that a handler supposedly forms a connection with the horse by chasing him around in circles. This is not a natural-horse technique. A horse won't connect with anyone who runs after him with a whip and forces him to run around in circles. If a trainer chases after a horse aggressively, like a predator would, he or she will only engage the horse's defensive instincts and make him feel he must get away from the trainer.

Instead, the round pen should be used to teach a horse to come to the trainer. The natural technique begins with the trainer pushing the horse away like another horse would, calmly moving him so he can think about and respond to the pressure the trainer is applying.

Establishing a Good Relationship

Part of a good relationship is having the confidence that your horse will not only perform how you want but also behave how you want. Conversely, the horse should also expect a harmonious relationship with his owner.

1. Learn about equine body language. Pay attention to what your horse is trying to say with his posture, ears, expression, and even tail position. If you understand how he is trying to communicate, you will have a better idea of how to speak to him in a language he understands.

2. Be considerate in your training. Endless circles, whether in the round pen at liberty, or in the arena under saddle, are counterproductive after a while. Don't drill. Just teach.

3. Learn how to better your riding. All the games and groundwork in the world will not help your horse if you have unsteady hands that jab him in the mouth, legs that constantly deliver the wrong cues, and a seat that bangs against the horse's back.

4. Make training enjoyable for your horse. Mix up your training day with things that your horse enjoys, whether that's grazing on the lead, a deep grooming, or a trail ride with a friend.

5. Leave your temper outside the arena. Your horse does; so should you.

6. Don't hold grudges. If your horse was spooky last time you rode, try to act like the last ride was perfect anyway. Don't take problems to heart; don't make them a bigger deal than they are.

Competition

Section 9

Riders who are not content to merely log in hours on the trail or in the arena are drawn to competition. There are as many different ways of competing with horses as there are riding disciplines.

Competitions range from western and English classes, such as pleasure and equitation, to sporting events, such as eventing and reining. Competitions test your riding skills and your horse's training against those of others; they also provide affirmation from a professional—a judge—that you are on the right track with your riding lessons and training. There are various levels of competition that suit riders of all experiences, so you can start out at entry level and then work your way up.

Chapter 22

Shows and Classes

Horse shows are divided into classes. There are classes for different abilities, age groups, and riding disciplines. One or more judges preside over the arena. Big shows and high-level ones have multiple judges. A judge either watches from a judge's booth, located adjacent to the arena, or observes from the arena floor itself, in the midst of the action. Based on the criteria of the class, the judge will select the individuals who best meet the class's ideal.

There is quite a bit of activity at a horse show, and the unfamiliar show ring often distracts riders and horses alike. Various factors such as readying the horse, packing up and getting to the show, and dealing with unfamiliar show grounds, different arena footing, and show-day jitters can affect a rider's performance. The key is to select a show that is appropriate for your level of riding. To start, many trainers suggest competing in classes that are slightly below the rider's schooling level at home, since green riders often go on autopilot and don't ride as effectively as they would in a familiar environment.

TYPES OF SHOWS

There are many types of competitions, from those for beginning riders, such as schooling shows and unrated local shows, to those for experienced riders, including rated A-shows and B-shows. Individuals who don't like arena riding can compete in other types of events.

Schooling Shows

Riders who are just getting started will more than likely enter schooling shows, which are specifically designed for beginners and green horses. These competitions are often put on by a boarding barn or an equestrian center for its clients. They feature classes similar to those at regular shows, but they are geared toward the novice rider or horse with the goal of educating riders on the basics of showing in a safe learning environment. At these shows, the judge is prepared to see a fair number of mistakes and botched rides.

Often, entrants are not required to wear show apparel—other than safety gear. Rules are less stringent, and sometimes the judge will take time to speak to each rider to offer friendly and constructive criticism just before the ribbons are announced and the class is "pinned." The schooling shows are usually held in a single day.

Unrated and Rated Shows

Local, unrated shows and 4-H shows are similar to schooling shows. They are community-based, affordable ways to compete without costly class fees. As is true of schooling shows, the caliber of horse and rider at unrated shows will be more entry level; they include weekend riders, riders looking to get experience for young horses, and those riding school mounts. Riders can get away with secondhand show clothes and tack that is merely serviceable. What counts in a horse at this level of showing is reliability rather than fantastic

This type of lead line class is commonly found at rated and unrated shows. It gives children a fun, safe introduction to competition.

The hunter over fences class at an A-rated show in West Palm Beach, Florida, is filled with fierce competitors.

performance. These shows are based more on fun and learning than on cutthroat competition. Unrated shows are also usually one-day events.

Rated shows are the highest level of competition. The United States Equestrian Federation (USEF) shows are rated according to the amount of prize money given, with AA being the highest and C the lowest. In addition to prize money, a rated show must offer a certain number of classes per division and be held over a certain number of days. Riders will go on the circuit, attending the various shows at their chosen level in the region. They must abide by USEF rules to comply with the rating. Although competition at local member shows can be fierce, the top horses and riders will be found at the AA shows. There are few C-rated shows in America—most lower-level shows are merely unrated or local.

B-RATED SHOWS

B-rated shows are a step up from local shows. These competitions offer divisions for novices, as well as for more experienced riders. Classes are generally geared toward green horses or junior and adult amateur riders, helping to create a supportive environment in which exhibitors are often heard encouraging each other at the back gate. The end of the show season culminates in either a members-only championship show or a banquet hosting the presentation of year-end trophies.

Although once considered merely a training ground for future A-circuit riders, or a limbo for horses and riders who somehow couldn't cut it in the big time, the B-shows are becoming increasingly popular as an end unto themselves. Lasting one day or over a weekend, these shows don't require the same commitment of time and money that nationally rated shows do. But they do require expertise. Errors that might still have earned a ribbon at a schooling show are simply not tolerated at B-shows.

A-RATED SHOWS

A-rated shows boast the crème de la crème of the equestrian world. There are no mediocre movers in the under-saddle classes and no hunters with uneven knees over a fence. The A-circuit is where the big money is. A solid junior or amateur hunter or western pleasure mount can easily cost $50,000 to $100,000.

In addition to the cost of the horse, the A-circuit requires a great investment of time and effort. Because competition is so tough at this level, there's no sliding by with just one riding lesson a week. Any cracks in a rider's foundation become glaringly apparent during the pressure of the multiday A-shows.

There are also the travel costs that are associated with attending an A-show. Because A-shows are multiday shows, junior riders must skip school, and adults must take time off from careers and domestic responsibilities. Besides paying for their own meals and hotel accommodations, exhibitors customarily share the cost of their trainers' meals and lodging during the show. This can be a substantial investment, especially if an entourage of barn managers, assistants, and groomers accompanies the trainer.

Open Shows and Breed Shows

An open show is just that: one that is open to all breeds. Today's open shows are usually discipline-specific (such as a dressage show), but occasionally you'll find an open show featuring a mix of different riding styles, with perhaps English classes in the morning and western ones in the afternoon.

Breed shows are held and sanctioned by their corresponding breed associations. Only horses registered with the association and in good standing can compete.

Ribbons

Ribbons are awarded for different placings. Most shows give out ribbons up to fifth place, although some larger events will award placings up to tenth. Each ribbon has a corresponding color with its place. The following are the colors for US shows:

- First place finish: blue
- Second place finish: red
- Third place finish: yellow
- Fourth place finish: white
- Fifth place finish: pink
- Sixth place finish: green
- Seventh place finish: purple
- Eighth place finish: brown
- Ninth place finish: dark gray
- Tenth place finish: light blue
- Championship: blue, red, and yellow
- Reserve Championship: red, yellow, and white

At these shows (which are often rated by the USEF as well), you'll see one specific breed competing in all kinds of classes, from the typical western and English classes to more uncommon events such as in-hand or halter classes, driving classes, and saddle seat. The points earned at breed shows will determine the horse and rider's ability to compete at the breed's year-end championship, often called a world show.

Other Events

For riders who do not enjoy arena riding, there are options for competing that don't involve horse shows. (*For a more extensive list see chapters 5 and 6.*)

Competitive trail rides (CTR): Trail rides offer an alternative to endurance racing. The sport is a blend of equitation, horsemanship, and trail manners. CTR takes place over several miles of course, with beginner rides starting around 15 miles and advanced rides around 50. In CTR, unlike in endurance, the first horse across the finish line isn't always the winner. From the arrival at camp to the ride's end, the riders are being judged. They receive scores on riding ability as well as horsemanship, which includes the fit and appearance of their tack, the horse's grooming, and the safety of their camp and trailer.

Endurance: This is the oldest sport recognized for competing on the trail. Endurance is essentially cross-country horse racing. At entry level, there are limited-

A young girl and her pony compete in one of the many fun and creative classes offered at gymkhanas.

distance rides of about 25 to 30 miles. Most of the main competitions, however, are for distances of 50 miles or more. The endurance elite cover 100 miles within twenty-four hours. Multiday rides cover 50 to 60 miles per day, for up to five consecutive days.

To cross the finish line first is one part of the competition, but another important aspect is earning the best-conditioned award. It is earned by the horse with the highest marks at the veterinary stations positioned throughout the competition route. The best conditioned horse may or may not be the race's overall winner.

Gymkhanas: Gymkhanas are contests that are primarily composed of several speed events; sometimes fun classes, such as best costume or egg on a spoon race, are included. At a gymkhana, riders and horses compete in a variety of events, including barrel racing, keyhole racing, flag racing, pole bending, and stake racing. Each event involves a starting line, a task (such as looping around the barrels, dashing through poles, or carrying a stake from one bucket to another), a timer, and a finish line. There are officiates, but no judges, and therefore no favoritism. The rider who completes the task and comes in fastest is the winner. Junior riders are the primary participants in play days/gymkhanas.

Vaulting: This team sport is basically gymnastics performed on the back of a horse. Vaulting involves three main participants: the vaulter (or team of vaulters), the horse, and the longeur. The longeur is the person who longes the horse in a circle. He or she controls the horse, so that the vaulter is free to perform a variety of movements as the horse canters along. Most riding disciplines showcase the performance of the horse, but in vaulting, the horse acts as the stabilizing force and the underlying impulse of the vaulter's artistic performance.

TYPES OF CLASSES

Enlist the help of a knowledgeable horse friend, riding instructor, or trainer to find the show that is right for you. These individuals can also tell you what level is the one to choose for your first outing. Your best bet is to start with a schooling show, although smaller open shows or C-rated shows are fine as well. Find upcoming shows by checking the bulletin boards at your local tack store, searching the classifieds in your local horse newspapers, or calling community equestrian centers. Some shows can be found online as well—you just

Junior competitors concentrate on proper rider position and control, both of which are judged in hunt seat equitation.

need to be savvy about where to look. Once you find a show, call for more information, and ask for a show premium—also called a prize list. The premium will contain a class list, as well as other pertinent information on the show such as starting times, parking and stabling availability, fee schedules, and entry forms.

When you look over the show premium, pick out some classes that sound interesting to you. Classes vary from show to show and from one part of the country to the next. A show that has western and English events scheduled on the same day will have a more all-around purpose feel to it than a hunter and jumper or western-only show. Here are the types of classes you may come across, as well as an explanation of what will be expected of horse and rider. Remember to choose a type of class that suits the riding you are accustomed to doing at home.

English Classes

There are several classes in English riding. Here are the primary ones along with details of what is expected of competitors in each class.

English pleasure: This class has been developed for breeds such as Morgans and Arabians, shown in full bridles and flat Lane Fox saddles. The horse is judged on manners, quality, performance, presence, and apparent ability to give a good pleasure ride. English country pleasure is a variation of the English pleasure class and is designed for horses such as Saddlebreds and National Show Horses.

This rider jumps her horse over a low fence at an unrated show. Unrated and schooling shows are good for new riders and horses.

SHOWS AND CLASSES

Learning the following terms will help you to talk horse shows and classes.

■ **Circuit:** A series of horse shows within the competition season.

■ **Lane Fox saddle:** Also called a cutback, this is a flat saddle used in saddle seat classes.

■ **Gymkhana:** Competition featuring timed speed events and fun mounted games.

■ **Open show:** All breeds can compete.

■ **Schooling shows:** Shows designed to help riders and horses gain experience.

■ **Show premium:** Printed information about the horse show.

Equitation over fences: This event tests the rider's seat, hands, and ability to control and show a horse while jumping over fences. Like riders in the working hunter class, riders in equitation over fences are judged on their ability to establish an even hunting pace. Judges evaluate the methods used by the rider and his or her effectiveness in influencing the horse throughout the event.

Hunt seat equitation (on the flat): In this event, the riders are judged on their ability to control their horses while showing them to their best advantage and demonstrating correct rider position and smooth aids. Riders show at the walk, the trot, and the canter in both directions, and then they stand quietly in a lineup. Occasionally, the judge may ask riders to drop their stirrups or to demonstrate a collected trot or the hand gallop; often they will be asked to back their horses in the lineup. In hunt seat, the horse doesn't have to be the prettiest mover in the class—it is the rider who is being judged.

Hunter hack: Hunter hack is the transitional English class between hunter under saddle and hunter over fences classes. The horse is judged alongside the other competitors in the arena at a walk, trot, and canter. Then, one at a time, each rider must complete a course of two fences. The riders may also be asked to gallop in hand and halt individually.

Hunters over fences: In this class, the horse is judged on his jumping style, even pace, flow of strides, balance, and manners, together with his way of going over a course of at least eight fences. There is no rail work, and each competitor is judged on his or her performance over fences alone in the arena. A winning horse must be forward but controlled, with a beautiful arc over each jump and even forelegs snapped up in front. The obstacles the horse jumps simulate ones found in a hunting field: fences, brush, and walls. Depending on the division, obstacles can range in height from eighteen inches to thirty-nine inches.

Hunters under saddle: In this class, the horse is judged on his manners, obedience, and responsiveness. The horse must move freely, demonstrating the walk, trot, and canter in both directions with only light contact from the rider's hand. Stellar movement will win over judges, as they are looking for a horse that closely meets the ideal, both in conformation and way of going.

Western Classes

There are several classes in western riding. Here are the primary ones, along with details of what is expected of competitors in each class.

Western pleasure: In western pleasure, a horse is judged at a flat-footed, four-beat walk; at a free-moving, easy-riding, two-beat jog; and at a three-beat lope. These movements are demonstrated in both directions around the ring on a reasonably looped rein without undue restraint. The judge may call for extended gaits. The rider's cues must be very subtle, almost imperceptible to the judge. The horse should appear calm, consistent, and responsive. Winning horses in pleasure classes are shown with a level topline and a slower collection during all three gaits. When performed correctly, western pleasure is one of the more challenging classes for horse and rider.

Trail class: In this class, riders must demonstrate how calmly and easily their horses can negotiate the sorts of challenges found out on the trail. The horse is required to work over and through obstacles, including negotiating a gate, carrying objects, riding through simulated water, stepping over logs or through simulated brush, finding a way across a ditch without lunging or jumping, crossing a bridge, backing through obstacles, demonstrating side passing, and allowing the rider to mount and dismount from either side. The winning horse approaches all challenges matter-of-factly and completes the course in a timely manner.

Horsemanship: This is a western under saddle pattern class, meaning that the horse and rider must perform a prescribed compulsory figure. The class is judged on the correctness of the rider's seat, hands, and feet; the finesse in riding; and precision as the competitor completes a series of maneuvers in a thirty- to sixty-second pattern. The pattern is posted on the day of the event, and competitors must note all instructions. Judges can ask for circles, spins, and other such figures and movements.

Western riding: This is a pattern class, but it is different from the horsemanship classes. It is designed to assess the skills and characteristics of a horse performing in a simulated ranch situation. Each pattern requires a horse to complete a series of lead changes at the lope around several different cone setups. Horses should be sensible, well-mannered, and free-moving. Tasks should be completed with reasonable speed. Horses are judged on the riding quality of their gaits and flying lead changes. Judges also observe the horse's response to the rider, manners, and disposition over the prescribed pattern.

Showmanship: This is an in-hand pattern class in which the handler works with the horse to execute a showmanship pattern selected by the judge. This class usually involves a series of movements, consisting of walking, jogging, stopping, backing, turning, and setting up. The top horses and exhibitors complete the pattern without the handler physically touching the horse—rather, the commands are given only through a lead shank.

Halter: Halter horse classes are judged specifically on conformation and movement at a trot. Halter is an in-hand class, but it is judged solely on how well the horse is put together. Halter horses are impeccably groomed and in flawless condition.

Handlers display their showmanship skills as a judge steps forward to make his decision during a western in-hand competition.

Chapter 23

Horse and Rider Preparation

Riders and their horses need to be turned out and equipped correctly to show. Like any sport, riding has its own uniform. There are rules for outfitting yourself and your horse with the right equipment, and then there are expectations for the correct fashion, which affect how riders see one another. At a schooling show, riders can get away with hand-me-downs or borrowed clothes—some shows may not even require full show attire—but those who are serious about competing should get the best equipment and clothing they can afford.

GROOMING AND TACK FOR THE SHOW

Now that you have entered classes for a show, you need to get ready in the days preceding the competition. Take the time to get your horse as slick and polished as possible for the show.

Clip and wash your horse a day or two before. Allow him to dry completely, using a clean horse sheet (a lightweight blanket) to help him stay warm as he dries. Many horses like to roll after they take a bath, so make sure you return the horse to a clean stall before letting him loose. Make sure to pull the mane to show length. At larger shows, riders braid (for English classes) or band (for western classes) manes, but for smaller shows or schooling shows, doing so is not required. See the grooming section starting on page 236 for specifics on mane braiding and show preparation.

Show grooming is an art unto itself. Many riders earn extra money as professional show grooms, and it is easy to see why. Preparing a horse for classes takes experience and patience—and a little bit of flair. Here are some of the basic classes and explanations of how the horse should be presented in them.

Western Rail Classes

These classes include pleasure, equitation, riding, horsemanship, and trail. Horses are groomed meticulously for the show ring—with much preparation for the show done year-round. The horses' coats are fine, tails are long and thick, and manes are pulled short and even. At the higher class levels, horses' manes are banded neatly. Use electric clippers to remove extra hair from your horse's legs from the knees to coronet bands. Clip the head, underneath the throatlatch, and around the muzzle; remove or trim long eyelashes, and shell out the ears—trim the hair from the edges and the insides of the ears. Scrub all white marks so that they are bright. The bridle path should be just over an ear's length long, and the mane should be pulled evenly to about four inches in length, or just long enough to lie flat. Scrub the hooves and apply hoof polish the day of show. Some competitors sand the hooves before polishing, but that removes the protective periople layer, causing damage over time.

If you are banding the mane, divide it into even quarter-inch sections, and secure them with small

Bathe your horse at least a day before a show. Clipping should also be done at least twenty-four hours before.

rubber braiding bands. The forelock can be braided and tucked under the browband or just left flowing. Wipe the muzzle, nostrils, face, and ears clean, and apply a thin sheen of baby oil to the eye and muzzle area.

The tail should be long and full, with a blunt, straight cut at the bottom. The tail should be long enough to reach the fetlocks. If your horse's tail is short, a good option is to tie an extension made of horsehair into the existing tail, although check your showing discipline's rule book to make sure this is allowed.

Most riders have a very ornate saddle, with many silver accents, that they use just for competition. The show bridle usually has a headstall that matches the saddle, often featuring ear loops instead of a browband and no noseband. Many riders use a leather or neoprene cinch. Under the saddle, the horse wears a Navajo blanket that coordinates with the rider's outfit. No leg wear is permitted.

Western Performance Horses

Western performance classes include reining, cutting, and working cow horse. The horses in these classes are not the hothouse flowers their show-ring compatriots

At the 2006 WEG reining competition, Hang Ten Surprize's mane is left natural (not banded) and the tail long and full.

appear to be, yet they are still presented impeccably. Clip and boot up the horse's legs, scrubbing all white marks so they are bright. Clip the head, underneath the throatlatch, and around the muzzle; remove or trim long eyelashes; and shell out the ears. Performance horses are shown with natural manes—the longer, the better. A bridle path is optional, but if you go with one, it should be at least six inches long. Wipe the muzzle, nostrils, face, and ears clean. Make sure the tail is long and full—at least long enough to reach the fetlocks—with a blunt, straight cut at the bottom. Scrub the hooves, and apply hoof polish the day of show. Again, a number of competitors sand the hooves down before polishing, but that removes the protective periople layer, causing damage over time. Use a discipline-specific saddle (reining, working horse, cutting), which will normally be less ornate than a pleasure saddle. Horses are permitted to wear protective leg wear. Reining horses should wear sport medicine boots on the front legs and skid boots on their back legs.

Hunters

Hunters have a robust, healthy appearance. The look is elegant and traditional. Closely clip the legs and hoof area, and add hoof oil just before going into the class. Clip the head, beneath the throatlatch, and around the muzzle; and shell out the ears. Make sure the bridle path is only about two and a half inches in length. Pull the mane to about four inches long and style on the right side of the neck. For larger shows, such as A-rated events, braid the mane into at least twenty-five hunter-style braids. For formal shows, the tail can be braided as well, but for most events, it is left natural and flowing.

The tack for hunters is conservative. No special show saddle is required; you may use your regular close-contact or jumping saddle. Place a shaped fleece saddle pad under the saddle; both saddle and pad should be impeccably clean. All girths should be leather, though some riders will place fleece covers over the girth. Horses may wear a standing martingale in hunters over fences. The bridle is simple and conservative, with only modest raised stitching on the nosebands and browbands—nothing else. Dropped, figure 8, flash, or any style other than a plain cavesson noseband is not permitted. No leg wear is permitted.

The Jumper Division

Although considered an English discipline, jumping is less conservative than the hunter division. A jumper can be groomed like a hunter, but there are a few differences. Normally, the mane is unbraided and kept a little longer and thicker than a hunter's mane, about five inches in length. If you wish to, braid the mane in the dressage style or with button braids. Leave the

This horse has a correctly braided mane for competing in a hunter competition.

Unlike the mane of the hunter above, the mane of show jumper Promise Me is unbraided.

tail either natural or banged (tail cut straight across at the bottom). The tail should hang at the cannon bone's midsection. Tails are generally not pulled at the dock (the horse's tailbone). The jumper wears an English jumping saddle and a girth of any material and color. Saddle pads can be any material and may be square or shaped to the saddle's outline. The bridle is the same color as the saddle, but a variety of nosebands are permitted, including figure 8, dropped, flash, and plain. Running and standing martingales are permitted. Horses with caulks (studs screwed into bored holes in the shoes to give the horse traction) will often wear belly guards for protection. A breastplate is sometimes worn to keep the saddle from slipping. Leg wear for horses, such as leather boots, is recommended.

Dressage Horses

Dressage horses are clipped and bathed similarly to the show hunter, with a few exceptions. The ears should be trimmed but not shelled out completely. Tidy up just the outer edges, leaving hair inside, so that insects do not irritate the horse and distract him from his performance (usually level off protruding tufts). Pull the mane to about four inches, and style it in dressage braids at least an inch apart. For breeds with long manes, leave the mane long, but use a French braid. Bang the tail at the bottom, usually just above the fetlock, and pull the tail at the dock. Show the horse in dressage tack. The saddle pad for dressage is rectangular in shape and is almost always white, although occasionally riders will be seen with black pads on gray horses. The bridle is black and is often accented with a jeweled browband and a white-piped noseband. The dressage bridle usually has either a plain or flash noseband. No leg wear is permitted.

RIDER APPAREL FOR THE SHOW

There's a saying that goes: life is a show—dress for it. When performing, you must look the part. At schooling or local shows, you can get away with simply pulling together an affordable outfit, but at rated shows, you must take the plunge and splurge on the latest, more expensive styles.

English Riders

English show apparel includes a hunt coat (also called a show jacket), and a show shirt (sometimes referred to by its old-school name, ratcatcher). The coat should

Turned out in proper style, Edward Gal and Moorlands Totilas compete in the European Dressage Championships in 2009.

This dressage rider is wearing a traditional four-button coat, a stock tie, and a safety helmet.

Hunter riders wear conservatively colored hunt coats in the show ring, such as this gray jacket.

be a dark color, such as navy or hunter green, although browns, olives, and charcoal grays are also popular. Teal coats are currently out of style, but other shades of blue are still trendy. The jacket should have a three-button front and two vents in the back that gracefully drape across the cantle of the saddle. The sleeves should be long, so your wrists remain covered when in the riding position.

The show shirt can display a little more individuality, with many style and choker embellishment options. Some shirts are plain white, while others have bold stripes or soft patterns. Men wear cotton dress shirts with a button-down collar and conservative tie. English boots are always black and tall, fitting snugly along the calf and coming up to the back of the knee so that they crease ever so slightly when the leg is bent. Hunt-seat riders wear field-style boots that lace up at the top of the

foot. Breeches are normally a beige color and feature a side-zip style with two slash pockets in front. However, various versions of beige styles are also suitable.

Riders should outfit themselves with an ASTM/SEI-rated (American Society for Testing and Materials/Safety Equipment Institute) safety helmet that has both a harness and a chin strap. Add gloves as a final touch. Jewelry should be kept to a minimum, and hair should be neat and kept above the collar (women should put their hair up under the helmet or in a hairnet).

Dressage riders wear a black jacket featuring a four-button front and a single vent in the back. They will wear a white show shirt, a white stock tie, white breeches, and tall dress boots. White gloves, traditional in dressage, should always be worn. Jumper riders are the most casual of the English set. They often wear polo shirts instead of jackets and show shirts.

Show Checklist

Have your show checklist handy to help you remember items often left behind. Here is a handy basic list; yours should be specific to your riding events.

RIDER EQUIPMENT:
- Belt
- Boots
- Chaps
- Gloves
- Hair accessories/ hairnet
- Hat or helmet
- Jeans or breeches
- Mirror and brush
- Money
- Safety pins
- Shirt
- Top or jacket

EQUIPMENT AND TACK:
- Bridle
- Chain shank
- Extra lead rope
- Extra towels
- Extra feed/grain
- Fly spray
- Girth
- Grooming kit (brushes, hoof pick, mane and tail brush)
- Hoof oil
- Hay net and hay
- Longeline and longe whip
- Manure fork/bucket
- Saddle
- Saddle pad
- Sheen spray
- Water bucket

Western Riders

Western riders have more options when it comes to riding apparel. Riders can opt for colors and styles that complement their own particular style and their horses' coat colors. At small schooling shows, you'll see a lot of western riders in casual dress, with crisp jeans, a long-sleeved western shirt, boots, a belt, and a western hat. While most riding clothes can be bought off-the-rack for smaller shows; for more advanced shows, competitive riders usually go for custom-made clothing.

Men have a simple clothing style in western riding. They wear crisp, starched shirts in earthy colors, jewel tones, or plaids. Women, however, usually try to put together outfits that set them apart from their competitors in the show ring.

For many years, cropped jackets and vests worn with tuxedo shirts were the favored uniform for rail classes. Anything with glitter and sparkle to catch the judge's eye made it into the show ring. Now, however, vests paired with bodysuit-style turtlenecks or tuxedo shirts are losing popularity in favor of vests with blouses and tail-out tunics. While there's still a bit of glitter to capture the judge's attention, it's done to dazzle, not blind. Trends continue to change with each show season, so riders should choose their outfits carefully. As they say, what goes around comes around, and no doubt, fashion in the show pen will come full circle.

As far as western pants are concerned, men tend to wear jeans in blue, black, or a neutral earth tone. These are starched and creased. Women have the same choices, but most choose show pants, which are similar to a polyester trouser. Underneath matching-color chaps, show pants give the rider a sleek look.

Western riders should spend as much as they can on two items: a pair of chaps and a hat. An ill-fitting or cheap hat will scream amateur. Hat quality is measured by a series of Xs. Most felt hats that are a blend of rabbit and chinchilla fur and wool are either 3X or 5X, while fur blends with beaver fur tend to be 10X or 20X. Pure beaver fur hats usually bear the symbol of

Children's western riding dress is much the same as that of the adults—hat, shirt, and chaps are key.

100X or higher. The shape of the hat varies by riding discipline and region.

Fur felt hats in light colors, such as silverbelly, sand, and buckskin, are popular, as is basic black. Finely woven straw hats (not the kind you see at the fair for five dollars) are also popular summer show hats. The hat is blocked—shaped along the crown and brim—and trimmed to frame the rider's face.

Chaps—suede, smooth leather, or artificial suede—need to match the rest of the rider's outfit and must fit properly. They must be high at the waist, snug through the leg, and cut long enough to drape across the top of the boot and cover the heel when the rider is mounted. The front should neither gape nor form a pouch.

As far as accessories go, women's neckwear rules have relaxed in recent years, so many female riders wear striking silver accents at the throat instead of a tie or a bolo. Jewelry can be pronounced, but should remain tasteful.

This closeup shows that the young rider's chaps correctly drape across the top of her boot and cover its heel.

BRAIDING EQUIPMENT

For braiding your horse's mane, you should have most of the items listed below (a few are optional):

- Yarn to match the color of the mane (cut to length by winding around thumb and elbow and cut top and bottom of the loop to make the pieces).
- Sponge and bucket to wet the mane (if desired)
- Braiding product (if desired)
- Step stool
- Pulling comb
- Pull-through or latch hook
- Small scissors (sewing scissors with tiny, pointed blades work the best)
- Braid aid (if desired)
- Hair clip or clothes pin
- Seam ripper to remove the yarn

A belt with a silver trophy buckle often finishes off the outfit (they can be purchased until you actually win one). Hair should be neat and stylish (with no hint of a "rodeo queen" hairdo). Most women pull their hair back into a tight bun. Gloves are optional, but currently the trend is to forego gloves in classes.

BRAIDING PRIMER

Braiding your horse's mane and tail shows respect for the judge and the competition. It is also a way of telling the judge that you have paid attention to every detail and have taken the time to go the extra mile to make your horse look as good as he can.

Step 1: Mane Preparation. The mane should be pulled to about six inches. Shampoo the mane a few days before you braid, but don't use conditioner or coat polish, otherwise the yarn will just slip out of the mane.

Step 2: Making Plaits. Assemble your equipment and secure your horse. Thread a hank of yarn (cut to about ten to twelve inches) through your belt loop or,

This style of braid includes a little bobble at the top of the plait. It is a variation on the traditional hunter braid and is more commonly seen in the dressage world.

if he doesn't mind, through the top ring of your horse's halter. Begin by either spraying water or a braiding product onto the mane. To make the braids equal in thickness use a braid aid or a comb and section off a small ponytail of mane. Make sure that the part is very straight. Once you've made your part, use a clothespin or hair clip to hold back the rest of the mane.

Grasp the ponytail and divide it into three sections. Always start braiding with the same strand on each braid. For instance, if you use the right strand of your ponytail first, continue using the right strand of each following ponytail. This makes the braids more uniform. As you plait the ponytail, clamp a thumb in the middle of each, turning to keep the pressure of the plait tight and even.

When you are halfway down the plait, hold the middle of the braid with one thumb and place a piece of yarn, folded in half, underneath your thumb. Continue plaiting, but this time incorporate the yarn pieces into the right and left strands of the plait. Remember to keep the pressure up. When you can't plait any longer, clamp off the braid with your thumb and forefinger. Take the two tails of the yarn, wind them around the end twice, and pass them through to make a knot. Tie the braid off again. You should have a tight braid that lies flat with about five inches of yarn hanging down. Continue plaiting the rest of the mane.

Step 3: Tying Off. Using your pull-through, begin pulling up the braids. Insert the pull-through into the center of the braid from the top to the base of the crest. Insert the yarn through the loop of the pull-through and pull the yarn back up through the top of the braid. Release the yarn, and with your hand, gently pull the yarn ends so that the knotted end of the plait is hidden within the base of the braid. Make sure that none of the prickly bits are sticking out of the top. Press the plait down flat. Cross the yarn ends behind the plait and knot them. Bring the pieces around to the front of the plait and knot it in the middle of the braid; clip off the ends of the yarn.

If you like, before you tie off the knot, you can push the plait up so that a little bobble of plait extends above the neck. Tie the knot just under this bobble. The trick is to keep all the bobbles even as you pull the other plaits up. This type of braiding is more commonly seen in the dressage world. Hunter braids usually lie flat against the neck. Hunters are shown with braids on the right side of the neck only. This is an

SHOW PREPARATION

Learning the following terms will help you talk about horse show preparation.

■ **Booting up:** To remove extra hair from the horse's knees to coronet bands.

■ **Bridle path:** A small shaved section of the mane, close to the poll, where the top of the bridle sits.

■ **Dock:** The top of the horse's tailbone.

■ **Mane banding:** Creating tiny ponytails in a western pleasure horse's mane.

■ **Periople layer:** The protective outer covering of the hoof.

■ **Shell out:** To clip out all the hair from inside the horse's ears.

old tradition that hails from the hunting fields. A rider mounts on the left, so braids must be on the right to avoid getting his or her hunting garb tangled in the braids. Dressage horses can be braided either on the right or left. It's the rider's choice.

You can repeat the same braiding regime on the forelock, or, if you're feeling creative, you can French braid the forelock. Begin by sectioning the top of the forelock into two strands. As you plait, gather up the third strand from each side of the forelock as you go down the forehead. As you reach the dangling end of the forelock, return to braiding normally. Tie the knot off, pull it through the forelock, and tie it off again as you did with the mane. Another nice finish is to forgo tying the plait off. Instead, pull the braid up through the forelock and bring the tail of the plait completely through. Fish that end back through the braid with your pull-through (from the bottom of the braid). Continue to pull the tail through the braid until the tail end is tucked completely under the plait.

Chapter 24

At the Show

Competing is not as easy as it looks; there are subtle nuances to performing. It may seem that all riders have to do is show up, ride around the arena for a few minutes, line up, and collect their ribbons. But taking a horse out of familiar territory into a foreign environment and expecting him to perform flawlessly requires practice and a proper mind-set.

Horse shows can be confusing for a novice rider. There is a whole litany of things to keep in mind, among them class times, warm-up etiquette, show-ring strategy, and ways to approach the judge. It's not just the ability to execute the perfect lope or a lovely, even trot that wins the prize—a judge can easily peg riders as beginners or as seasoned campaigners by watching how they handle themselves in the heat of the moment. The choices made in the warm-up arena, entering the ring, and during the class help give you and your horse an even chance in the competition.

ARRIVING

Horses tend to be a little apprehensive when they are in unfamiliar surroundings. The trailer ride is enough to tell the horse something is up, and he may arrive at the show grounds feeling nervous and high-spirited. These feelings usually go away with experience, but it is important to be prepared and develop the skills necessary to help your horse keep a calm head during the show's commotion.

Once you arrive at the show grounds, your horse will usually be eager to get off the trailer. Remember, he is a very strong animal in unfamiliar surroundings; he has the ability to be dangerous, even if he doesn't mean to be. Attach a stud chain to the halter by running the chain through the near (left) ring of the halter's noseband, from the outside to the inside, looping it once around the noseband, and running it out through the off (right) ring of the halter's noseband. This will give you a little more control over the horse and make him more respectful.

If your horse is behaving himself, walk him around the grounds to let him see, smell, and adjust to his new environment. He may whinny, and other horses may respond to him.

When you think he has satisfied his curiosity and seems comfortable, take him to the stable. If you are only there for one day, take him back to the trailer instead, tie him, and allow him to eat hay from a full hay net (tied high enough so he cannot get a leg through).

Now is the time to check in at the show office. The payment procedure varies by type of show—some will require that the entry fee be sent in ahead of time. Pick up your assigned number, an updated class list, and any other information you may need (such as contact information for the show's photographer or videographer). Then head back to your trailer, and find out when your first class will be.

WARMING UP

At the bigger shows, there are usually two warm-up arenas: one to longe in and one to ride in. At smaller shows, however, space is often an issue, and there may be only one warm-up arena, with the rail reserved for riding and the middle reserved for longeing. This can be tricky because the riders on the rail need to watch out for those horses longeing in the middle. Invariably, you will get somebody who tries to ride between the horse and the longeur. You must realize that the handler has little control of the horse on the end of the longe-line; the horse can buck or kick, so give him at least a horse's length of space. Another point to bear in mind is that the farther away you are from a longeing horse, the more your own horse will listen to you.

Remember that the show is not a place to train your horse. If your horse doesn't know his work, you're not going to teach him in the warm-up arena. So treat the warm-up as just that: a place to let your horse loosen his muscles and to get him focused and ready to work.

Some horses need to keep walking around in the warm-up arena to remain calm and relaxed; however, you do need to leave the warm-up arena in time for last-minute grooming before the class. Wipe down your horse and his mouth, clean his hooves out again, and comb out his tail. Plan to do this close to the show arena, around the area of the ingate.

Rules of the Warm-Up Arena

You need to know the rules and proper etiquette for the warm-up arena. Here are some important points to bear in mind as you warm up.

Passing: Pass left shoulder to left shoulder. However, keep your eyes up. There may be a rider who may not be able to pass on the left side, or perhaps there is a trainer on a really green horse that cannot be maneuvered in time. If you keep your head up and make eye contact with the other riders, you will be able to tell which way the traffic is moving.

A call for "rail": Someone may call out "rail, please" in an effort to get you to give way. You don't have to move out of the way, but most likely that rider is having difficulty and needs space, so try to respect the request. A good rule of thumb is to always give the right-of-way to the inexperienced person, the young horse, or the child.

Slowpokes: Slowpokes have the right-of-way on the rail. Just like on the freeway, faster riders should

Show competitors wait in the warm-up ring, also known as the collecting ring, preparing for their upcoming class.

Riders warm up over fences at a hunter and jumper event. When warming up, be courteous and watch for the inexperienced.

move into the far lane to pass. Faster riders do not get to monopolize the rail—much as they would like to. The slower horse stays on the rail and keeps his position. However, rail riders should remain aware of those riders attempting to pass and make sure not to swing out into the faster traffic.

Trainers: If you see a trainer working a group of horses in a show pattern in the center of the arena, give them a little consideration and don't ride into their pattern. But don't feel intimidated by the trainers and their riders either. You have just as much right to warm up your horse as any other rider or trainer does. Expect that more experienced and established trainers may try to push you around and bully you out of their way. If they do, simply stand your ground and let them know you exist. Hold your space and say "excuse me."

The Call

In some shows, such as dressage shows, a ring steward keeps competitors aware of their class times. This may not always be the case in your particular discipline. Stay close to the show environment so you can hear your number called. Don't go wandering off to the trailer two minutes before your class starts. If you miss the call for your class, and they've closed the gate by the time you get there, you're out. Some

THE THREE-CLASS STRATEGY

Always pay attention to the show's announcer, who can serve as your guide for when to tack up, mount up, and warm up. He or she will not tell you these things directly, but by using a three-class strategy, you should have ample time to get ready. Three classes before yours, you should have your horse groomed and saddled. Two classes before, you should be mounted up and in the warm-up ring, using the twenty minutes or so during those classes to warm up your horse. By the time the class before yours is called, you should be mentally focusing on your task at hand.

If your horse is really being fresh, you may have to longe him to shed some of his energy. This will add time to the three-class strategy. Plan instead to start preparing for your first class at least an hour early.

smaller shows will give the class a last-minute call and sixty seconds to enter the arena. But if you're late, you're making a bad impression on the judge.

If you're showing in another arena or if you have to make a tack change, that's a different matter. You may be allowed to be late. You must stay in contact

DON'T BES

When you are competing, don't be:

Late: If show nerves tend to paralyze you, ask a friend to keep you on track.

Unprepared: Don't search for your spurs at the last minute or try to learn your pattern just before a class.

Rude: Be polite to other exhibitors, volunteers, and office staff. Discuss issues calmly.

An arena hog: Respect everyone's place in the arena, but don't give in to more experienced riders or trainers that may try to bully you out of their way.

A poor sport: Judging is one person's opinion; it may not mesh with yours. At day's end, you can talk to the judge about what he or she thinks you should work on.

with the show staff, however, to let them know what you need. Try to be quick with your changes because nobody wants the show held up.

COMPETING

A grand entrance is everything for rail classes. It's the judge's first impression of you, so put your best foot forward. Try to get in early to find a good spot on the rail, but keep in mind that it may not always be possible. When the class is called, there's a mad dash to the arena, and everyone bunches up to get into the gate. If you get caught in the melee, don't worry; you can still get a good space. Simply pause at the gate, and let the person in front of you go ahead before you enter.

To get the best position, it also helps to know your competition. If you know that your horse has a long stride and you have a slow-moving horse in front of you, for example, you may not want to be directly behind him. Don't put yourself in a bind right away by having to pass right from the start. The ideal situation is to get a spot on the rail and keep it. You're going to be noticed more if you're on the rail and in the clear, and your horse will also be steadier if you're away from the other horses.

If you get surrounded by a pack of horses, ride deep into the next corner. This makes your route longer and allows the other riders to pass you more easily on the inside. If you've been stuck in the midst of a group the entire first half of the competition, take your time when reserving, and you'll break free of the group.

Show-Ring Strategies

The best position is to be alone on the rail. During gait transitions, if you have the extra room, when asked to lope or canter, you can wait a moment if you need more space, or take more time to reverse. Try not to pass in western classes, or you'll make your horse look fast and possibly not well controlled. English riders can pass without worry, but they still need to keep a solo

Western pleasure riders wait in the lineup for the judge's results. Give yourself plenty of time to reach your class on time.

position on the rail. If someone is having trouble, you can pull off the rail and pass. A judge will not fault you for pulling off the rail to get away from a fracas.

Watch a class prior to yours to tell where the judge tends to stand in the arena. Most riders will use the time the judge is turned to make subtle corrections. Keep an eye on the judge, but don't stare. If you've determined that he or she doesn't watch a certain area, use that space to prepare for the area he or she does watch.

Acknowledge the judge when you enter the arena, give him a quick smile, and move off with an attitude that you have a great horse and you know what you are doing. The judge will see your confidence. Don't use a phony smile, but do have a confident expression.

If you make a mistake, move on as if it was what you intended. Don't give up; fellow competitors may have made the same or worse mistakes. React to gait change requests as quickly as possible, but make sure the horse in front is changing at the same time.

During the final lineup of the class, the judge will ask certain riders to cue their horses to back up. This shows the rider's control over the horse. So make sure you position yourself in the lineup with enough space on either side of you, so if the judge asks, you can back up without banging into other competitors. (If there's been a horse in the class who has been misbehaving, try not to stand next to him.) When it comes time, glance behind you and back up smooth and steady.

Judges look at the entire picture: a well-groomed horse in great condition, wearing clean, quality tack; an English rider in up-to-date, conservative hunter clothing; or a western rider in a well-shaped hat and fitted chaps that coordinate with the horse's color. You don't have to spend a lot of money to look good. Well-tailored, neatly pressed, stylish clothing is most important. Outdated clothing shouts to a judge that this is your first horse show. See what's being worn at shows before you purchase your outfit. One stylish, coordinated look is all you need for the show season.

The Judge's Decision

You will know the judge is done judging when he or she turns in the scorecard. If you're in any doubt, hold your position, and don't start talking to the person next to you. The judge will be walking down the line making a final decision, and you don't want to be caught slouching.

If your name is called for a ribbon, step out of the lineup, but let any rider who placed higher than you

International Competition

Eight riding disciplines shine under the international spotlight: eventing, vaulting, dressage, para-dressage, show jumping, combined driving, reining, and endurance. Although the Fédération Equestre Internationale (FEI) now governs all of these sports, in the past each discipline held its own championship in the previous winner's country. There was no chance for everyone to compete together, not even in the Olympics, which includes only show jumping, eventing, and dressage. In 1983 it was proposed that all world championships take place under one roof. Every four years, the event would be held in horse-centric cities throughout the world, two years before the next summer Olympics. In 1990, former FEI president Prince Philip (husband of Queen Elizabeth II) organized the first World Equestrian Games (WEG) in Stockholm. Since then, the WEG has been held since in The Hague; Rome; Jerez, Spain; Aachen, Germany; and Lexington, Kentucky.

Many countries have their own events, too; Malaysia, for example, holds the semi-annual Premier Cup Series (*above: a June 2009 jumping competitor*).

accept his or her ribbon first. Once you've collected your ribbon, continue riding out of the gate.

If you didn't win, be a good loser. You should never discipline or punish your horse while you are in the arena. After the competition, ask your trainer or a friend for feedback about why you didn't do well. Go home and work on the areas you need to improve; chances are you'll do better next time. After all, that's what horse shows are all about—discovering where your weaknesses, as well as your strengths, lie. If you and your horse did the very best you could, be happy with your performance. Awards are sometimes arbitrary and subjective.

AFTER THE SHOW

Once your classes are over, you still have a few more duties before the day is through. Take your horse's tack off, store it, and give your horse a sponge down (or if the grounds have wash racks available, a rinse from a hose). Scrape him dry and let him unwind while you quietly pack up so as not to distract the other competitors. Deposit any manure or trash in the proper places. Close out your check at the show office. Load your horse calmly, and leave the show grounds slowly, so you don't spook other horses or create large clouds of dust. As you drive home, you can recall the day's events with satisfaction and contentment.

Horse talk

COMPETITION

Learning the following terms will help you talk about horse competitions.

■ **Barn blind:** Someone who sees only the best in his or her horse and fails to see the flaws.

■ **Final lineup:** Riders wait in a line as the judge makes the final decision.

■ **Makeup ring:** The arena or marked space where riders gather prior to their class.

■ **"Rail, please":** A request for the right-of-way.

■ **Ring steward:** The competition official who runs the competition arena.

■ **Warm-up arena:** The arena where riders warm up or longe their horses before competition.

Two riders from Culver Academies in Indiana clean and replace their tack after a show.

The winner of the Supreme Hack Championship, Hickstead, 2010, wears the red-white-blue championship sash.

Breeding Mares and Raising Foals

Section 10

Breeding your own mare certainly sounds like an excellent idea. Some owners dream of getting the alchemy just right so they can produce the horse that will take them to the very height of competition. Breeding also seems like a much more affordable option than purchasing a trained show horse.

But there is a great deal you need to think about before pursuing a breeding program, including the question of whether your mare is a good prospect for producing superior offspring. Unfortunately, there exists a surplus of average horses in the world—the by-products of some owners' determination to breed their mares before carefully considering everything that is involved in breeding.

Chapter 25

Breeding

Breeding your mare should be considered carefully. Yes, you get to pass along your beloved horse's genes to another generation, and foals are indeed an appealing prospect. But will those foals turn out to be superior horses, worth all the time, money, and commitment involved in raising them? A great deal of preparation, research, and prenatal care must take place before a baby is born.

DECIDING WHETHER TO BREED

The first step in the process is considering whether your mare is worth breeding at all. A mare with poor conformation, a genetic predisposition to particular diseases, or a terrible temperament is more than likely to pass these faults on to her foal. Do not expect a stallion with good genes to correct the shortcomings of your horse. Unless you have no problem with getting a replica of your current horse, you should not attempt to breed her.

A horse worthy of breeding should have a calm and willing temperament, sound conformation (no physical weaknesses), and the athletic ability for the discipline you prefer.

If your horse seems to be a good candidate for breeding, you must next select the right sire. The stallion needs to have not only the right pedigree but also a good performance record. He should also have the physical traits that will be a good match for

your mare. Understanding what those traits are will take time on your part. Consult trustworthy sources, including your veterinarian, experienced breeders, and knowledgeable friends.

When deciding whether to breed your mare, you must also factor in your ability to raise a foal. Do you have the room on your property necessary for a growing foal to run and get strong? If you board your horse, do you have permission from the stable to raise a baby on the premises? Do you have the funds necessary to board your foal at a farm or ranch that will care for a baby while he is growing?

Anyone who wishes to raise a foal from birth also needs to have endless amounts of patience. A foal is not mature enough to carry a rider until about three years of age. Waiting for the foal to come of age is one type of patience a handler must have.

Another is the patience to deal with the foal's young brain as he learns what is expected of him. And if you're not the one doing the training, you'll have to factor in the expenses of hiring a trainer to start your colt or filly under saddle.

It may turn out that your affordable alternative to buying a quality show horse will be just as expensive as purchasing a horse already trained and ready for the show ring. (And with the latter, you won't have a waiting period to begin riding.)

However, if you have done your research carefully and thoroughly, and you have concluded that your

Foals, like this Morgan, need plenty of room to grow and mature.

mare and you as a handler are ideal candidates for raising a foal, then congratulations—the hard work can now commence.

CHOOSING YOUR STALLION

Make a list of the pros and cons of your mare—her best attributes as well as any shortcomings. Using this list, search for a stallion with attributes and traits that will balance your mare's positive characteristics and minimize her faults. Remember, you can't correct her faults, but you can minimize them. If your mare has a short back, for instance, don't breed her to a stud with a short back; you are most likely going to end up with a short-backed foal. Look instead for a stallion with good conformation over his topline to minimize the fact that your mare is short backed.

If your mare is registered with a breed association, you may want to breed her to a stallion of the same breed to obtain a purebred foal. Or maybe you are more interested in breeding for height, performance, or color. Whatever you decide, you can explore a number of different avenues to find the ideal stallion.

They include, among other sources, breed association magazines, national publications, breeding farm listings, and the Internet. You should also speak with your veterinarian or local trusted trainer for advice. There are millions of stallions worldwide, and you should consider as many as possible when choosing your mare's mate.

Do you know how to pick the right stallion? It takes some know-how to get past the slick ads in

STALLION QUALITIES

When selecting your ideal stallion, keep in mind these five elements: conformation, gaits and movement, performance ability, attitude, and mature offspring. Then think of your own mare, and see how she fits into the equation. Determine what your ultimate goal is for your foal. What do you want to do with the foal? How do you want him to look and behave? If the stud you are considering seems likely to give you what you desire, then go with him.

This Paint stallion is beautiful, but be sure to evaluate temperament and performance as well as looks.

breed magazines and those on colorful websites. Keep in mind that stallion ads are made to sell the stud's services. In this type of advertisement, the stallion is presented at his best (or better than his best), perfectly groomed, touched up, and posed. There is often a description that lists his many accomplishments. Take everything you see and read with a grain of salt. Investigate the stallion and breeder elsewhere to discover what kinds of reputations they have. If possible, see the stallion in person to get a better feel for what he is like.

Bear in mind that the stallion's temperament is crucial. A sire that is too aggressive or ill-tempered can make for a bad-tempered baby. Take a look at the stallion's existing offspring, his get, to see if they have any undesirable behavioral traits. Never pick a stallion because he's "at a farm nearby" or because "there's a great deal on his stud fee." Remember, you are creating a life, and you should not skimp when it comes to choosing the right sire.

PREPARING FOR BREEDING

You have made up your mind to breed your mare, and you have an eye on the stallion you want to use. Those are merely the first steps, however. There is much more to contemplate, decide upon, and set in place before boy meets girl. At times it will feel overwhelming. Don't worry: with forethought and planning, you can ensure that the process will go as smoothly as possible. There will be hitches along the way, of course, but as long as you're prepared, you should be able to deal with them.

BREEDING

If you plan to travel down the breeding path, here are some terms to help you talk horse breeding.

■ **Anestrous:** The time when the mare is not cycling, the ovaries are not active, usually in late fall and winter, although this can vary due to location.

■ **Caslick's procedure:** This veterinarian procedure sutures the lips of the vulva together after the mare has been bred so that no contaminants can enter the uterus.

■ **Diestrus:** The time in her cycle when the mare is not fertile. She will not be receptive to the stallion. This time usually lasts fourteen to sixteen days and occurs twenty-four to forty-eight hours after ovulation.

■ **Estrous cycle:** Usually a twenty-one-day cycle, it begins the first day of estrus and ends on the first day of the following estrus period. Also called the heat cycle.

■ **Estrus:** The fertile time of a mare's cycle. It lasts five days and coincides with the egg leaving the ovary. Since the egg is developing in the beginning of estrus, the best time to breed is mid- to late estrus.

■ **Get:** Offspring.

■ **Inflammatory cytokines:** A function of the immune system that helps protect the body from foreign substances. In this case the stallion's semen.

Vaccinations

Before breeding, the mare will need the yearly core vaccines: rabies, tetanus, West Nile virus, and eastern/western encephalomyelitis, plus any risk-based vaccinations recommended in your region. For instance, Kentucky breeders vaccinate against botulism, which is caused by toxins produced by soil-borne bacteria called

A clinician performs an equine blood test, screening for diseases that may be harmful to a mare or her foal.

Clostridium botulinum. As a precaution, many people will also give booster vaccines to their mares prior to the breeding season. Mares should also have a Coggins test, which screens for equine infectious anemia (EIA). Always consult your veterinarian for advice on vaccinations prior to breeding. (*See section 6 for more on health care.*)

Nutrition

Many veterinarians suggest that mares go into the breeding process on an upward plane of nutrition. This means that the mare should start slightly thinner than her ideal weight and increase in weight through her pregnancy rather than starting on the heavier side, as fat mares are more difficult to get in foal. Overweight mares don't want to move about as much, and movement is important with currently barren mares because it helps them clear the uterus of fluid and also decrease inflammatory cytokines (a function of the immune system that protects against foreign substances, in this case stallion semen). It is very beneficial to have your mare out moving around as opposed

BEING IN SEASON

Most mares are bred in late spring so that their foals will be on the ground in early spring the following year, since the gestation period is eleven months long. Mares begin to cycle, or go into season, as early as February and can cycle all through summer and even into early winter. Estrus, which starts when an egg is released from the ovary, occurs about every four weeks. Be aware: Being in season can affect the mare's personality. She may become more temperamental when ridden or exhibit a willful attitude while in the barn.

to being stalled all the time. Luckily, most horses are naturally thinner coming out of winter.

The type of feed you choose to give your mare will depend on her. Some mares are harder keepers than others, meaning they have a difficult time putting weight on. Feed also depends upon your type or quality of pasture, and whether your mare is kept outside in a field. Work on balancing her diet with the kind of hay you

Proper nutrition is important in gestating mares. Good-quality roughage should make up a good part of the diet.

have. If you only have grass hay, for instance, you may need to supplement for extra protein. Roughage is the major source of nutrition, and supplemental feed can balance what is missing. The extension officer at your nearest land grant university can help with analyzing your hay and recommending supplementation.

(*Note*: If your mare is at pasture, tall fescue, a type of drought resistant grass, has an endophyte that causes problems in gestating mares. Consult with your extension office for advice.)

Reproductive Preparation

Reproductive preparation for the breeding season depends on the age of your mare and her reproductive history. These factors will determine how much veterinary attention your mare will need prior to the breeding season. Regardless of her age and history, a mare must have a thorough reproductive examination prior to the breeding season, preferably when she is cycling. Schedule this exam at the end of the breeding season or in the fall when your mare is still cycling; don't wait until the beginning of the year, when she is anestrous (not cycling). It's more difficult to completely evaluate her reproductive tract—such as the functionality of her cervix, uterine tone and content, and ovary size—when she is not cycling.

The exam will include rectal palpation and an ultrasound. The veterinarian will also obtain an endometrial

CHANGING CYCLES

Most people want to have an early foal, which will be more mature when it comes to being broken, raced, or sold. But mares are long-day breeders. This means they cycle when there is increasing daylight and don't cycle in shorter daylight hours.

To get around this, breeders employ phototropic stimulation. In other words, they place the mares under lights at nighttime to bring the natural breeding season in earlier. Savvy breeders will bring their mares into the barn during the late afternoon and leave the lights on until 11 p.m., thus simulating a longer day. There should be enough light so a newspaper can be read in all corners of the area.

You can also work with your veterinarian to use different combinations of compounds and hormones, such as estrogen, domperidone, and sulpiride, to accomplish a similar outcome.

culture and/or cytology with a uterine swab, brush, or low-volume lavage. This ensures that the mare doesn't have inflammation or infection in her uterus (endometritis). Studies reveal that cytology is a better indicator of inflammation than is a culture—a culture only identifies infection. Cytology looks at the type of cells evaluated: inflammatory, bacterial (plus urine crystals), and

A veterinarian performs an exam on a mare. Ultrasound and endoscopic exams help determine whether a mare should be bred.

fungal elements. Just because a culture doesn't show infection doesn't mean there's no inflammation.

The cervix will be evaluated to make sure it is fully functional in closing and opening. Again, depending on the age and history of the mare, your veterinarian may then do an endometrial biopsy, which may be accompanied by a hysteroscopic examination. This is an examination of the mare's uterus with an endoscope. During this examination, the veterinarian can evaluate what the endometrium looks like and if there are any foreign bodies, infections, or other abnormalities. This exam needs to be done while the mare is in diestrus and her cervix is closed so her uterus can be distended and viewed.

An endometrial biopsy tells you more about the deeper layers of the uterus and the uterine environment. It can tell you the degree of inflammation, the degree of scar tissue, the integrity of the vessels, and whether there are uterine drainage problems. It also acts as a predictive index for carrying a foal. When the tests are done, you will be better able to judge the likelihood of the mare's ability to carry a foal to term. If the mare is a good candidate, you will also have a better idea of what you need to do to get her in foal and keep her in foal.

The veterinarian will also look at your mare's perineal conformation; this is the easiest way to determine if she has any problems that will affect the breeding or the foal. Perineal conformation has a huge impact on whether the mare is more likely to become contaminated and infected by her own bodily functions, develop endometritis, or aspirate air or urine into the vagina. If she does have poor perineal conformation, many of these issues can be resolved by a Caslick's procedure, in which the lips of the vulva are sutured together after the mare has been bred so contaminants cannot enter the uterus. The vulva will be opened thirty days to two weeks before the mare is due to foal.

Resolving Problems

Maiden mares (mares that have never carried a foal) often can't relax the cervix properly. Once the mare is bred, the normal inflammatory response of the uterus to the spermatozoa is to create fluid. This fluid can't leave the uterus if the cervix isn't properly relaxed. This also occurs in older mares for other reasons. If a mare is bred with this fluid, the spermatozoa may not survive. Retained fluid will also create a poor environment for the embryo as it develops and moves into the uterus. Your veterinarian can use different methods to get rid

Pasture breeding is not as prevalent as it once was. Most live coverage is supervised by a breeder or specialist.

of the fluid, such as administering oxytocin or cloprostenol (a prostaglandin), flushing the mare's uterus, or using acupuncture. If a cervix does not open correctly in maiden mares, it may trap debris, mucus, and dead cells within the uterus. In this case, using compounds to relax the cervix in conjunction with uterine lavage helps remove the mucus and debris.

LIVE COVER VERSUS ARTIFICIAL INSEMINATION

Horses used to be bred only through the method of live cover, which is when the stallion is actually present to impregnate the mare. Today, however, through artificial insemination (AI), horse owners can choose a stallion from the other side of the world to impregnate the mare. This provides a greater opportunity to select the ideal mate for your mare, even a deceased stallion whose semen was frozen before he passed away.

Covering

If you are going to have your mare bred the traditional way, contact the farm where the stallion is located and arrange to have your mare delivered to the farm around the week she goes into season. She will stay on the premises until she is bred.

The mare shows that she is in season by urinating, raising her tail, winking her vulva, and often neighing. When a stallion is present, he will approach the mare with an arched neck, prancing and nickering, and will nuzzle and sometimes even bite the mare. The pheromone the mare releases tells the stallion she is ready for mating.

Occasionally, a teaser gelding will be used to determine if the mare is indeed ready to cover. The mare's reaction to the teaser will indicate when it's time to bring out the stallion. The stallion is the only horse that will actually mount the mare.

When it is time for the mare to be bred, she will be placed in breeding hobbles, which fit on her hind legs and prevent her from kicking out at the stallion and injuring him. Her tail will normally be bandaged up so as not to interfere. The stallion will be led out and will mount the mare.

As the stallion mounts, his front hooves will rest near the mare's front shoulders. Once the stallion inserts his penis, he will rest heavily on the back of the mare.

The act is over quickly; the stallion ejaculates soon after entering the mare (horses do not thrust repeatedly). Once he has done so, he will be removed. The mare will have the hobbles taken off, and she'll be back in her stall. Because a mare's body will actually try to expel the semen, she may have to be covered more than once before an egg is fertilized.

Artificial Insemination

Although live cover is fairly straightforward, artificial insemination (AI) requires a host of equipment on the stallion's side.

The artificial vagina (AV) is the most common tool used for collecting stallion semen for AI. It is a tubular handheld canister with a latex liner. A water jacket, filled with warm water, surrounds the liner. There is a collection apparatus on one end, usually with a filter to keep out the gel part of the ejaculate. Once the stallion mounts the breeding mare or dummy (a type of mounting device), the AV is held in front of the mare,

This hard-shelled insulated container is used to ship semen long distances.

the stallion's penis is diverted into it, and the semen is collected. Afterward, the AV is held in an upright position so no semen leaks out, and the water is drained away. There are several makes and models of AV to choose from, depending upon lightness, ease of use, and stallion preference.

Shipping semen is possible through cooling and shipping devices. These containers lower the temperature of the semen to keep it viable for twenty-four to forty-eight hours—basically the time it takes for next-day delivery and insemination of the mare.

There are two types of containers to choose from. The hard-shelled, insulated container is the first, base shipping container. The semen is shipped in a cooling container, which is then placed in a bag with freezer packs. This container is the best for optimum cooling and has a lead shield to protect the contents from airport security X-ray machines. The drawback is that it is so expensive that the stallion's owner has to pay not only to ship it to the mare's owner but also to have it shipped back.

Alternatively, semen can be sent in disposable cooling and shipping systems, which are made from

Once a horse has been successfully bred (as this Welsh Pony obviously has), crucial prenatal care begins.

cardboard and Styrofoam. The enclosed semen is cooled with a special freezer pack. The advantages of using a disposable system is that it is inexpensive and, unlike a hard-shelled container, does not need to be returned. The other convenience is that the semen is shipped in a syringe instead of a bag. This makes it a little bit easier for the practitioner, because it removes the extra step of filling a syringe. The one drawback is, on hot days, the disposable containers do not keep the semen as cold during shipping.

Once it arrives, the semen is warmed in the most natural way possible: in the mare's uterus. The veterinarian passes a disposable pipette through the cervix and deposits all the semen into the body of the uterus.

DETERMINING PREGNANCY

Your veterinarian will be able to tell if your mare is pregnant by palpation of the uterus. Feeling for the growing embryo via the mare's rectum does this. A veterinarian puts a hand in the rectum and feels the uterus for signs of pregnancy. These signs include the uterus's tone, its shape, and the presence and size of the small sac containing the fetus, called the amnionic vesicle. This method can be used from as early as sixteen to nineteen days of pregnancy, but it does have to be done by a very skilled veterinarian.

Generally however, a pregnancy exam is usually performed by ultrasound at fourteen to sixteen days. It's important to have the mare examined early because you can plan to rebreed her if the breeding didn't take.

The next important exam should be performed about thirty days after breeding. If the mare miscarries and loses the embryo within this window, you can have her rebred. If she aborts anytime after thirty-five days, she won't be fertile for at least another three months. Normally, this will mean that she can't be bred for that season.

Chapter 26

Pregnancy, Delivery, and Foal Care

The most important thing you can do for your pregnant mare is to give her correct prenatal care. You can ensure a healthy pregnancy and robust foal by planning a complete program of good nutrition, deworming, and vaccinations, and by alerting your veterinarian to the slightest sign of trouble. To ensure a successful delivery, you need to prepare carefully and be ready for any complications. Of course, that's just the beginning. Once the foal is born, you have a whole new little equine to care for.

CARING FOR THE PREGNANT MARE

Caring for a pregnant mare includes two critical components: proper nutrition and proper health care. Because the foal only develops modestly during the first two trimesters of pregnancy, the mare's nutritional needs stay fairly constant during this time. The mare should receive sufficient grain ration to make sure she keeps good weight and muscle and good-quality roughage. If the mare started out the pregnancy underweight, these months should be used to bring her weight up to good condition. In the last trimester, a mare's nutritional needs change greatly. Administering vaccinations and dewormers at the proper time is also crucial.

Nutrition during Pregnancy

Near the end of the second trimester, at seven months, the foal starts growing rapidly. Begin to increase your mare's feed so she has good milk production once the foal arrives. The mare's feed should be adjusted so that her energy requirements (grains and concentrated feeds) increase by 20 percent and her protein requirements increase by 30 percent. At the beginning of the eighth month, she will need to have her diet supplemented with trace minerals. During this time, the fetus begins to store iron, zinc, copper, and manganese in its liver because the mare's milk will not supply adequate amounts of these minerals to meet the newborn foal's needs. Adding a supplement that contains copper, for example, has been shown to reduce the incidence of bone formation problems in foals.

If the mare begins to put on fat late in her pregnancy, she needs to be switched to a feed that supplies a larger amount of protein and minerals per pound so less feed can be given to her per day. Overweight mares have more complications during foaling. Thus it is important to monitor the mare's body condition throughout pregnancy so the appropriate condition is maintained. As the foal grows larger, the mare has less and less capacity to digest large-concentrate meals. By the time she is ready to foal, she should be eating 70 percent forage so her nutritional needs are met without overloading her digestive system with grain—or making her obese.

After foaling, the food requirements increase dramatically for the mare, as she is producing milk high in energy content, protein, calcium, phosphorous, and vitamins. After giving birth, mares usually require up to fourteen pounds of grain a day, depending on the

Journey from Egg to Foal

A fertilized egg stays in the oviduct for about six days and then travels to the uterus. During the first part of pregnancy, the embryo is free-moving in the uterus for slightly longer than two weeks, until it fixes itself to the uterine wall. At about a month, the embryo's heartbeat can be seen on ultrasound. Soon after, the placenta forms over the embryo, and the embryo becomes a fetus.

At about seventy days, the sex of the fetus can be determined via ultrasound. By about five months into the pregnancy, the fetus is the size of a medium-size dog. Most of the foal's growth occurs during the last trimester, when it grows the last 60 percent of its birth size. The gestation period of a horse lasts about eleven months (340 days approximately, although it can also be as few as 320 days or as many as 370 days). A colt is carried a few days longer than a filly.

By the time this pregnant mare is ready to foal, 70 percent of her diet should be forage.

quality of the hay being fed. She will need a feed or a supplement that contains calcium and phosphorous, as well as other minerals, protein, and vitamins. Once the foal has been weaned, the mare's milk production reduces. The mare's grain ration should be reduced as well to make sure she does not gain weight.

Health Care during Pregnancy

Use dewormers that are safe for a pregnant mare, and check with your veterinarian to determine a deworming program during pregnancy. Keep her pasture and stall clean of manure. Be sure to reduce the possibility of her transferring internal parasites to her foal by deworming her with ivermectin the day before she foals.

Discuss with your veterinarian the vaccinations your mare will need for your region. She should have been vaccinated before she was bred (see previous chapter). Once she is pregnant, do not give her any booster shots during the first two months. To ensure that the foal gets the most protection as soon as he is born, vaccinate the mare in the last three to six weeks of pregnancy.

The vaccination program has several objectives: to keep the mare healthy through pregnancy, to keep the mare from aborting her foal, and to protect the foal by improving his immunity via the mare's first milk.

FOALING

An average pregnancy lasts 335 to 342 days, but some pregnancies have varied from as few as 315 days to more than 400. About a month before foaling, many mares start to develop swelling (edema) close to where their udders are on their belly. The mare's teats will begin to enlarge, particularly about two weeks before the foal is born. During this early development, the udder remains firm, but a few days before foaling, it gradually softens and fills with fluid, which slowly changes in appearance from watery to viscous colostrum. The colostrum is generally present a couple of days before birth.

Before the foal arrives—and he often does so without warning—you need to prepare the foaling area to help the foaling process go smoothly. Although plenty of foals arrive safely without human intervention, things can go wrong during the foaling process or shortly thereafter, and tragedy can ensue. Thus someone needs to be in attendance during the foaling process. That's why most major farms and ranches have special staff on-site during foaling season.

If you're going to function as the nursery attendant yourself, then you can expect to lose some sleep until the foal arrives. It's a good idea to team up with an experienced pal, who can provide guidance and alternate night duty with you.

A woman administers a dewormer. Consult a veterinarian to establish a deworming program during your mare's pregnancy.

Horse talk

PREGNANCY AND FOALHOOD

Learning the following terms will help you talk about equine pregnancy, delivery, and foal care.

■ **Amnion:** Also called the water sack, this is a tough, liquid-filled sack that covers the embryo or fetus.

■ **Colostrum:** The mare's first milk, which is filled with antibodies that will protect the foal against diseases. This first milk is also higher in protein and lower in fat.

■ **Creep feeding:** A small feeder placed at foal level and is restricted so that adult horses can't get in and eat the foal's food.

■ **Foal heat:** A mare's first fertile cycle after foaling.

■ **Foal scours:** A common diarrhea, also called foal heat diarrhea, which occurs in foals at one to two weeks of age.

■ **Malposition:** The foal is misaligned or twisted in the birth canal, which impedes the foaling process.

■ **Red bag delivery:** An extreme emergency. Rather than seeing the translucent white amniotic bag and the foal, you'll see the red placenta emerging from the birth canal.

■ **Weaning:** The foal is no longer dependent upon his mother for nutrition; at this time, he is removed from his mother's side. He is now referred to as a weanling.

■ **Wobbler Syndrome:** Where the spine compresses from abnormally developed vertebrae in the neck. Provide nutrition for a moderate growth rate to keep the foal healthy.

RIDING DURING PREGNANCY

As long as your veterinarian hasn't told you otherwise, you should still continue to ride your mare during pregnancy. Keep in mind that while she isn't a hothouse flower, her heavy work should taper off after five months. She can still benefit from light, easy rides up until her tenth month, although she may be too round to fit her usual tack. Allow her plenty of time to be turned out—preferably in a pasture that she can walk around on her own—for a good portion of each day.

You can also install a closed-circuit video camera in the foaling stall. This way, you can remain indoors and not disturb the mare until she definitely demonstrates that it's time for the foal's arrival.

Foaling Area

At least a month before your mare's due date, prepare the area where she will foal. Most horse owners will create a foaling stall. It should be substantially larger than a standard box stall—usually the size of two stalls.

This allows you to better observe your mare and allows her space to lie down during the final stage of labor without being crammed against a wall. The stall should be well ventilated, cleaned, and sterilized. All adjacent stalls should be empty, but she should have companions nearby for company. Place a thick layer of straw for bedding, rather than shavings, which can become stuck in a foal's air passages. Any feed and water buckets should be placed high so that the mare doesn't deliver the foal into them, drowning him inadvertently.

If your mare will deliver outside, make sure there is a separate area for her fenced away from other horses and place straw there. She may not use the area, but she should have the option. Look for any object that could injure a newborn, such as a loose board or broken fence wire. Other horses on the property may become excited over a new foal, and your mare may become fiercely protective of her baby, so double-check the security of every fence panel. Ultimately, whether your mare will foal indoors or out, keep the area as meticulous as possible. A clean, safe environment will prevent infections and injuries.

Preparation for foaling also includes making sure you have everything you need to aid the delivery. (*See Foaling Kit sidebar, opposite, for more information.*)

A pregnant mare stands in a foaling stall; this large stall will give her and her foal the room needed during and after the birth.

The foal usually rests for up to twenty minutes with his hind legs in the mare's vagina. Note the amnion at the back.

Foaling Process

The time from the mare's water breaking to the completed delivery of the foal is usually about twenty minutes, though occasionally it may take an hour. The mare generally rolls onto her chest and stands within fifteen minutes of delivery, and the umbilical cord breaks one to two inches from the foal's abdomen. With normal, vigorous foals, the amnion (or water bag) usually ruptures by the time the chest is passed so that the foal can breathe. If it doesn't break, it should be torn open and cleared away from the head.

The next stage of labor is the passage of the placenta, which usually occurs within an hour after the foal is born. The mare may be in slight pain with cramps while this is occurring. The placenta should be saved so it can be checked for completeness and for any signs of infection. Even though it's estimated that foaling problems occur in only 1 percent of all births, tragedy can happen if you own that particularly unlucky mare and you aren't on-site. Although you should not be intrusive during labor, you should be a silent, patient observer, ready to call your veterinarian immediately at the first sign of trouble.

Foaling Emergencies

There are four main categories of foaling emergencies. One is a foal in malposition. Once your mare's water breaks, the foal's front legs and head should be visible within five minutes. If not, it's probably due to the foal being misaligned or twisted in the birth canal, which impedes the foaling process.

Another emergency situation is premature separation of the placenta, known as a red bag delivery.

FOALING KIT

Prepare a foaling kit. It should include the following items:

- **Cell phone:** With vet's number on speed dial.

- **Clean towels:** For rubbing down the foal or mare on a cold night.

- **Enema:** A baby-size human one in case you have to help the foal pass its first stool, called meconium.

- **Flashlight:** Unless your stable has sufficient overhead lighting.

- **Illuminated watch or clock and checklist:** So you can jot down what time the final stage of foaling began, when the foal stood up, and so on. Your veterinarian can give you a general timeline for each of these events. Any major departures from this timetable should elicit a call to your veterinarian for advice.

- **Latex gloves:** To keep your hands clean when handling the birth.

- **Nolvasan (a strong antibacterial solution):** For application on the navel stump.

- **Plastic bag (large) and tub:** Keep on hand for stashing the afterbirth and placenta so your veterinarian can inspect it for anything abnormal.

Once you've assembled your foaling kit, allow your veterinarian to inspect the contents and make suggestions for any additional items.

Rather than seeing the translucent white amniotic bag and the foal, you'll see the red placenta emerging from the birth canal. Since the foal receives his oxygen supply from the placenta's attachment to the uterine wall until he's completely delivered, a red bag delivery is an extreme emergency.

A third emergency situation is if the new foal does not stand and nurse within two hours of delivery. If he does not, he may become weak. It may also be a sign of other problems, including septicemia (infection).

A Quarter Horse foal nurses from his mother. The mare's first milk is packed with nutrients for her newborn.

The fourth major foaling emergency is gastrointestinal problems in either the mare or the foal. If the mare does not seem interested in eating after foaling, or if the foal strains to defecate and pass the first stool (the meconium), then the veterinarian needs to intervene.

Since only horsepeople experienced in foaling out mares should provide hands-on assistance during emergencies, as a novice you should have your veterinarian's phone number on speed dial when you expect your mare to deliver. Otherwise, your meddling may do more harm than good. Your veterinarian can advise you on what to do until he or she arrives on scene.

Of course, you can also decide to entrust experts other than your veterinarian with the foaling process. About a month before your mare is due, transport her to a local ranch, farm, or clinic that specializes in foaling out mares. Major breeding farms, with their own annual crop of foals, usually offer such services. Though they do charge fees for assuming the roles of birthing coach and midwife, your peace of mind may be worth it.

FOAL CARE

Have your veterinarian check your foal within twelve to twenty-four hours after birth to assess his overall health and to see if he has received adequate colostrum, which contains important antibodies. Have your mare checked as well. Your newborn's first instinct will be to get on his feet and have his first meal. He may not be able to stand right away, but don't interfere with his attempts. The foal will gain more strength as the blood begins to circulate throughout his body, so as long as he is in a safe, hazard-free environment, leave him to find his own feet. As discussed above, however, a foal that shows little or no urgency to get up and nurse is cause for concern.

The foal's initial attempts to nurse will often be clumsy, as well. When first searching for his mare's bag, a foal will often latch onto a hock or an elbow instead. Allow the newborn to figure it out on his own. (If he is still unsuccessful after a couple of hours, ask your veterinarian what he recommends.) After the foal has nursed successfully from the mare, he will want to rest and even sleep, though he will get up occasionally to test his legs.

It's rare, but there are some instances of mares not bonding with their foals. Sometimes this is because the mare hasn't had enough private time with her baby. Occasionally, she may be sensitive to nursing or a first-time broodmare with little experience. Allow the hormones of motherhood to kick in—it is better if you stay out of the way for your own safety, too. But be sure to let your veterinarian know if she is aggressive or dangerous around the foal.

When your foal is about a week old, he should be nursing, sleeping, and playing vigorously. At this point, you may see that the foal has watery diarrhea, but seems otherwise healthy. This common diarrhea is called foal heat diarrhea, or foal scours, which usually occurs at one to two weeks of age. No need for alarm. The time mares usually have their foal heat (the first heat after foaling) usually coincides with the foal's diarrhea.

Your veterinarian will likely ask you to just watch the foal and make sure he has an appetite and is still active. Rub Desitin or A&D ointment under his tail to help prevent skin sores caused by diarrhea. Usually the diarrhea resolves within four to six days. If it lasts too long, the foal will become dehydrated, so call the veterinarian if it intensifies or is heavy for more than ten days.

Keep the foal separate from other horses on the property for the first few days. They will recognize that there is a newcomer soon enough and will be very excited. But there needs to be a period of adjustment for all equines stabled close together. Allow your mare time to get to know and dote on her baby, and allow the foal time to gain strength with each passing day.

It's important that while you are molding your equine child to be a good partner down the road, you also need to remember at all times that foals are powerful and agile. Use your head around them. While they look sweet on the outside, they are creatures of instinct and self-preservation.

Foal Nutrition

After about a week, the foal will begin to test out his mother's feed. Foals can be fed some solid food at ten to fourteen days. A good option is creep feeding: a small feeder is placed at foal level and is restricted so

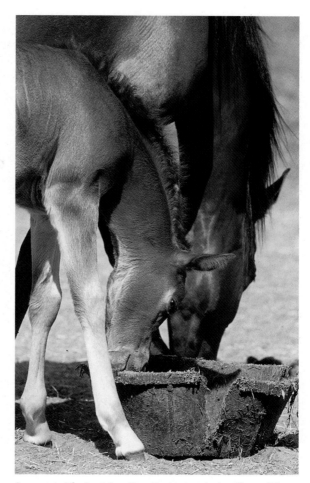

A mare and foal eat together. Foals can start eating solid food around fifteen days of age.

the mare can't get in and eat it. The foal is free to nibble at a grain ration nutritionally ideal for foals. He will continue to nurse often, however. You will also notice that he will eat his dam's manure—this is nature's way of giving him essential intestinal bacteria.

Up to six weeks of age, most of the foal's nutrition comes from milk. At two to three months of age, the foal may need more sustenance than the mare can provide, and supplemental feeding will probably be necessary. It is important not to overfeed the foal in an attempt to increase his normal rate of growth, as is the practice among some owners who wish for an early sale or for better performance in the arena.

To do so can lead to skeletal weakness and possible deformity. These problems are all grouped together as developmental orthopedic disease (DOD). Physitis is a disease that affects the growth of the horse's lengthening long bone at the growth plate. Osteochondrosis (OCD) involves the growth of the bone underlying the joint. There is also a condition called wobbler syndrome, in which the spine compresses from an abnormally developed vertebrae in the neck. Provide nutrition for a moderate growth rate in order to keep the foal healthy.

A foal can become severely lame, and when this happens, it's time to call the veterinarian. A foal can get septic arthritis, a dangerous condition in which his joints become infected. This is usually the result of bacteria that has entered the foal's system through an infection—perhaps of the umbilical cord or the intestines. The bacteria travel through the blood and settle within the joints, becoming a life-threatening condition. Check the foal's abdomen for any unusual swelling, which could point to an infection.

Mares and foals, like these Quarter Horses, have a strong bond. Weaning must be done right to minimize stress on both sides.

Health Care

Foals don't need vaccines until they are about four months old, as long as their dams were on good vaccination programs. (If their dams weren't vaccinated, the foals should be vaccinated at two months of age). They should receive their first shots for encephalitis, tetanus, and West Nile virus, and receive boosters for them. Many veterinarians suggest holding off on a flu-rhino vaccination until the foals are at least six months old. Check with your veterinarian on the specific vaccines for your area.

It's important not to vaccinate too soon. Since a foal receives antibodies from the mare's colostrum, he is already protected from disease, so giving vaccinations at that point is like doubling up. Vaccines, which are doses of dead viruses, become bound up with the antibody that is already present from the colostrum. When this happens, the foal ends up unprotected against disease.

It's very important to put foals in a deworming program, as they are very susceptible to internal parasites. A foal should be dewormed with a product safe for foals and should be given small doses frequently, such as monthly. This should continue for a year. After the foal is a yearling, the schedule can be adjusted.

WEANING AND WEANLING CARE

Eventually, your foal has to graduate to a weanling—which can be traumatic or not, depending on your resources and how you decide to handle it. The foal needs to move closer to adulthood and independence, and the mare needs to recover from the physical stress of being a mother.

Foals should be weaned when they are between four and six months old. Usually by the third month, the mare will be gradually producing less milk, and the foal will start eating his own grain and quality forage ration to make up for this decrease. By the time the foal stops nursing, he should be eating about 2 to 3 percent of his total body weight per day of a ration that is made up of 14 to 16 percent protein by weight. Good-quality roughage should compose at least 30 percent by weight of the total ration.

When to wean your own foal is a personal decision. Make sure that these prerequisites are in place before weaning:

- The foal is healthy and illness-free.
- He is showing signs of independence.

Warmblood mares and foals share a pasture. Before weaning a foal, make sure he is well socialized with other horses.

• There is another foal or companion that can keep him company and distract him.

• He is used to and comfortable with the pasture or paddock that will become his new home away from his dam.

• He is well socialized with other horses.

Most foals are born in the spring, so weaning time is often in the heat of summer, adding to the stress of the situation. If the area that you live in is extremely hot at the proposed six-month window of weaning, try to delay it until more hospitable weather. It is easy to lose a foal due to stress in these circumstances.

The abrupt separation method, in which the foal is dragged from its dam in the midst of panic, pacing, and whinnying is a method that is falling out of favor. While some experts say that it is like ripping off a bandage—painful, but only for a short time—proponents of gradual separation think that abrupt weaning leads to health and behavioral issues, including gastric ulcers, delayed reproductive function, increased levels of cortisol (a hormone secreted by the adrenal cortex), which can weaken the immune system, poor sleep, and injury due to panic.

Abrupt weaning removes the mare to another location, leaving the foal in his familiar surroundings and with the other weanlings or with an older babysitter horse. When the mare is moved out of sight and sound of the foal, the action causes a great deal of panic for both parties. A better method, if you have the room, is to keep the mare in one pasture and the foal in an adjoining one. This way, they can still be together, yet the foal cannot nurse. The mare's milk will eventually dry up, and the foal can still be close to the mare. Usually what happens is that the foal will gradually get used to being away from his mother's side and will go off on his own. You can eventually move the mare away from the foal.

What hasn't proved effective, however, is separating the foal and mare and then putting them back together after a certain time period. It's actually counterproductive to separate the mare and foal during the day but then place them in the same stall at night, as this only forces them to relive the stress of separation over and over again each day.

Once your foal is weaned, his need for human contact will increase, as young horses can revert to their instincts—bonding with the herd rather than you and developing the desire to flee once again. Make sure that your relationship is solid with the foal and remains so after he becomes a weanling.

Now that your horse is a weanling, he will still have foal characteristics, continuing to eat, sleep often,

Sharing a pasture and a stall with each other eases the difficulty of separation from their dams for these weanlings.

and nap frequently. Continue to spend time handling the weanling, and make your time with him short but frequent, to match his attention span. Make sure you get the foal accustomed to standing for grooming, following when leading, and even lifting up his feet for the eventual hoof care he'll receive.

Make sure that your baby has other babies his age to socialize and play with. Many people recommend that the horse be placed in an environment where he can be around other weanlings, such as a breeding farm. Others think that too many babies together make for mischief and instead recommend a well-mannered, older, calm horse. A quiet, older brood-mare can work very well because she will be able to teach the weanling social skills without being mean or vicious, and he'll learn good pasture manners (if he doesn't have them already). An older mare's calm temperament also helps to keep weanlings from getting too excitable. When a foal is weaned, he needs to begin a careful regimen of hoof trimming from a farrier who is an expert in dealing with youngsters. Not only will he know how to handle a young horse, but he will also help ensure proper hoof growth and limb angulation. Corrective trimming can often help with some abnormal leg issues. Weanling hoof care can easily be too little or too much.

Weanlings also need the proper vaccination and deworming schedule for the areas in which they live.

Don't expect your weanling to be perfect, and don't demand that he be on his best behavior, because it isn't going to happen. Correct dangerous behavior, as you did when he was a foal, but expect that he will misbehave occasionally when his attention span gives out, when he gets annoyed with his lessons, or when he is fresh and feeling good and doesn't want to settle.

HANDLING AND IMPRINTING

Beginning your foal's lessons early and keeping his training sessions positive and short will make him easier to handle as he gets larger and older. Gentle your foal by getting him accustomed to being touched and handled. He should learn to wear a halter as soon as you can get one on him. At the very least, he should be wearing one within his first week. It is easier to do when he is a few days old rather than a few months old, when he is harder to handle. He should learn to lead soon after, but not by his face, as the foal's neck is delicate. A rope should be pattered in a loop around his back and buttocks, and he should be guided around by his body instead. (*For more information on training your foal to lead, see page 323, chapter 18.*)

Make sure to keep handling your foal daily, many times a day, stroking and rubbing him all over, making the human touch a happy experience. Remember that while he is cute, he grows stronger every day, and foals generally roughhouse when they play. Therefore, don't treat your foal like a dog. You need to be firm with him but not mean. You have to protect yourself from getting hurt by a rambunctious youngster. Make sure that you always expect him to behave well around you by setting the rules early and being consistent with him.

There are different schools of thought on whether to imprint. In psychobiology, imprinting is a form of learning in which a very young animal fixes its attention on the first object with which it has visual, auditory, or tactile experience and thereafter follows that object.

In nature, the object is almost invariably a parent; when training horses, the handler imprints the foal by exposing him to human touch and some of the stimuli he will encounter as an adult, making him confident and bonded to people. Some experts wish to leave the baby alone on its first day, while others say to immediately begin imprinting: touching and rubbing the foal, rapping on his feet to simulate shoeing, and running clippers over him. An experienced horseperson should do this; imprinting the foal roughly or interfering with the mare could result in injury to the foal or to yourself. (*For more information on imprinting, see page 320, chapter 18.*)

THE YEARLING

A yearling is at least one year old, but less than two years old. Yearlings are out of the baby stage, but they are still nowhere close to their mature stage. Hormones are barely emerging, and yearlings are still very suggestible and easily influenced—in positive and negative ways. As a human, the yearling would be a tween.

They are awkward and ungainly, but they are still a joy to be around and to train. How you handle young horses as yearlings will determine how they turn out as mature equines. The growth rate will slow as the foal approaches two years, and his food ration should be reduced as necessary to about 1½ to 2 percent of his body weight.

The grain supplied should provide half the diet, and good-quality roughage should provide the other half. Remember that the goal is to raise a well-nourished foal that is not overweight. Excess weight taxes the skeleton and may compromise the foal's later athletic career.

This group of Warmblood yearlings band together while at pasture. In human years, a yearling would be a "tween."

The Senior Horse

Section 11

With good nutrition, care, and exercise, horses can live longer and, more importantly, have productive lives into their later years. Thanks to science and continual research, horses are staying healthy and are enjoying careers at ages people never thought possible a few decades ago. New developments in pharmaceuticals have helped the older horse thrive into his twenties and beyond. Understanding when a horse is truly geriatric and should be retired is about more than knowing his age.

Perhaps the most difficult decision you will come to is deciding when it's time to let your old horse go, when there is only suffering left for him. Then it will be up to you to ease his passage.

Chapter 27

Senior Horse Facts

Tufts University, in Medford, Massachusetts, conducted a study in New England in 2007 and discovered that out of 1,000 equines, 14 percent of them were between ages twenty and twenty-nine. Many of those seniors were still being ridden, and 10 percent were still competing. The study also found that the oldest horses and ponies were around forty-five years of age.

Since the late 1980s, a great deal of money has been poured into funding research to find out how the older horse ticks. The appropriate diet and types of exercise for older horses, as well as a better understanding of their immune response and physiology, are now becoming common knowledge.

The question is, does all this information prove that buying an older horse is a worthwhile investment? What are the advantages of choosing an older horse over a younger one? Are there good reasons for hanging on to that aging partner who's been with you through thick and thin?

SIGNS OF AGING

That horse has "been ten." Years ago this statement meant that a horse had already seen the age of ten and was now on the downward slide. Ten years old is no longer seen as the dividing line between young and old in a horse, but experts say it's difficult to pinpoint at exactly what age a horse should be considered elderly. In past research studies, horses that had reached the age of twenty were considered geriatric, but many twenty-year-olds today are still useful and many are still performing. Although horses start to show the signs of aging by twenty, nowadays it's not until twenty-five or even thirty that most people think of a horse as truly geriatric.

You can tell that your horse is aging by obvious physical changes. One of the most apparent changes is a swayed back, otherwise known as lordosis. A spinal deviation, swayback is caused in part from a loss of tone in the back and abdominal muscles, as well as a weakening and stretching of the ligaments. Lordosis doesn't cause the horse pain, however, and with the right saddling and padding, the horse can still be used for light work.

Older horses are also likely to be thinner, as they have a harder time keeping on weight. Even when older horses eat the same amount as before, their bodies are not able to utilize the feed as well. Horses' coats also change with age. With each winter, their coats will grow longer and thicker, and with each spring, they will take more time to fully shed out. You will see gray hair appear as well, especially around the eyes, the muzzle, and the ears.

The pasterns on the horse's hind legs will begin to slope more. Hollows above the eyes may become more prominent. Eyes may develop cataracts or get cloudy. Incisor teeth will become elongated. An older horse will often have a droopy lower lip.

An aging Andalusian trots around his pasture on a Michigan farm. You can still enjoy many productive years with your older horse.

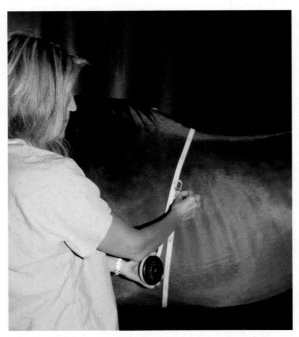

Use a weight tape as shown here to measure your aging horse to track significant weight loss or gain.

These signs do not mean, however, that the horse has only a few years left to live. Many old—and very old—horses continue to benefit from regular use. It is up to the owner to determine how much work the horse can do and monitor the amount as the horse ages.

NUTRITION AND CARE

Proper nutrition is crucial to keeping the aging horse healthy. As noted above, his digestive system isn't as efficient at processing foods, and there are a host of hormonal and metabolic changes that may have some effect on digestion and absorption of his food's nutrients. Older horses must have food that provides energy and is high in protein, vitamins, minerals, and fiber. If you have an older horse, give him food that is highly palatable and easy to chew and swallow. It should be dust free to prevent coughing and chronic obstructive pulmonary disease (COPD) flare-ups.

There are several types of complete feeds designed specifically for the needs of the older horse in ingredients, palatability, and how easy it is to chew. A number of these feeds are also designed to break down easily in water to make a mash for a horse suffering from dental problems. If a horse is having trouble keeping weight on, he may benefit from a probiotic, a supplement that contains healthy microflora to help him digest his feed better. Many senior feeds have this extra ingredient added in.

Because of the older horse's challenges with nutrition, you should check his weight on a regular basis. Seeing your horse every day, you may find it difficult to tell if he's losing weight by merely looking at him. Use a weight tape—a measuring tape designed to give a rough estimate—to check every couple of months.

Don't just check that he isn't losing weight; make sure he isn't gaining too much weight, either. In fact, it can be more dangerous for horses to be overweight than under. Excess weight places strain on muscular and skeletal structures, and makes the horse more susceptible to conditions such as laminitis and the development of fatty tumors. Make sure you are not providing your horse with more energy than he can work off.

Older horses also have less resistance to parasites, so periodically have your horse checked for worms by performing a fecal test and also set up a regular deworming program.

Dental Problems

Since older horses may have problems digesting their food and utilizing the nutrients it contains, make sure that your horse's teeth are well maintained. Be aware of the condition of his teeth in his teens, and don't wait until he is twenty to fix a problem. It's hard to correct something that has already gone wrong in the mouth. Good dental maintenance will allow a horse to age well.

There are many steps you can take to maintain an older horse's dental health; among them is having your veterinarian or equine dentist check your horse's teeth twice a year. You should also be on the lookout for signs that your horse is having dental problems. One sign is weight loss; another is dropping most of his feed on the floor as he eats. If your horse has stopped eating altogether, have your veterinarian check for a gum or tooth infection.

Sometimes a horse's teeth are too worn to chew the feed correctly. Unevenly worn teeth can develop sharp points, which can make eating painful. Your should have your horse's teeth by your veterinarian or an equine dentist at least once a year. If the horse

Life Expectancy

With good care, good nutrition, and the right genetics, many horses will live into their thirties. A horse's age doesn't match exactly to human years. He matures rapidly in his youth, but then the aging process slows once he is fully grown. Aging also depends on the individual horse's genetics, as well as what type of life he has led. If a horse is handled and cared for properly when he is young, he will undoubtedly be in better condition than an older horse that didn't get as good a start. Like people, a hard life as a youngster will affect the horse's elder years.

ESTIMATED EQUIVALENT AGE IN HUMAN YEARS		
HORSE		HUMAN
1	=	6.5
2	=	13.0
3	=	18.0
4	=	20.5
5	=	24.5
7	=	28.0
10	=	35.5
13	=	43.5
17	=	53.0
20	=	60.5
24	=	70.5
27	=	78.0
30	=	85.5
33	=	93.0
36	=	100.5

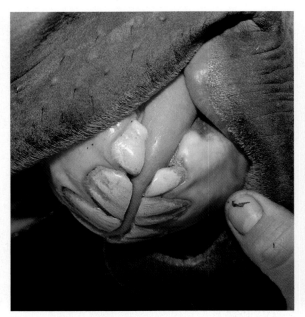

These are the teeth of a five-year-old horse. By this age, all permanent teeth are through.

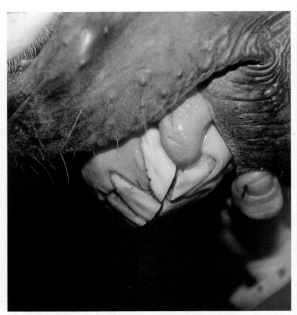

These are the teeth of a twenty-year-old horse. Notice how the jaw has become oblique, which can make it difficult to eat.

is having trouble chewing because of tooth loss, you may want to give him a hand by creating a mash of his feed. Alfalfa pellets and complete feeds can be soaked in water and broken down into a gruel. Grain formulated for seniors can also be broken down into a mash that he won't need to chew—and it still provides all necessary nutrients and roughage.

Other Ailments

As a horse gets older, he experiences his share of aches and pains. Poor conformation, poor nutrition, hard use, injury, obesity, and improper hoof care can all increase problems for the aging equine. Here are some ailments that afflict the older horse.

Arthritis: Nearly all older horses are plagued by arthritis to some extent. Work with your veterinarian to manage arthritis. He may suggest using a nonsteroidal anti-inflammatory such as phenylbutazone (often simply called bute), or by using joint supplements or injections.

Joint injections can provide a great deal of relief because they either mimic the synovial fluid that lubricates the joints or they bind to the cartilage components within the joint itself, depending on what type of medication is used. Any oral pain relievers, such as bute, should be given as sparingly as possible—for

example, only on a bad arthritis day—since regular use of this type of medication can lead to stomach ulcers.

Hormonal imbalance: Problems with hormone imbalances are common for both horses and ponies. These imbalances stem from a deterioration or disease of the endocrine system and may result in the development of pituitary pars intermedia dysfunction (PPID, more commonly known as Cushing's disease), Addison's disease, or pituitary adenoma. Your veterinarian

Unshedable coats are a sign of Cushing's disease, a disorder this thirty-five-year-old pony has.

should check your horse's endocrine function if he becomes weak or sweats excessively, is constantly thirsty, or grows a long coat of hair that he can't shed. If he has a drastic personality change or seems lethargic, let your veterinarian know this as well.

Hypothyroidism: If you do notice that your horse is continuing to gain weight regardless of how much you cut back on his feed, he may be suffering from hypothyroidism (an underperforming thyroid). This condition is treatable, so seek advice from your veterinarian. Remember to make any changes or additions to your horse's diet gradually so that he doesn't develop colic.

Ask your veterinarian to help you set up a wellness program specifically designed for your aging horse. This may include extensive physical examinations, blood work, and dental care, all performed on a schedule.

FITNESS

People now understand that horses are good athletes and can continue to be ridden into their teens. Studies have compared heart rates and cardiovascular changes between young horses and middle-age horses (horses in their teens up to age twenty). The results have shown that there is actually little difference between these age groups.

SENIORS

Learning the following terms will help you to talk about senior horses and their issues.

■ **Bute:** Short for phenylbutazone, which is a nonsteroidal anti-inflammatory.

■ **Colic:** Gastrointestinal illness.

■ **Laminitis:** Inflammation of the laminae of the hooves; known also as founder.

■ **Lordosis:** A spinal deviation; known also as swayed back.

■ **Probiotic:** A dietary supplement that contains healthy microflora to help the horse with digestion.

■ **Weight tape:** A measuring tape, placed around the horses barrel, that estimates his weight.

Vaccinating

Ongoing research projects at many veterinary institutions study the response of aged horses to vaccination. Aged horses seem to have a less robust response to vaccinations and may require vaccinations more frequently than their younger counterparts. It is advisable to keep showing or traveling horses separate from senior horses, as their immune systems may be less able to protect them from exposure to pathogens, particularly new ones.

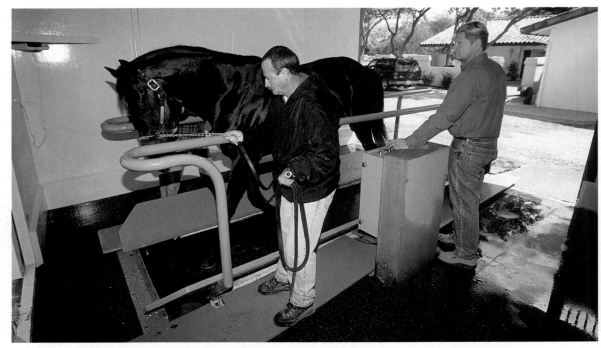

Equine treadmills can help rehabilitate an elderly horse or an injured horse or aid in a sport horse's training.

A good example of an excellent older equine athlete is the 1976 Olympic team bronze medalist, Keen, a Thoroughbred gelding ridden by American Hilda Gurney. He dominated dressage in the 1970s and into the 1980s; he was well into his late teens, and he was still competing and winning. At the age of eighteen, at the 1984 Los Angeles Olympics, Keen placed fourteenth individually. At nineteen, he was second at the FEI North American Championships and won the West Coast Olympic Selection Trials. In 1985, the gelding won the U.S. Equestrian Team National Championships for the sixth time.

Older horses can adapt to exercise and increase their fitness level. Treadmill studies at universities have concluded that older horses, once they became accustomed to the machine, were eager to move faster after one or two sessions, with the exercise actually helping loosen up joints and muscles. Fitness in the older horse must be gradual, with many short but frequent sessions of exercise. The horse should have a break at the moment he needs to recover, and then continue with the session. These sessions can be lengthened as the horse's condition improves.

One of the issues for horses in their twenties concerns thermoregulation. Older horses have less plasma volume than younger horses. Older horses don't have a large enough plasma volume in their blood, so they won't have a reserve large enough to sweat efficiently. This means they have less fluid in their bloodstream, and perspiration won't cool the body down quickly enough. In addition, the amount of fluid within the body is lower in an older horse, so a horse in his twenties runs the risk of getting dehydrated faster than he would have when he was younger.

The older horse may be able to handle the exercise, but he may have trouble handling the effects of trying to stay cool. For this reason, the intensity and the duration of exercise should be reduced—most likely when the horse enters his late teens or early twenties.

Physiology, Not Age

Should a horse of a certain age be retired? It has been proven overwhelmingly that horses do much better if they are left in work, as long as they are still comfortable. It's just as it is with people. The fitter you are, the better you'll feel. Judge the horse's aging based on his physiology, not numerical age. If he is doing well, he should continue to be ridden and enjoyed.

Instead of going at a fast canter for an endurance ride, for example, he may have to do the ride at a trot.

Expect the older athlete to heat up faster and to have a heart rate that increases more quickly. He will also breathe harder. The good news, however, is that he should recover easily if he isn't pushed too far; in fact, ten minutes later he should be back down at his usual level and be able to continue. As a rule of thumb: don't push as hard with an older horse, and take a longer amount of time to cool down between sessions. And of course, all this is contingent on whether the horse is orthopedically sound.

INVESTING IN AN OLDER HORSE

If you buy an older horse, are you simply buying a big vet bill? Not if you buy the right horse. In fact, there are some real pluses to buying an older horse. Some advocates of the sound, older athlete believe he is a fabulous investment, for the simple reason that he has already proven he can do the job and stay healthy. Many horses, for example, can't do high-level dressage work; they break down. If you are considering purchasing a fifteen-year-old Grand Prix horse, you'll likely get three to seven more years of work from him. Once a horse has made it to the upper level, and he is sound, meaning he can go to work drug-free, chances are he will continue to perform well. But the odds of a four-year-old staying sound and doing Grand Prix work are actually slimmer because it's difficult to tell at such an early age whether a horse will stay sound throughout training.

Although you may have to give the older horse some preventive medicine or some joint supplements to maintain him comfortably, there is still great opportunity for learning and competing, and he can be a real pleasure to own. At fifteen, a horse still has many competitive years left, depending on how well he has been cared for and used during his life.

Don't automatically assume that an older horse will fall apart and pass him by. Understand that the animal may still give you many years of pleasure riding and learning. It depends, however, upon what you are going to ask a horse to do. Compare the horse with a human. For example, you take a risk when you hire a forty-eight-year-old pitcher, but then there are the Roger Clemens of the world who have played for twenty-three seasons. Each horse, like each person, is different. You have to assess the history of the horse, how many owners he has had, and what he has been doing.

Prepurchase Exam

Can an older horse pass a prepurchase exam? As with all prepurchase exams, you have to recognize that you are not going to find a blemish-free horse, regardless of age. With an older horse, for example, there will probably be some joint changes visible in the radiographs. Work closely with your veterinarian to determine which changes are significant and which are not likely to hinder the animal's performance. Look at which changes are progressive and which are not.

WHAT AN OLDER HORSE OFFERS

Does an older horse have anything left to offer a rider? If you want a trained horse, obviously you'll need to have an older horse. If you want an untrained horse, with plans to train him yourself to a high level of competition, you'll want to start with a fairly young horse because, if you choose an unschooled older horse, you will run out of years to train him.

For some trainers, a good horse will be even better when he is old. What older horses lose in physical ability, they make up for in knowledge. For example, try jumping a liverpool (an open-water obstacle) on a green five-year-old. It will be a difficult task because the youngster will not want to go near the water, and if by some chance he does make an effort, it will be clumsy and wild since he hasn't learned to use his body correctly. Do the same jump on an older Grand Prix horse, and he will easily take you up to—and over—the jump. Eventually, one day your older horse will be too old for competitive work, and then you will need to pay him back for what he has given you by taking care of him for the rest of his life.

Chapter 28

Retirement and End of Life

Retirement for both human and horse is a fact of life. It comes to us all, but unlike humans, horses can't tell us when it's time to throw in the towel. So how do you know when your equine partner needs to slow down and take life easier?

Age alone won't tell you. Like people, horses age at their own rates. There are plenty of examples of twelve-year-old horses that have been worked hard or suffered a career-ending injury, thus forcing them into retirement, and then there are the ancient thirty-year-olds that are still gamely carrying dudes through the mountains along the Montana-Idaho border. The bottom line is: age should never be the deciding factor when considering retirement.

Much more difficult than determining when your horse should retire is determining when your horse has reached the end of his life, when all that may be left to him are days filled with pain. Letting a beloved animal go is never easy, but it is your responsibility to make the best decision you can for him at all stages of his life.

RULES FOR RETIREMENT

Knowing your horse goes a long way toward determining when it's time to slow down. Since aging is basically bodily damage and breakdown, if the horse's activity is a contributor to this, then it's time to take it easy and transition into retirement.

If the horse can only perform and compete with medications, is it right to keep him in competition? For most age-related conditions, the medication needed is a painkiller, usually an NSAID (nonsteroidal anti-inflammatory drug) such as bute. Although bute may relieve the pain, it does nothing to relieve the underlying cause, which means the problem is going to get worse irrespective of the fact that the horse acts like he feels better. Bute and other NSAIDs also accelerate the aging process, since the original condition causing the discomfort hasn't been resolved. If a horse can't compete unless he's "buted up," the horse's activity needs to be amended. And most show organizations prohibit or control the use of drugs in competition.

Listen to your horse. He will give you subtle—and overt— clues about his comfort. There may be a change in his attitude, a hesitation in his response to a cue, or a failure to take a lead that wasn't a problem in the past.

THE RETIREMENT PROCESS

Normal aging occurs one cell at a time. It's a gradual, gentle process that has surprisingly little accompanying discomfort because, when it occurs a little at a time, your horse can adapt to it day by day and continue to do his job. But if the aging starts accelerating, his workload needs to be adjusted immediately; otherwise, it will contribute to the accelerated breakdown. Look for distress signs such as loss of weight, loss of visual acuity, loss of athletic ability, recurrent impaction colic, dry or mucus-

This thirty-nine-year-old Thoroughbred partbred is not that unusual. With good care, many horses are living longer lives.

coated manure, dry coat, degenerative joint disease, heart arrhythmia, or shortness of breath. If your horse exhibits some or all of these signs of accelerated aging, you need to immediately lessen his workload. If you don't, you are setting him up for further age acceleration.

Lessening Activity

A complete shutdown of activity may not be necessary. Many older horses still do well with light exercise to keep their muscles toned, their joints lubricated, and their minds active. You might consider allowing an older child to ride and care for your horse part-time. This will not only give the horse something to do, but the child will also benefit from experience with a placid equine. The child's lighter weight will not cause as much stress on the horse's aging bone structure. If your horse has lost a lot of muscle over his back, try using an extra saddle pad to make him more comfortable. A liniment rubdown can also soothe his tired muscles after a workout. Keep in mind that, while riding sessions for older horses are beneficial, they shouldn't be too rigorous. Keep the sessions brief and

Retired horses still have much to give, especially to a child who will benefit from the horse's experience.

manageable. The tight circles and the fast gallop he used to perform should be a thing of the past. Instead, replace them with slower work and easier patterns.

If he's earned his retirement, you can still keep him active and moving by ponying (leading) him off the

A group of retired Arabians at Jordan's Royal Stables provide companionship for each other—so important for older horses.

back of another horse. If you do decide to retire your horse, you might also consider making him a companion to another horse. Companionship is important to all equines—especially to retired horses, who don't have many distractions during their day. What you do for your horse in his last years is repayment for all he has done for you during his prime.

Retirement Caveats

Out of sight, out of mind. Some horse owners simply forget their horses once they send them off to pasture. A senior horse still needs all the care described in the previous section, "Special Care of the Senior Equine." He can have his shoes pulled off if he doesn't have any serious lameness issues, such as laminitis. Choose his retirement home well. He will still need protection from the elements, which usually means a run-in or loafing shed in his pasture. He will also need to be blanketed during the coldest part of the year.

Pasturemates should get along, and the horse should not be chased away from his food or bullied to the point that he is injured. Do not just throw him in a pasture and let the new pasturemates fight it out. A horse that is not used to being with others, particularly in a new place, will have to first establish a place in the existing pecking order. Select herd mates that are placid and friendly toward one another. Age has little to do with it; go for the right personality. There should always be someone who will check on him every day and alert you if any emergencies arise. Finally, make it a point to see your old friend from time to time. Don't just write the check for his boarding and leave it at that.

RETIREMENT/END OF LIFE

Learning the following terms will help you talk about retirement and the end of a horse's life.

■ **Accelerated aging:** Includes loss of weight, visual acuity, and athletic ability, recurrent impaction, colic, dry or mucus-coated manure, dry coat, degenerative joint disease, heart arrhythmia, or shortness of breath.

■ **Degenerative joint disease:** Known also as DJD or osteoarthritis, this condition of lameness is caused by a breakdown of cartilage in the joints.

■ **Euthanasia:** Ending a life to relieve suffering.

■ **NSAIDs:** The abbreviation used for nonsteroidal anti-inflammatories.

■ **Pasturemates:** Equine companions in a pasture.

END OF LIFE DECISIONS

When a horse gets old, he has similar issues to those that geriatric people face. He may become more cantankerous in personality and may even display some signs of senility. His attention span may shorten, and his reaction time may diminish as well.

There will eventually come a time, however, when all of your best efforts to manage your horse's aging processes cannot protect his health and well-being. He will have all but stopped eating, he will take no interest in you or his other companions, and he will be in pain whenever he tries to get around. When this happens, it is time to let your old horse go.

Although death is a part of life, it is still a time full of sorrow for a horse owner. Having a vet send your horse to his final rest is the most humane way to give him a fitting end to his life—quickly and painlessly. While this is a sad time in the owner's relationship with his or her horse and one of the most difficult decisions to make, it is one of the kindest things one can do for an ailing horse.

Death is different for animals than it is for humans. Their reality is different from your own. They live very much in the present, without a concept of the future. Although it is difficult for people to understand, they must always put their horses first. Allowing a horse to linger as his body begins to betray him in earnest—simply for the sake of not wanting to make that call—is insensitive and selfish. Death by natural causes is hardly pleasant, despite what one might think.

Death for a horse is simply an end. He doesn't have lost dreams or sadness—just dignified acceptance. Veterinarians have reported that a peacefulness comes over a horse at the moment of passing.

In the United States, horses are humanely euthanized, or put down, by administering an overdose of barbiturate that acts on the central nervous system. The barbiturate is delivered intravenously. Within a heartbeat, the drug travels to the brain to depress its functions. The horse first loses consciousness, like he would if he were undergoing surgery. The parts of the brain that control breathing and heartbeat are more protected, so it takes another minute or two for these functions to cease. It is important to remember that the horse is unconscious and can't feel any pain.

Euthanasia is not always a quiet event. When a horse is given the initial drug to end his life, he may fall to the ground awkwardly and may groan. Take advice from your vet before deciding whether to be present during the procedure.

Euthanasia is one of the most difficult decisions to make. If your inner voice tells you that euthanasia is the best decision, yet you can't seem to let go, ask yourself why. You may be trying to put off the pain of losing your best friend. You may be feeling like you could have done more for him. You may even hope that natural death will take place so you won't have to make that hard decision yourself. Put aside your own feelings, and do what is best for the horse that has done so much for you.

Grieving

The death of a horse can be every bit as tragic as the death of a human loved one. For many people, their horse is their best friend and a constant being in their lives, one that provides unconditional love. When the horse passes away, the owner experiences profound

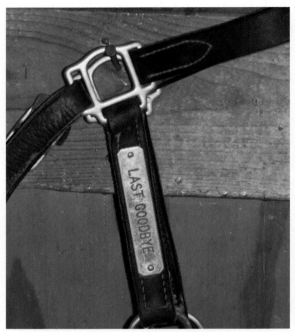

Memorialize your horse in a way that's meaningful to you, such as a specialized halter (*shown here*), a plaque, or a photo.

loss. It is never easy to put a horse down. You may feel overwhelming sadness, anger, and depression—all of these emotions are normal. If friends and relatives don't understand why you are so upset, don't feel that there is something wrong with you. People expect the bereaved individual to return to their normal work and social life, yet for the person who has sustained the loss, life is not as it was before. Your horse was a special friend you lost, and it's natural for you to grieve and mourn his passing.

1. Here are some tips you can use to help ease the pain: Accept the fact that very intense, often unpredictable feelings of grief are normal, especially during the first two weeks. Some people are shocked that they feel more upset when they lose a companion animal than they did when their close friend or relative died. Owners experience a roller coaster of emotions. This is just part of normal grief. You are not going crazy.

2. Horse owners are likely to feel guilty or concerned that they made the wrong choices regarding their horse's care. These feelings should lessen after a couple of weeks. Remember that the decision to euthanize the horse is not the same as killing the horse. Euthanasia provides a good, peaceful end for an animal that would otherwise have suffered and passed anyway.

3. Memorialize your horse in a way that includes others who also cared for him. These individuals can become a support system when family and friends don't understand the depth of the attachment. Websites such as www.petloss.com offer online memorials that can also be helpful in the grieving process.

4. Only return to the horse world when you are ready. Some people love to ride—it is their passion, and without the activity in their lives, they are lost. Others can't bear to be around horses if their own has just died. If it helps, spend some time with a friend's horse. The act of grooming a horse, smelling his coat, and feeling his muzzle on your hand can be very therapeutic. But don't push yourself if you aren't ready. You will know when it is time to get back in the saddle.

Coping

Do whatever helps you to heal. If you hope to reunite with your horse in heaven or want to have a memorial service and invite friends, ignore people who discourage these ways of comforting yourself. Some owners like to publicly take out an ad in equestrian magazines in tribute to their horse, sharing a photo, poem, or story. Others write in journals or compose letters to their horses expressing gratitude for the love and friendship that they shared and the emotions that they are feeling once their companion is gone. Many people also collect small mementos and favorite photos of their horses in an attractive container and keep it in a special place. You can even honor your horse by helping other animals and making a donation to an animal shelter, rescue group, or sanctuary in his memory.

Understand that we all grieve in our own way and in our own amount of time. Your emotions over a horse's death—sadness, guilt, anger, denial, and hopelessness—are as intense and varied as over any other type of loss.

It is said that time heals all wounds, and undoubtedly this is true for the death of your equine companion. It may take some time to get over the loss, but remember that your grief is a reflection of how much he meant to you. You were lucky to have a special relationship for the time you shared together. What you are left with is a wealth of wonderful memories to help you adapt to life without him.

In the sunset of life, it's important to slow down and enjoy each moment you get to share with your horse.

Moira C. Reeve

In Memoriam

Remembering
Moira
1964–2010

SHARON BIGGS: AUTHOR

In 2009, I was on my way to Chicago to write a story about the equine spectacular *Cavalia* when Moira called me on my cell phone and told me she had been approached to write the *Original Horse Bible*. She wanted to know if I'd coauthor it with her. I jumped at the chance! We had always wanted to write a book together and Moira was in remission from breast cancer then, so we thought this would be a great way to celebrate her remission. Although her cancer returned in the middle of writing this book, in perfect Moira form, she persevered, and we were able to finish the book before she became too ill to write.

My association with Moira began in 1996, when I started writing for *Horse Illustrated* and she became my editor. When I finally met her in person, at a big show in Del Mar, California, it was "friends at first sight." We bonded through *Seinfeld* humor, and just knew we'd be best buds forever.

We shared a passion for England, as well as horses, and when a trip there dropped into my lap, she supported my idea for a *Horse Illustrated* story about horses that lived in London. (It was through this assignment that I met Mark, my husband, then a mounted police constable.) Moira also cheered me on when I decided to run away to England in 2000, and even offered to keep my cat, Lira. Moira turned out to be the best foster mom ever! And I was lucky enough to see Moira in England many times during the six years I lived there.

Moira changed my life in so many ways, personally and professionally. I would not be the writer I am today if I hadn't had her guiding hand and understanding. She was the one who taught me how to write a breed profile and how to make a dull subject—such as how to give a horse a bath—remarkable and fun to read. She could look at every subject with a creative eye. Moira also knew when to compliment me and when to correct me. In both situations, she always had encouraging words, and I always grew as a writer. I don't think I write one word today without having her comments in the back of my mind. I know many writers and photographers who feel fortunate to have worked with Moira and learned so much from her.

I will always remember drinking pints of beer at the Bag O' Nails pub down from Buckingham Palace in London with Moira and laughing about the silly things in life. I will always remember her support during some difficult times in my life. And I will always remember her courage as she fought breast cancer. I'm proud to have written this book with her, and I'm proud to have called her my friend.

BOB LANGRISH: EQUINE PHOTOGRAPHER

I worked with Moira for many years and in many places. I came to California on three or four occasions to do day photo shoots with her, particularly for western and English riding. Moira always had a great eye for the finer details, and I was always guided by her expertise. She also had a great sense of humor and could always see the funny side of anything.

I can remember driving her around in California and her joking about whether I knew which side of the road I should be on. At one point, when we were stopped for a red light, I was so busy chatting that I didn't notice when the light changed. Moira finally said to me, "Bob, these lights are *not* going to get any greener. Go!" I use that expression to this day.

Moira and I also worked together compiling a few coffee-

table books, which were sold all over the world. Her knowledge of horses was vast, and if she didn't have the information, she would research it to the nth degree and then write in a way that was light, informative, and jovial, as though she had spent her whole life with some of the rarer breeds of the world. The last published project we worked on before this one was *Wild Horses of the World,* for which the Princess Royal wrote the foreword.

After a hard day's work on shoots, Moira and I always had to sample the local grape juice. When she came to visit us in the UK, we had a wonderful time as we were able to fete her and her husband and shower them with some of our British hospitality. She was due to come over again in October 2010. We were looking forward to her visit tremendously, and I had stocked my wine cellar especially with the best of everything that she would love. Even though she was an incredibly brave person, she was too weak to make the journey, which I will always regret. I will sadly miss my Chardonnay drinking partner.

Moira was a very talented and special person, one I had the honor and privilege to know, work with, and call friend.

Acknowledgments

We have so many people to thank for helping us to make *The Original Horse Bible* a reality.

Our thanks to Sandy Kintzele, Jessica Kutch, Phaedra Valencia, Mark Waller, Terri and Russell King, and Ashley Biggs for lending their amazing help and support to this massive project.

We are also extremely grateful to the following professionals; a good deal of the information in this book would not have been possible without their expertise: Dr. Brian Nielsen, Penny Jargella, Kass Lockhart, Rachael Cox, Ty Cannon, Tom and Neva Scheve, Dr. Lisa Lancaster, Dr. Melyni Worth, Eleanor Blazer, Dr. William A. Schurg, Dr. Lynn Taylor, Dr. D. Paul Lunn, Dr. David W. Horohov, Dr. Pete Sheerin, Dr. John Steiner, Susannah Hinds, Kate Norris, Dr. Julia Wilson, Mark Pranckus, Mitch Taylor, Joel Sheiman, Dr. Craig Reinemeyer, Dr. Dickson Varner, Dr. John Donecker, Dr. Tom Kennedy, Dr. Mary Scollay, Dr. Karen Wolfsdorf, Dr. Jenifer Nadeau, Dr. Thomas Klei, Sherie Grant, Robert Malmgren, Sue Harris, Mary Gatti, Christine Livingston, Sidley Paine, Leslie O'Neal Olsen, Kent Mullenix, Paula Gore, Dr. Thomas Gore, Captain Richard Waygood, and Mike Winter.

We also owe a big thank you to our world-renowned photographer Bob Langrish. Thanks as well to photographers who supplied extra photos, among them Lesley Ward, Elizabeth Moyer, Shawn Hamilton, and Trent Miles. Our thanks to models Katie Lifto, Lori Gabrellie, and Laura Forrester.

Thanks to Jarelle Stein: we couldn't have asked for a more understanding or harder working editor. Our thanks to all the hardworking staff at BowTie Press, especially Lindsay Hanks, Karen Julian, and Andrew DePrisco, and to Elizabeth Moyer, Lesley Ward, and Holly Werner.

Appendix:
Equine Ailments

COMMON CONDITIONS AND DISEASES

Like any animal, the horse is susceptible to a number of conditions and diseases. The following are the ones that are most commonly found in horses. Always talk to your veterinarian for advice about how to treat these conditions and diseases.

Colic

Colic is the generic term for abdominal pain that can be caused by several different things. The attacks and pain can be mild to severe, since every horse reacts to pain differently; if you suspect your horse is having a colic attack, phone your veterinarian right away. There are four common types of colic.

1. Spasmodic colic: This is a spasm of the intestines and the most common of the four types. Spasmodic colic has many causes; one of the most frequent ones is intestinal damage caused by migrating parasite larvae.

Symptoms: The horse may look or bite at his sides, show distress, sweat, lie down, and roll. Guttural sounds may be heard.

Treatment: The veterinarian will administer a relaxant, which usually does the job.

2. Impaction colic: The second most common type, this colic can be caused by an impaction of food, sand, or round balls of mineral deposits, called enteroliths, in the large intestine. Impaction can also be caused by parasites.

Symptoms: The horse may look or nip at his sides or show signs of distress, such as circling or sweating. Your veterinarian will conduct a rectal exam to find out whether the problem is impaction.

Treatment: Your veterinarian will dose the horse, via a stomach tube, with mineral oil or salt water and will administer painkillers.

3. Gas (Tympanic) colic: Less common than the first two types, this colic is caused by fermenting food material or an impaction in the intestine.

Symptoms: Pain is severe, and the horse may roll violently.

Treatment: Gas colic can be relieved either by tapping the stomach with a needle to release the gas or by surgically tapping the intestines.

4. Strangulation: This is the worst type of colic a horse can experience. It occurs when the intestine twists, misaligns, or telescopes onto itself.

Symptoms: The horse will exhibit violent signs of pain, including rolling and thrashing. Shock and death (usually within eight hours) are almost a certainty if a horse is not treated immediately. If the blood supply is cut off to the intestine, a part of the intestine can die off.

Treatment: Surgery is required to correct the twist or to remove the affected intestine. Many times euthanasia is the only option.

Further information: A horse with chronic colic needs to be managed daily. Horses are designed to eat little and often, so they should have access to hay or forage at all times or be given several small meals throughout the day. Any changes in diet should be done gradually. Feeding yeast, probiotic, or prebiotic supplements long-term can help reduce the occurrence of digestive disturbances. Regular deworming is important. Talk to your veterinarian about managing a horse with chronic colic.

People used to believe that, if you let a horse roll when he had colichis intestines would twist. But a twisted gut is not going to occur with rolling or thrashing. A horse can, however, injure himself from rolling or cast himself (get himself in a position where he can't get up). And people can be endangered too. If a horse is thrashing around, you can get caught up and injured. While you're waiting for the veterinarian to arrive, leave your horse in his stall, but remove objects he could get caught up in, such as buckets and hay nets. Don't try to walk your horse. Only a small number of colic cases are eased with exercise. Leave your horse in his stall while you wait for your veterinarian to arrive.

If your horse has a colic attack bad enough to warrant surgery, don't panic. The vast majority of horses undergoing colic surgery make a full recovery. If your horse has colic surgery, be sure to follow the postoperative care instructions closely. A lengthy rehabilitation period is necessary to allow the incision to heal. An eventual return to regular exercise and work is important to build and maintain abdominal muscle.

Encephalitis
Sometimes called sleeping sickness, encephalitis (Western, Eastern, and Venezuelan strains) is a terrible equine disease that is spread by mosquitoes. Although some strains are deadlier than others, all strains infect the brain and spine. If the horse survives, he may have permanent brain damage.

Symptoms: Horses exhibit strange behavior, walking in circles or into walls and other objects. Severe depression, weakness, fever and difficulty with eating are other symptoms.

Treatment: Treatment includes the administration of broad-spectrum antibiotics. Relapses are common, so treatment may have to be repeated.

Further information: Although mosquitoes are not prevalent in dry areas, the American Association of Equine Practitioners (AAEP) continues to list Eastern and Western encephalitis on its core vaccination guidelines due to the grave nature of the disease. The Venezuelan strain is not a problem at the moment, so it is not a core vaccine.

Equine Gastric Ulcer Syndrome (EGUS)
Ulcers are commonly found in mature horses and in foals, particularly in those confined in stalls. In humans, eating stimulates the production of hydrochloric acid. Horses, however, have evolved to eat often, so hydrochloric acid is produced constantly to aid in digestion. If a horse doesn't eat, acid builds up in the stomach, irritating it. Ulcers are created when this stomach acid comes into contact with the acid-sensitive lining at the top of the horse's stomach. Ulcers can affect horses from all disciplines. Studies show that an ulcer can be can be induced in recreational horses in as quickly as a week simply as a result of changing their housing and routines.

Symptoms: In foals, symptoms include lying on the back, colic after nursing/eating, teeth grinding, salivation, and diarrhea. In mature horses, symptoms include poor appetite, weight loss, colic, dull hair coat, attitude changes, and poor performance. Horses with ulcers often crib or windsuck as a way to alleviate the discomfort.

Treatment: In 1999, the FDA approved the use of the acid pump inhibitor omeprazole for horses. This medication was found so effective that little research has been conducted to find new medications. Instead, efforts are turned toward honing dosages, finding how quickly ulcers are formed, and discovering how stress and nutrition affect EGUS. Constant treatment with omeprazole can be costly, but putting an animal on the medication for a week during any change in routine can be effective in helping to prevent ulcers. Dietary change can be helpful. Decrease the amount of grain; instead feed fat, such as corn or vegetable oil, or rice bran in lots of small meals. Up the roughage by allowing constant access to grazing or forage. Alfalfa, with its high calcium content, acts as a buffering agent. Five to eight pounds of alfalfa fed in small meals throughout the day can decrease the pH in the stomach and intestine. Probiotic supplements are also beneficial.

Equine Infectious Anemia (EIA)
EIA is caused by an equid-specific lentivirus in the retrovirus family, equine infectious anemia virus (EIAV). Mosquitoes were once thought to carry it, but horseflies and deer flies appear to be the most common vectors.

Symptoms: High fever, swollen limbs, listlessness, and anemia are symptoms of EIA. Sometimes no symptoms appear at all. EIA is extremely contagious and can infect a herd almost silently. Most countries keep EIA at bay through a test that checks for EIA antibodies in the blood. In North America, this is called the Coggins test and is often required when taking your horse to shows, to a new boarding facility, or across state lines. When purchasing a new horse, ask if the horse has a negative Coggins test.

Treatment: There is no vaccine or cure for EIA, and it is usually fatal. Horses that do recover will be carriers for life and will usually be euthanized.

Equine Influenza

Equine influenza, caused by the orthomyxovirus equine influenza A type 2 (A/equine 2), is a very common contagious disease that affects the upper and lower respiratory tracts. It is spread from horse to horse via droplets expressed through coughing.

Symptoms: Sneezing, fever, runny nose, cough, loss of appetite, and muscle stiffness.

Treatment: As with humans, the flu must run its course. Rest is the best medicine.

Further information: Infection of the herd can be avoided by quarantining all new horses for fourteen days. All horses should be vaccinated for the flu unless the facility is closed to visiting horses. Vaccination usually provides coverage for up to six months. It is an AAEP risk-based vaccine.

Exertional Myopathy: Tying-Up Syndrome and Azoturia

Tying-up is a metabolic syndrome causing muscles to spasm. It is usually seen in the first minutes of exercise, typically during the trot and the canter. Azoturia and the tying-up syndrome are degrees of exertional myopathy, with azoturia being the more pronounced form.

In the past, exertional myopathy was called Monday morning disease because workhorses would show these signs when they went back to work on Monday, after a day off. Exertional myopathy is linked to excess glycogen (sugar) in the system caused from eating grain ration while resting. The sudden return to exercise brought on spasms for those horses.

Symptoms: With azoturia, the horse halts within 15 to 60 minutes into exercise and refuses to move forward. If he is forced to move, his gaits may be very short and stiff. Because azoturia is painful, the horse may sweat and have a rapid pulse. Finally, the horse becomes immobilized, and the muscles in the hindquarters stiffen. The horse may even collapse. The other form of exertional myopathy (tying-up) happens after vigorous exercise when the horse is cooling down. Symptoms include stiffness, trembling, and sweating. The muscles in the hindquarters and loin are hard and sensitive to touch.

Treatment: If the horse shows signs of any form of exertional myopathy, you must seek veterinary care immediately. Leave the horse where he is if it's safe to do so; forcing him to move can cause muscle-tissue damage. If you need to move him to a place of safety and help, walk him slowly and carefully.

Your veterinarian will administer medication to ease the spasms and pain. She or he will also draw blood to look for muscle-tissue damage. Eliminating or restricting grain and molasses (starch) in the diet is key because this reduces glycogen in the system. Horses prone to exertional myopathy have difficulty metabolizing glycogen, which causes the symptoms. There are many excellent low-starch feed products, created to help with this problem. Daily exercise is also important; days off only exacerbate the problem.

Further information: Horses that have chronic issues with tying-up may have an inherited disease called polysaccharide storage myopathy (PSSM). PSSM causes a horse's system to overreact to the insulin created when consuming carbohydrates and starches, which raise blood sugar levels. This is most common in American Quarter Horses and Paints. A muscle biopsy can diagnose PSSM. Treatment is easy, and many horses never display any symptoms if they are exercised regularly and fed a diet low in starch and high in fat, such as vegetable oil or rice bran. You may supplement the diet with vitamin E and selenium.

Potomac Horse Fever (PHF)

Potomac horse fever is a disease caused by the organism *Neorickettsia risticii*. It is prevalent near waterways throughout the United States from late spring to early fall, with most cases occurring in the summer. The organism uses water snails as its first intermediate host, which then infect aquatic insects (caddisflies, mayflies, damselflies, dragonflies, and stoneflies) as its secondary host. These insects hatch in great numbers in the humid months of summer and are attracted to horse facilities by artificial lighting. PHF infection occurs when insect carcasses get into stalls and hay supplies where they are ingested by horses.

Symptoms: Mild colic, fever, diarrhea, and (in pregnant mares) abortion. Severe cases can lead to laminitis.

Treatment: PHF can be treated with the antibiotic oxytetracycline, fluids, and NSAIDs (nonsteroidal anti-inflammatory drugs). If PHF is caught early enough, horses generally respond quickly. Laminitis is usually severe and difficult to treat.

Further information: There is strong evidence to suggest that turning lights off at facilities will go a long way toward preventing Potomac horse fever. Motion detector lights are a good alternative to floodlights. PHF vaccine is an AAEP risk-based vaccine.

Rabies

Rabies is not frequently found in horses, but it is possible for one to catch rabies through a wild animal. Rabies is always fatal. The rabies virus is transmitted via saliva, usually from the bite of an infected wild animal such as a raccoon, a fox, a skunk, or a bat. Since rabies is zoonotic (meaning it spreads to humans), anyone who has had contact with a rabid animal may be at risk. Fortunately, the vaccine is extremely effective.

Symptoms: Lack of coordination, depression, and aggression.

Treatment: There is no treatment; euthanasia is required.

Further information: Public health consequences are significant if a horse is found to be rabid, therefore the AAEP has placed the rabies vaccine on its core vaccination guidelines, which means all horses should have the vaccine regardless of region or circumstances.

Strangles

Strangles (also known as distemper) is caused by a bacterium called *Streptococcus equi* subspecies *equi* (*S. equi var. equi*). Although strangles is a highly contagious disease that usually affects young horses (weanlings and yearlings), all other ages can be infected.

Some horses, called "shedders," may continue to release the virus well after infection. The organism is transmitted from contact with infected horses or shedders. Transmission can also happen through indirect contact with water buckets/troughs, hoses, feed mangers/buckets, pastures, stalls, grooming equipment, sponges/cloths, human hands and clothing, and even insects who have come into contact with nasal discharge or pus.

Symptoms: Strangles affects the respiratory tract. Although usually not fatal, strangles can cause permanent damage to the lungs. Coughing, fever, weepy eyes and nose, loss of appetite, a fever of 102–106 degrees Fahrenheit are all signs of strangles. Swollen lymph nodes under the chin are hallmarks of the condition. These abscesses will eventually burst and release pus.

Treatment: Isolation and therapeutic doses of antibiotic, such as procaine penicillin. Everything the horse has come into contact with must be disinfected.

Further information: If strangles is a common problem in your area or if your horse is exposed to many new horses, you should consider vaccination. However, vaccination in the face of an outbreak should be carefully considered, as there is a significantly increased risk of adverse reactions to the vaccine in already exposed horses. Strangles vaccination is considered a risk-based vaccine by the AAEP. This is because many otherwise healthy horses, especially mature horses, may develop adverse responses to the vaccine. Consult your veterinarian for advice about vaccinating before you administer it.

Tetanus

Tetanus is caused by the spore-forming bacterium *Clostridium tetani*, and it all too frequently proves to be fatal for the infected horse. The bacterium is common in the intestinal tract and feces of animals and humans; the bacterium is also abundant in the soil. Spores can live in the environment for many years, which presents a constant risk to horses and people. It is not contagious, but infection can occur through puncture wounds, incisions, and exposed tissue, such as the umbilicus in foals.

Symptoms: Muscles become stiff and rigid.

Treatment: Treatment includes injections of tetanus antitoxin and administration of antibiotics, tranquilizers, and sedatives. The affected horse will be sensitive to light and noise, so he must be kept in a quiet, darkened stall.

Further information: Tetanus is on the AAEP's core vaccination guidelines.

West Nile Virus (WNV)

West Nile virus causes arbovirus encephalitis in horses and in humans. Horses represent 96.9 percent of all nonhuman mammalian cases of WNV disease.

The virus is prevalent in all of the United States and most of Canada and Mexico. WNV originates in birds and is transmitted to horses and humans (the dead-end hosts) via several species of mosquitoes. The virus does not spread from horses or humans; it must originate in birds.

Symptoms: The virus affects the central nervous system, and symptoms mimic encephalitis. Symptoms also may include the following: loss of appetite, depression, aimless wandering, walking in circles, weakness/paralysis of hind legs, and inability to swallow.

Treatment: There is no specific treatment for WNV; supportive veterinary care is recommended. Although death can occur, the majority of horses infected with the West Nile virus recover.

Further information: Vaccination is not a guarantee against infection. Since mosquitoes carry WNV, it's important to rid the stable yard of any breeding sites. Mosquitoes breed in stagnant water, so dispose of or turn over any vessels, such as cans, buckets, and pots. Remove tires from your property, as this is a favorite breeding site. Fogging the area with a commercial insecticide at night when mosquitoes are active is helpful. Report any dead birds found near the stable to your state's Department of Health. West Nile virus is on the AAEP core vaccination guidelines list. This is a mosquito-borne virus vaccination and timing of administration is important. It must be given before mosquitoes come out. Check with your veterinarian for the optimal timing to vaccinate your horse.

EQUINE HEAT ILLNESSES

In general, horses handle cold weather much better than they do hot weather. Of course, the desert breeds, such as the Arabian and the Akhal Teke, can manage heat much better than can cold-blooded draft breeds, such as the Percheron and the Clydesdale. There are a few heat illnesses to be mindful of during hot weather, especially in older horses that have problems with perspiration and body temperature regulation. Humans and horses are the only animals that perspire through their skin to regulate the body temperature, but that can become more difficult for both species as we age.

Anhidrosis

Anhidrosis is the inability to sweat during exercise or hot weather. The cause is thought to stem from problems with epinephrine receptors.

Symptoms: Your horse does not sweat in typical situations. He will also pant in a rapid pattern in an effort to cool himself. He might develop a fast pulse, dry skin, and a high fever.

Treatment: Anhidrosis is rare and usually affects horses in hot and humid areas of the country. Recommended treatment is to keep the horse in an air-conditioned stall or to move him to a temperate state.

Heat Stroke or Heat Exhaustion

When a horse can't sweat efficiently enough to cool his body, heat exhaustion or heat stroke can occur.

Symptoms: Depression, weakness, appetite loss, hot dry skin, and immobility. Pulse/respiratory rates increase and temperature may rise to 106-110 degrees Fahrenheit. Convulsions, coma, and death can result.

Treatment: Lower the horse's temperature quickly. Get him into shade and a breeze (such as a fan), and hose him down with cold water. Ice packs are also helpful. Call your vet immediately.

Synchronous Diaphragmatic Flutter (SDF)

Commonly known as thumps, SDF stems from low serum calcium due to electrolyte depletion.

Symptoms: A twitch or spasm in the flank not related to normal breathing. You'll feel or hear a thumping sound in unison with the horse's heartbeat

Treatment: SDF will usually go away on its own. But sometimes thumps are serious enough to require IV treatments. Phone your veterinarian if you suspect SDF.

FOOT DISEASES

Although the horse's foot may look unremarkable at first glance, it's an intricate work of engineering that can be compared to the bones of the human finger. *Foot* is a broad term encompassing the hoof and its internal structures. The hoof refers to the outer covering of the internal structure. Here are four of the most common diseases that affect the foot.

Laminitis (Founder)

Laminitis, or founder, is a metabolic and vascular disease that causes a breakdown of the laminar tissue. It is a painful and debilitating affliction that often leads to euthanasia. It is more common in ponies and stallions than in mares, and it will usually affect only the two front feet, although it can occur in all four. Laminitis can happen when bacterial endotoxins and lactic acid release into the bloodstream. This causes the digital arteries in the foot to dilate at the same time causing the small capillaries that feed the laminae to constrict. The resulting swelling causes the damage to the tissue.

Acute laminitis can be caused by several things, such as a retained placenta in a broodmare, but it is often caused by an over consumption of carbohydrates, namely grain and other concentrated feed. This changes the delicate bacterial population within the cecum and leads to endotoxin and lactic acid release.

Symptoms: These include fever, sweating, diarrhea, rapid pulse and breathing. The digital artery at the fetlock will throb. The hooves will be extremely painful to the horse, and he may rock back or shift his weight from side to side to relieve pressure in his feet.

Treatment: Call your veterinarian immediately, and remove all feed. Your veterinarian will coat the stomach with mineral oil via a stomach tube and administer pain relievers such as nonsteroidal anti-inflammatory medications (NSAIDs). Cold hosing or packs to the hooves can help relieve the pain. Other medications, such as vasodilators to improve blood flow to the hoof, may also be prescribed. Removing the shoes and providing a soft place for the horse to stand is also beneficial.

Further information: If your horse continues to show symptoms for more than two days, he may have chronic laminitis. This can lead to permanent damage in the hooves, where the coffin bone detaches from the hoof wall and rotates within the hoof. The tip of the bone can penetrate the sole. A low-starch diet and corrective hoof trimming and shoeing will be important if your horse has chronic laminitis. In severe cases, a horse will be unable to return to previous work and may only be sound and comfortable enough for a life at pasture.

Navicular Disease (Navicular Syndrome)

Navicular disease is an inclusive term that denotes problems with the navicular bone and surrounding structures in the foot. It is found in horses older than five years of age and causes front leg lameness. Horses that start and stop at speed, such as racehorses, barrel racers and jumpers, are particularly prone to navicular disease because this type of work puts stress on the front feet.

The navicular bone is a small, flat bone found at the heel of the foot beneath the frog. It is attached to the pedal bone by the impar ligament and to the pastern joint by suspensory ligaments. The navicular bone's job is to provide support to the coffin and short pastern bones. The surrounding structures include the deep digital flexor tendon (a pulley-type device that runs around the bone and attaches to the back of the coffin bone) and the navicular bursa (which acts as a cushion). Navicular disease can affect horses with feet that provide poor shock absorption, such as with poor foot conformation, underslung and contracted heels, and bad or infrequent hoof trimming. The primary cause of navicular disease isn't well understood, but trauma to the navicular bone, the interference of the blood supply, and damage to the tendon, ligaments, and bursa are factors that can lead to the disease.

Symptoms: These include a shortened stride when moving on a downhill slope or on hard ground or small circles and mild lameness that comes and goes. As the disease progresses, lameness becomes more common. The pain occurs in the heel, and a sign of this pain is toe-stabbing, in which the horse puts his toe down first as he moves; sometimes, he'll stumble over his toe. One foot may be affected more than the other, and the horse will often stand with the afflicted foot pointed forward.

Your veterinarian will use tongs called hoof testers to test for pain, applying pressure with these to the frog area. He will also conduct a diagnostic called a nerve block, in which various areas in the foot around the nerves are blocked with a local anesthetic to see if lameness is alleviated. If the pain is alleviated, this is a sign that the lameness lies within the navicular area. Your veterinarian also may take radiographs, although this method of diagnosis may not be accurate, as many professionals dispute what is normal and what is not.

Treatment: Navicular disease cannot be cured, only managed. It is very important to employ corrective trimming and shoeing, such as with a rolled toe eggbar shoe, which removes pressure on the heels and encourages correct toe breakover. NSAIDs can help with inflammation, and medications, such as isoxsuprine hydrochloride, can help with blood flow to the navicular bone. The navicular bursa can be eased with injections of corticosteroids and hyaluronic acid. A last resort treatment is a surgical procedure called a palmar digital neurectomy to desensitive the foot, which removes both digital nerves in the back half of the foot.

Further information: Because navicular disease can be severely debilitating to a horse (and indeed cause the end of a career), make sure that you ask your veterinarian about navicular disease during a prepurchase exam, especially if your potential horse is more than five years of age.

Thrush

Thrush is a common bacterial infection of the frog, the spongy fleshy portion in the center of the sole of the foot. There are many bacterial species that cause thrush, but the usual culprit is *Fusobacterium necrophorum*. Thrush is caused by lack of care, namely stabling the horse in a dirty stall or muddy pasture, or failing to clean the hooves of mud and debris each day. Wet and dirty environments and a buildup of debris in the hooves create the perfect environment for bacteria to thrive. Regularly scheduled farrier visits will also help keep thrush at bay; paring back the frog will stop debris from becoming enclosed in flaps of tissue. In bad cases, thrush can cause lameness.

Symptoms: Signs of thrush include a greasy black and foul-smelling substance, usually seen when cleaning the clefts of the frog.

Treatment: Move the horse to a dry environment. Pick out the hooves daily, and clean around the frog and heal bulbs with a Popsicle stick. Apply a topical anti-thrush solution, such as Thrushbuster. Talk with your farrier about shoeing the hoof to help promote a healthy frog.

Further information: Thrush caught early can be fairly easy to clear up; however if left untreated, thrush can eat into the inner sensitive structures of the foot.

White Line Disease

White line disease, which is also called seedy toe, affects the hoof wall. The white line is the connection between the sole and the inner part of the hoof wall. White line disease begins in the outer portion of the hoof wall. It starts at the bottom of the hoof at the ground surface. The damage can work its way up, in some cases moving all the way to the coronary band.

The disease may more correctly be called outer hoof wall disease because that is the location of the infection that has been found in published studies. This infection is usually fungal. Some studies have documented bacterial invasion, although the fungus is probably the first invader on the scene.

Symptoms: To check for white line disease, you have to look at the sole of your horse's bare foot. You cannot see it in the shod hoof because the shoe covers the hoof wall at the ground surface. To check for white line disease, clean out your horse's feet and brush off any dirt or shavings covering the outer edge of the bottom of the foot. In black feet, you will see the division between the inner and outer hoof wall easily because the inner hoof wall has no pigment. Look for cracks in this zone where the outer dark hoof wall joins the inner unpigmented horn. (In white feet, the color of the inner and outer hoof wall looks the same, but if there is white line disease you will still see damage.) If you see cracks in this region, white line disease may be starting. You may see some dirt or pebbles stuck in the cracks. Sometimes you can see white flaky horn or black ooze in the cracks.

In some feet, there isn't much to see in the cracks, just open space. In bad cases, the cracking and infection can spread from the outer wall to the white line, which connects to the sole. If the white line is involved, you may see separation, stretching, or small pockets of missing white line where the infection has eaten away at the horn. If the damage extends up the hoof wall, you can sometimes hear a hollow sound when tapping a hoof pick against the outer hoof wall.

Treatment: Since a weak hoof wall predisposes the foot to this disease, the treatment involves removing the underlying problem. For example, if it was long toes or imbalanced shoeing that weakened the hoof wall, a farrier may need to reshoe at more frequent intervals and/or change hoof balance. Sometimes antifungal medication is needed if microbial invasion is extensive.

Hoof horn tissue does not heal like skin, because hoof horn is dead tissue, like our hair and nails. New healthy horn must replace it. White line disease can take weeks to months to clear up, depending on how far up the hoof wall the invasion has caused damage.

Further information: Anything that weakens the outer hoof wall may lead to the disease. Some common causes of weak hoof walls include long toes, laminitis, unbalanced trimming, harsh environmental changes such as excess dryness alternating with wet conditions, shoes left on too long, or overworking a horse that is not conditioned for a hard ride. Although fungal infection is identified in this disease, it is not usually believed to cause it. The organisms cultured from diseased hoofs are common in the soil and can be found on the surface of healthy hoofs too. But if the hoof wall becomes weak, the organisms can invade and cause damage to the wall, and eventually to the white line if left untreated.

Resources

ASSOCIATIONS AND ORGANIZATIONS

Here are associations and organizations that can help answer questions on equine health, welfare, and breeds.

Akhal-Teke
Akhal-Teke Association of America;
    www.akhal-teke.org

American Cream Draft
American Cream Draft Horse Association;
    www.acdha.org

American Indian Horse/Horse of the Americas
American Indian Horse Registry;
    www.indianhorse.com
Horse of the Americas Registry;
    www.horseoftheamericas.com

American Paint Horse
American Paint Horse Association;
    www.apha.com

American Quarter Horse
American Quarter Horse Association;
    www.aqha.com
Foundation Quarter Horse Registry;
    www.fqhrregistry.com
National Foundation Quarter Horse Association;
    www.nfqha.com

National Quarter Horse Registry;
    www.nqhr.com

American Saddlebred
American Saddlebred Horse Association;
    www.saddlebred.com

American White and American Creme
American White and American Creme
    Horse Registry;
    www.cheval-creme.com/engawachr.html

Andalusian and Lusitano
International Andalusian & Lusitano Horse
    Association;
    www.ialha.org
The Foundation for the Pure Spanish Horse;
    www.prehorse.org

Appaloosa
American Appaloosa Association;
    www.amappaloosa.com
Appaloosa Horse Club;
    www.appaloosa.com
Foundation Appaloosa Horse Registry;
    www.foundationapp.org
International Colored Appaloosa
    Association;
    www.icaainc.com

Appaloosa Sport Horse
Appaloosa Sport Horse Association;
> www.apsha.org

Appendix Horse (Quarter Horse X Thoroughbred)
American Appendix Horse Association;
> www.americanappendix.com

Arabian
Arabian Horse Association;
> www.arabianhorses.org

Pyramid Society/Egyptian Arabian;
> www.pyramidsociety.org

Azteca
American Azteca Horse International Association;
> www.americanazteca.com

Banker Horse
Corolla Wild Horse Fund;
> www.corollawildhorses.com

Barb
Spanish Barb Breeders Association;
> www.spanishbarb.com

Belgian Warmblood
Belgian Warmblood Breeding Association/
> North American District;
> www.belgianwarmblood.com

Belgian
Belgian Draft Horse Corporation of America;
> www.belgiancorp.com

Buckskin
American Buckskin Registry Association;
> www.americanbuckskin.org

International Buckskin Horse Association;
> www.ibha.net

Canadian Horse
Canadian Horse Breeders Association;
> www.lechevalcanadien.ca

Canadian Sport Horse
Canadian Sport Horse Association;
> www.c-s-h-a.org

Carolina Marsh Tacky
Carolina Marsh Tacky Association;
> www.marshtacky.org

Caspian
Caspian Horse Society of the Americas;
> www.caspian.org

Chincoteague Pony
National Chincoteague Pony Association;
> www.pony-chincoteague.com

Cleveland Bay
Cleveland Bay Horse Society of North America;
> www.clevelandbay.org

Clydesdale
CIydesdale Breeders of the USA;
> www.clydesusa.com

Colorado Ranger
Colorado Ranger Horse Association;
> www.coloradoranger.com

Connemara Pony
American Connemara Pony Society;
> www.acps.org

Curly Horse
American Bashkir Curly Registry;
> www.abcregistry.org

Curly Sporthorse International;
> www.curlysporthorse.org

International Curly Horse Organization/
> North American Curly Horse Registry;
> www.curlyhorses.org

Dales Pony
Dales Pony Association of North America;
> www.dalesponyassoc.com

Dartmoor Pony
Dartmoor Pony Registry of America;
> www.dartmoorpony.com

Dutch Warmblood
Dutch Warmblood Studbook in North America;
> www.kwpn-na.org

Fell Pony
Fell Pony Society and Conservancy of
  the Americas;
  www.fellpony.org
Fell Pony Society of North America;
  www.fpsna.org

Florida Cracker Horse
Florida Cracker Horse Association;
  www.floridacrackerhorses.com

Friesian
Friesian Horse Association of North America;
  www.fhana.com
International Friesian Show Horse Association;
  www.friesianshowhorse.com

Georgian Grande
International Georgian Grande Horse Registry;
  www.georgiangrande.com

Gypsy Vanner Horse
Gypsy Cob and Drum Horse Association;
  www.gcdha.com
Gypsy Horse Registry of America;
  www.gypsyhorseregistryofamerica.org
Gypsy Vanner Horse Society;
  www.vanners.org

Hackney Horse
American Hackney Horse Society;
  www.hackneysociety.com

Haflinger
American Haflinger Registry;
  www.haflingerhorse.com

Hanoverian
American Hanoverian Society;
  www.hanoverian.org

Holsteiner
American Holsteiner Horse Association;
  www.holsteiner.com

Hungarian Horse
Hungarian Horse Association of America;
  www.hungarianhorses.org

Icelandic Horse
United States Icelandic Horse Congress;
  www.icelandics.org

Irish Draught Horse
Irish Draught Horse Society of North America;
  www.irishdraught.com

Kentucky Mountain Saddle Horse
Kentucky Mountain Saddle Horse Association;
  www.kmsha.com
Spotted Mountain Horse Association;
  www.kmsha.com/smha/

Kerry Bog Pony
American Kerry Bog Pony Society;
  www.kerrybogpony.org

Kiger Mustang
Kiger Mesteño Association;
  www.kigermustangs.org
Steens Mountain Kiger Registry;
  www.kigers.com/smkr

Lipizzan
Lipizzan Association of North America;
  www.lipizzan.org
United States Lipizzan Registry;
  www.uslr.org

Mangalarga Marchador
United States Mangalarga Marchador Association;
  www.usmarchador.com

Miniature Horse
American Miniature Horse Association;
  www.amha.org
American Shetland Pony Club/
  The American Miniature Horse Registry;
  www.shetlandminiature.com

Missouri Fox Trotter
Missouri Fox Trotting Horse Breed Association;
  www.mfthba.com

Morab
International Morab Breeders' Association;
  www.morab.com

Purebred Morab Horse Association and Registry;
www.puremorab.com

Morgan
American Morgan Horse Association;
www.morganhorse.com
Lippitt Club;
www.lippittclub.net
Lippitt Morgan Breeders' Association;
www.lippittmorganbreedersassociation.com
Morgan Single-Footing Horse Association;
www.gaitedmorgans.org

Mustang
American Mustang and Burro Association;
www.ambainc.net
Cloud Foundation;
www.thecloudfoundation.org
National Mustang Association;
www.nmautah.org
National Wild Horse Association;
www.nwha.us
Spanish Mustang Registry;
www.spanishmustang.org

National Show Horse
National Show Horse Registry;
www.nshregistry.org

New Forest Pony
New Forest Pony Association and Registry;
www.newforestpony.net

Nokota
Nokota Horse Conservancy;
www.nokotahorse.org

Nez Perce Horse
Nez Perce Horse Registry;
www.nezpercehorseregistry.com

Norwegian Fjord
Norwegian Fjord Horse Registry;
www.nfhr.com

Oldenburg
Oldenburg Horse Breeders' Society;
www.oldenburghorse.com

Oldenburg Registry N.A. and Intl. Sporthorse Registry;
www.isroldenburg.org

Palomino
Palomino Horse Association;
www.palominohorseassoc.com
Palomino Horse Breeders of America;
www.palominohba.com

Paso Fino
Paso Fino Horse Association;
www.pfha.org
Pure Puerto Rican Paso Fino Federation of America;
www.puertoricanpasofino.org

Percheron
Percheron Horse Association of America;
www.percheronhorse.org

Performance Horse
Performance Horse Registry;
www.phr.com

Peruvian Horse
North American Peruvian Horse Association;
www.napha.net

Pinto
Pinto Horse Association of America;
www.pinto.org

Pony of the Americas
American Quarter Pony Association;
www.aqpa.com
Pony of the Americas Club;
www.poac.org

Racking Horse
Racking Horse Breeders' Association of America;
www.rackinghorse.com

Rocky Mountain Horse
Rocky Mountain Horse Association;
www.rmhorse.com

Shackleford Horse
Foundation for Shackleford Horses;
www.shacklefordhorses.org

Shagya-Arabian

North American Shagya-Arabian Society;
www.shagya.net

Shetland Pony

American Shetland Pony Club/American Miniature
Horse Registry;
www.shetlandmini.com

Shire

American Shire Horse Association;
www.shirehorse.org

Sport Horses of Color

International Sport Horses of Color;
www.shoc.org

Spotted Saddle Horse

National Spotted Saddle Horse Association;
www.nssha.com
Spotted Saddle Horse Breeders & Exhibitors Association;
www.sshbea.org

Standardbred

Standardbred Pleasure Horse Organization of Mass.;
www.standardbredhorse.com
United States Trotting Association;
www.ustrotting.com

Suffolk Punch

American Suffolk Horse Association;
www.suffolkpunch.com

Swedish Warmblood

Swedish Warmblood Association of North America;
www.swanaoffice.org

Tarpan (Modern)

North American Tarpan Association;
www.tarpanassociation.com

Tennessee Walking Horse

Friends of Sound Horses;
www.fosh.info
National Walking Horse Association;
www.nwha.com
Part Walking Horse Registry;
www.hiplainswalkers.com

Tennessee Walking Horse Breeders'
and Exhibitors' Association;
www.twhbea.com
Walking Horse Owners Association;
www.walkinghorseowners.com

Thoroughbred

Jockey Club;
www.jockeyclub.com

Trakehner

American Trakehner Association;
www.americantrakehner.com

Walkaloosa

Walkaloosa Horse Association;
www.walkaloosaregistry.com

Welara

American Welara Pony Registry;
www.welararegistry.com

Welsh Pony and Welsh Cob

Welsh Pony and Cob Society of America;
www.welshpony.org

OTHER EQUINE ORGANIZATIONS

Included below is a listing of associations for various
riding activities, professional horsepeople, and welfare issues.

Activities

American Driving Society;
www.americandrivingsociety.org
American Endurance Ride Conference;
www.aerc.org
American Polocrosse Association;
www.americanpolocrosse.org
American Ranch Horse Association;
www.americanranchhorse.net
American Trail Horse Association;
www.trailhorse.com
American Vaulting Association;
www.americanvaulting.org
Carriage Association of America;
www.caaonline.com
Cowboy Mounted Shooting Association;
www.cowboymountedshooting.com

Extreme Cowboy Association;
www.extremecowboyassociation.com
Intercollegiate Dressage Association;
www.teamdressage.com
Intercollegiate Horse Show Association;
www.ihsainc.com
International Hunter Futurity;
www.inthf.org
International Performance Horse Development
Association;
www.iphda.com
International Side Saddle Organization;
www.sidesaddle.com
Masters of Foxhounds Association of America
and the MFHA Foundation;
http://mfha.org
National Barrel Horse Association;
www.nbha.com
National Cutting Horse Association;
www.nchacutting.com
National Hunter Jumper Association;
www.nhja.org
National Reined Cow Horse Association;
www.nrcha.com
National Reining Horse Association;
www.nrha.com
National Snaffle Bit Association;
www.nsba.com
National Team Roping Horse Association;
www.ntrha.com
National Versatility Ranch Horse Association;
www.nvrha.org
North American Trail Ride Conference;
www.natrc.org
Professional Rodeo Cowboys Association;
www.prorodeo.com
United States Dressage Federation;
www.usdf.org
United States Equestrian Federation;
www.usef.org
United States Equestrian Team Foundation;
www.uset.org
United States Eventing Association;
www.useventing.com
United States Hunter Jumper Association;
www.ushja.org
United States Polo Association;
www.us-polo.org

United States Team Penning Association;
www.ustpa.com

Care and Welfare
American Humane Association;
www.americanhumane.org
American Horse Defense Fund;
www.ahdf.org
American Society for the Prevention of
Cruelty to Animals;
www.aspca.org
Communication Alliance to Network
Thoroughbred Ex-Racehorses;
www.canterusa.org
Foundation for the Preservation and Protection
of the Przewalski Horse;
www.treemail.nl/takh/
Hooved Animal Humane Society;
www.hahs.org
Humane Society of the United States;
www.hsus.org
Unwanted Horse Coalition;
www.unwantedhorsecoalition.org

General/Policy Organizations
American Horse Council;
www.horsecouncil.org
American National Riding Commission;
www.anrc.org
British Horse Society;
www.bhs.org.uk
Kentucky Horse Park;
www.kyhorsepark.com
North American Riding for the
Handicapped Association;
www.narha.org

Professionals
American Association of Equine Practitioners;
www.aaep.org
American Riding Instructors Association;
www.riding-instructor.com
Certified Horsemanship Association;
www.cha-ahse.org
Master Saddlers Association;
www.mastersaddlers.com
United Professional Horsemen's Association;
www.uphaonline.com

Photo Credits

Photographs in this volume are copyright Bob Langrish, except for the images that are listed below. These photographs and illustrations are copyright/courtesy of the following:

PREFACE—**6**: Shutterstock.
CHAPTER 1—**14 (top, right)**: Library of Congress, LC-DIG-npcc-27022; **14 (bottom)**: Painting by Heinrich Harder, courtesy David Goldman, www.search4 dinosaurs.com; **16**: Library of Congress, LC-DIG-ppmsca-14367; **18, 20**: Shutterstock; **21**: Library of Congress, LC-DIG-jpd-00811; **22**: Custom and Border Protection, photo by James Tourtellotte; **24**: Library of Congress, LC-DIG-ppmsca-06607; **26**: Library of Congress, LC-USZ62-49241; **28 (top)**: World Horse Welfare, www.worldhorse welfare.org; **28 (bottom)**: Nevada Historical Society.
CHAPTER 2—**32, 33, 36, 37**: Robin Peterson, www.fernwoodstudio.com; **38**: Moira C. Reeve.
CHAPTER 4—**60 (left)**: American Indian Horse Registry; **68 (bottom)**: Library of Congress, LC-DIG-ppmsca-18429-00020; **72 (top)**: Main and State Stud Marbach; **77**: Jackie McFadden, Carolina Marsh Tacky Association; **92**: International Georgian Grande Horse Registry; **93 (bottom)**: Library of Congress,LC-DIG-ppmsc-00031; **109 (bottom)**: Library of Congress, LC-USZ62-76953; **111 (bottom)**: American Kerry Bog Pony Society (Linda Ashar); **112**: Courtesy of Chris Bredeson, Board Member, Hungarian Horse Association of America, photo by Diane McGregor; **114 (top)**: Tereza Huclova, www.terezahuclova .com; **116 (bottom)**: Library of Congress, LC-DIG-ggbain-17082; **122 (bottom)**: James R. Spencer; **125 (bottom)**: Library of Congress, LC-DIG-ggbain-00350; **135 (bottom)**: Association for Recuperation and Development of Pleven Horse and Gidran Breeds in Bulgaria; **146 (top)**: Lynda Konrad, North American Tarpan Association; **148 (top)**: Library of Congress, LC-DIG-cwpb-01997; **149**: Walkaloosa Horse Association; **152 (top)**: Baden-Wurttemberg Breed Registry, photo by Olav Krenz; **153 (bottom)**: Library of Congress, LC-USZ62-118299.
CHAPTER 8—**198, 199 (bottom)**: Shutterstock; **202**: Moira C. Reeve.
SECTION 5 OPENER—**212–13**: Shutterstock.
CHAPTER 9—**221**: Shutterstock.
CHAPTER 10—**225**: Shutterstock; **230**: Adrian G. Stewart, Adrian@mulberry alpacas.com.
CHAPTER 13—**274**: United States Department of Agriculture, Agricultural Research Service, K7505-1, photo by Scott Bauer; **278**: Shutterstock.
CHAPTER 14—**280, 284 (bottom)**: Shutterstock.
SECTION 7 OPENER—**288–89**: Lesley Ward.
CHAPTER 15—**292**: Lesley Ward; **297–99**: Giovanna Tarantino.
CHAPTER 16—**300**: Lesley Ward; **303, 306, 307 (top)**: Moira C. Reeve.
CHAPTER 17—**312**: Shutterstock.
SECTION 8 OPENER—**316–17**: Moira C. Reeve.
CHAPTER 19—**324, 334**: Moira C. Reeve.
CHAPTER 21—**344**: Library of Congress, LC-USZ62-45060.
CHAPTER 22—**353**: Shutterstock; **355 (right)**: Moira C. Reeve.
CHAPTER 23—**358**: Moira C. Reeve; **363 (left), 366**: Shawn Hamilton, CLiX Photography, www.clixphoto .com; **363 (right)**: Lesley Ward.
CHAPTER 24—**371**: Moira C. Reeve; **373**: Shutterstock; **374**: Trent Miles, Culver Academies.
CHAPTER 27—**407**: American Veterinary Medical Association.
CHAPTER 28—**410**: Shutterstock; **414**: Elizabeth Moyer.
IN MEMORIAM—**416, 418, 419**: Steve Zepezaur; **417**: Charles Mann.
INDEX—**460**: Elizabeth Moyer.

Index

airs above the ground movements, 158, *159*
Akhal-Teke (Turkmenistan)
    about, 48, *59,* **59–60**
    and Karabakh, 107
    and Lokai, 119
    and Nez Perce Horse, 122–23
ALBC (American Livestock Breed Conservancy)
    and Carolina Marsh Tacky, 77
albino-type horses, 53
alfalfa hay, 237, 239, 426
all-around horses, 179–80
alternative medical care, 281, 282, 284–86.
    See also holistic horsemanship
Altèr-Real (Portugal)
    about, **58,** *58*
    and Lusitano, 112
amble gait
    about, 37
    Banker Horse, 71
    Florida Cracker Horse, 96
    Kentucky Mountain Saddle Horse, 110
    Narrangansett Pacer, 64
    Walkaloosa, 149
American Association of Equine Practitioners
    (AAEP), 263, 428, 429
American Cream Draft (US), *60,* **60–62**
American Cream Draft Horse Association, 61, 62
American Farriers Association (AFA), 262
American Hanoverian Society (AHS), 102
American Horse Council, 27
American Horse Defense Fund (AHDF), 27
American Horse Slaughter Prevention Act (2006), 27
American Humane Association Film and
    Television Unit, 29
American Indian Horse (US), **60,** *60*
American Indian Horse Registry (AIHR), 60
American Livestock Breed Conservancy (ALBC)
    and Carolina Marsh Tacky, 77
American Paint Horse (US), 54, *61,* **62–63**
American Paint Horse Association, 62
American Quarter Horse (US)
    about, *40, 42, 54, 63,* **63–64**
    and American Paint Horse, 62
    and Appaloosa, 67
    and Azteca, 69
    and Barb, 71
    crop-outs, 62
    muscle fiber of, 32

    and Pony of the Americas, 131
    speed of, 33
    and Thoroughbred, 149–50
American Quarter Horse Association, 62, 64
American Saddlebred (US)
    about, *64,* **64–65,** *161*
    and Canadian Horse, 75
    and Georgian Grande, 92
    and Missouri Fox Trotter, 114
    and National Show Horse, 120
    in saddle seat competition, 161, *161*
    and Thoroughbred, 149–50
American Saddlebred Horse Association, 64
American Veterinary Chiropractic Association, 285
American white and creme colors, 53–54
Americas
    about, 19–20
    Azteca, *69,* **69–70**
    See also Canada. South America; United States
Amish-made barns, 226
amnion (water sack), 391, 393
amnionic vesicle, 387
anatomy and physiology
    about, 31, 211
    body language, 36–37, 347
    bone structure, 31–32, *32,* 69
    digestive system, 32
    evidence of vestigial toes, 16
    hoof-pastern axis, 259
    of hooves, 259
    intelligence, 40
    life cycle, 38–40, 41, 390, 405, 413–15
    ligamentum nuchae, 253
    muscles and tissue, 32, *33*
    nervous system, 33
    points of the horse, *37*
    senses, 33, 36
    of western horses, 312
    See also gaits and movement
Anatomy of a Horse, The (Stubbs), 25
ancient types of horses, 47
Andalusian (Spain)
    about, *65,* **65–66,** *240, 243, 295, 404*
    and Altèr-Real, 58
    and Azteca, 69
    and Friesian, 96
    and Lipizzan, 109–10
    Lusitano vs., 65–66, 111, 112

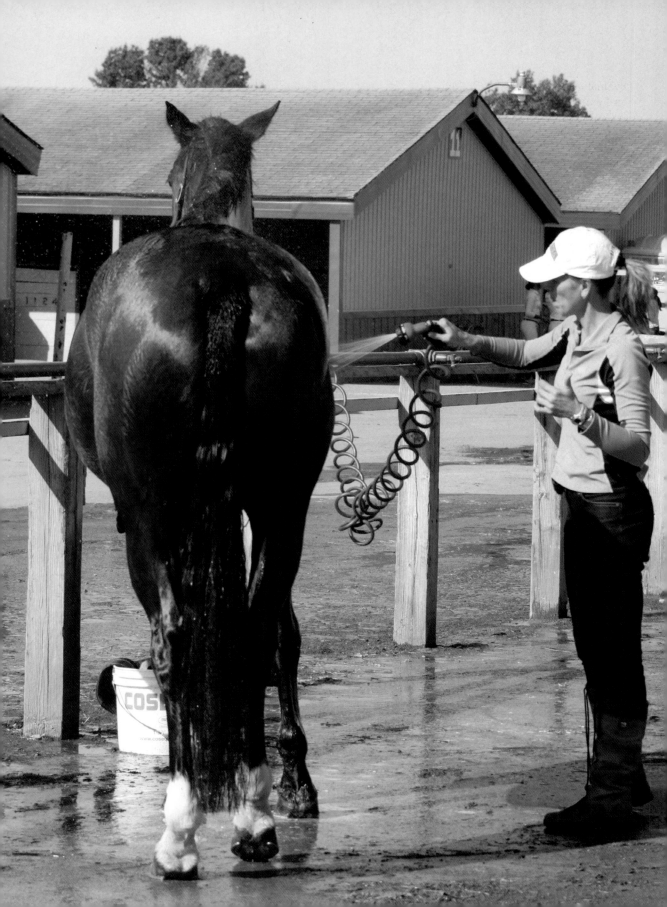

hunters over fences class, 356
hunters under saddle class, 356
Huntington, Rudolf, 82–83
Hunt, Ray, 344
hunt seat equitation class, 356
Hunt, Sharon, *51*
hypothyroidism, 407
Hyracotherium or *Eohippus,* 13
hysteroscopic examination, 385

I

Iberian Horse. *See* Andalusian; Sorraia
Icelandic Horse (Iceland), 52, **104–5,** *105*
ice tail, 93
illnesses and diseases
    anhidrosis, 429
    arthritis, 406, 411
    developmental orthopedic disease, 395
    encephalitis, 426
    equine gastric ulcer syndrome, 426
    equine infectious anemia, 381, 426–27
    equine influenza, 427
    exertional myopathy, 427
    foal scours, 391
    heat stroke or heat exhaustion, 429
    hormonal imbalances, 406–7
    hypothyroidism, 407
    laminitis, 407, 428, 429–30
    navicular disease, 430–31
    osteochondrosis, 396
    and overweight condition, 405
    physitis, 395–96
    polysaccharide storage myopathy, 427
    Potomac horse fever, 427–28
    rabies, 263, 428
    septic arthritis, 396
    signs of, 251
    strangles, 428
    tetanus, 428
    thrush, 259, 266, 431
    vaccinations for preventing, 263–65, 381–82,
        396, 407, 428
    West Nile virus, 428–29
    white line disease, 431
    wobbler syndrome, 391, 396
    See also colic; swelling
ILPH (International League for the Protection of
    Horses), 26

immune system and acupuncture, 286
immunization vs. vaccination, 263
impaction colic, 425
imprinting, meaning of, 398
imprint training, 320–23, 398–99
independent seat (rider), 303
India
    Kathiawari, **108,** *108*
    Marwari, 108, **112–13,** *113*
Indonesia
    about, 102
    Batak (Indonesia), **67,** *67,* 132
    Java, **102,** *102*
    Padang, **132,** *132*
    Sandalwood, *141,* **141–42**
    Sumba and Sumbawa, **144,** *144*
inflammatory cytokines, 381, 382
in-hand pattern, 357
injections
    administering, 255–57, 264–65
    for arthritis, 406, 411
    for collagenolytic granuloma, 196
    for navicular disease, 430
    for treatment of tetanus, 428
    vaccinations, 263–65, 381–82, 396, 407, 428
injuries
    boots and wraps for preventing, 204–8
    disaster planning for preventing, 234–35
    eye injuries, 255
    from fences, 230–31
    first aid, 254–59
    first aid kit, 255, 285
    from girth and cinch, 197
    from loose horseshoes, 257
    overreach injury, 211
    from rolling with colic, 426
    from saddle pads, 196
    swelling and heat in hooves and legs, 254
    from training halters or chains, 326
    tying safety for preventing, 295, 328
    from weather while trailering, 217
    wounds, 254–55
    See also swelling
in season, 382
instructors. *See* riding instructors
integrative medicine, 281, 282, 284–86. *See also*
    holistic horsemanship
intelligence, 40

Lord Darney (Clydesdale), 82
lordosis (swayback), 403, 407
Lusitano (Portugal)
    about, **110–12**, *111, 284*
    and Altèr-Real, 58, 112
    Andalusian vs., 65–66, *111, 112*
lysine, 239

M

macrocyclic lactones, 273
Magdalenian horse, 132
maintenance and fire safety, 232–35
Making Your Horse Barn Safe booklet, 235
malposition of foal, 391, 393
manes
    banding, 359–60, 367
    conditioners, 243, 244
    hogged, 53
    pulling, 243–44, 359
Mangalarga Marchador (Brazil), **120**, *120*
manger trailers, 216
manure
    in arenas, 228
    in barns, 224–25
    and compost temperature, 233
    fecal egg count reduction testing, 274
    of healthy horses, 251
    in pastures, 231
    pests and parasites, 271, 272, 276
Maremmana (Italy), **121**, *121*
mares
    about, 17
    bonding with their foals, 395
    foaling, 389, 390–94
    gestation period, 390
    herd behaviors, 41–42
    and natural breeding process, 39
    nurse mare industry, 28
    pregnancy, 382, 387, 389–90
    See also breeding
mark out (bronco riding), 166
Marks, Kelly, *342, 345*
martingales, 203–4, *204*
Marwari (India)
    about, **112–13**, *113*
    Kathiawari vs., 108
massage therapy, 250, 285
Masuren, and Wielkopolski, 151

mature offspring and breeding, 380
Max #2 (Colorado Ranger Horse), 83, 84
maximum trailer weight rating (MTWR), 216
medications
    NSAIDs, 406, 407, 411, 413, 428, 430
    and purchasing your horse, 184, 189
    See also veterinarians
medium pace, 302–3
mesh fencing, 230–31
Mesohippus, 13, *14,* 15
Messenger (Thoroughbred), 144
metacarpal and metatarsal bones, 16
Metis Trotter (Russia), **123**, *123*
Mexico, Azteca, *69,* **69–70**
microbials, 239
Middle Ages, 18–19
Middle East
    Bedouins, 68
    Caspian, 48, *48,* 78, *78,* **78–79**
    Plateau Persian, **134**, *134*
    See also Arabian
migration, 15–16
military. *See* warhorses
milkweek *(Asclepias spp.),* 278–79
Miller, Robert M., 320
mine ponies. *See* pit ponies
Miniature Horse (Europe), **113–14**, *114*
Miohippus, 16
Missouri Fox Trotter (US), 52, **114–15**, *115*
Misty of Chincoteague (Henry), 80
mobile phones, 295, 393
modern types of horses, 48–50. *See also* breeds
modular barns, 225–26
Mohr, Erna, 134
Mongolian (Asia), 100
Morab (US), **116**, *116*
Mores, Marquis de, 123
Morgan (US)
    about, *117,* **117–18**
    and Canadian Horse, 75
    and Missouri Fox Trotter, 114
    and Morab, 116
    and Paso Fino, *240*
    and Pony of the Americas, 131
    and Standardbred, 144
    and Tennessee Walking Horse, 147
mouthing behavior, 36
moving close behind characteristic, 176, 179–80

moxidectin, 273, 275, 276, 277

Mullen mouth snaffle bit, 199

Mulvihill, John, 111

Muraközi (Hungary), **126,** *126*

Murgese (Italy), **127,** *127*

muscles and tissue, 32, *33*

Mustang (US)
    about, *118,* **118–19**
    and Chincoteague Pony, 80
    and Nokota, 123

mustangs (US)
    about, 18, *238*
    and BLM, 29, *29*
    Kiger Mustang, 71, **107–8,** *108*
    Pryor Mountain Mustang, 71
    Spanish Mustang, **143,** *143*
    See also Mustang

mutton withers conformation defect, 39

Muybridge, Eadweard J., *24, 25*

mythologies, 17, 23–24, 25

N

Namib (Africa, Namibia), *119,* **119–20**

Narrangansett Pacer (US)
    about, 50, 64
    and American Saddlebred, 64
    and Kentucky Mountain Saddle Horse, 110
    and Standardbred, 143–44
    and Tennessee Walking Horse, 147

National Breeding Association of Warmbloods, 70

National Cutting Horse Association, 163

National Museum, 14

National Pony Society, 92

National Resource Conservation Services
    (NRCS), 226–27

National Show Horse (US), **120–21,** *121, 228*

Native Americans
    American Indian Horse, **60,** *60*
    and American Paint Horse, 61
    and Appaloosa, 66
    Dakota Indian tribe, *153*
    and Mustang, 118–19
    Nez Perce Horse, *122,* **122–23**
    Nez Perce tribe, 66, 122–23, 149
    Pueblo Indians, 19
    Seminole tribe, 95

natural healthcare choices, 282, 284–86

natural history, 13–16

natural horsemanship, 343–47

navicular disease, 430–31

Neapolitan Horse (Italy)
    about, 50, 109
    and Oldenburg, 126
    and Salerno, 140

Near East, 20–21

neck-reining, 315

needle and syringe disposal, 265

nervous system, 33, *36*

Netherlands
    ban on rodeo, 165
    Friesian, 83, 94, 95, **96–97,** *97,* 98, 126
    Gelderlander, 70, 91, **91,** *91,* 99
    Groningen, 91, **95,** *95*

New Forest Pony (UK, England), **121–22,** *122*

Nez Perce Horse (US), *122,* **122–23**

Nez Perce tribe, 66, 122–23, 149

niacin, 239

nipping and biting issues, 339, *340*

Nokota (US), *123,* **123–24,** *241*

nolvasan (antibacterial solution), 393

Nonius (Hungary)
    about, **128,** *128*
    and Furioso, 89

nonsteroidal anti-inflammatory drugs (NSAIDs),
    406, 407, 411, 413, 428, 430

Norfolk Trotter/Roadster (UK, England)
    about, 50, 74
    and Breton, 74
    and Connemara Pony, 84
    and Dales Pony, 87
    and Frederiksborg, 86
    and Freiberger, 87
    and French Trotter, 88
    and Morgan, 118
    and Nonius, 128
    and Norman Cob, 130
    and Selle Français, 137
    and Standardbred, 144

Noriker (Austria), **129,** *129*

Norman Cob (France), **130,** *130*

North Swedish Horse (Sweden), **131,** *131*

Norway
    Døle Gudbrandsdal, **80,** *80,* 131
    Døle Trotter, **81,** *81*
    Norwegian Fjord, *124,* **124–25**

nose and health, 250

S

sabino coat/color pattern, 55, 62, 82

Sable Island Horse (Nova Scotia), *135,* **135–36**

sacrifice area (paddock), 225, 232

saddle bronc (rodeo), 167

saddles

    about, 193

    care of, 200, 202, 295

    for competition, 360, 361

    English saddles, 193–95, *194*

    girth and cinch, 197

    jumper saddle, *194*

    Lane Fox saddles, 355, 356

    for longeing, 329

    saddle pads, 196–97

    safety straps, 296

    sidesaddles, 160, *160*

    western saddles, *195,* 195–96

saddle seat competition, *160,* 160–61

saddle seat riders, 303, 304, 306

saddle tree, 195, 211

safety around horses, 294–95, 337

safety drill in case of fire, 235

safety of horse trailers, 215–16, 220

safety tests before purchasing, 180

Salazar, Ken, 29

Salerno (Italy), **140,** *140*

salting hay, 233

samurai warriors, 21, *21,* 97

Sandalwood (Indonesia), *141,* **141–42**

sanding hooves, warning about, 359, 361

Sardinian (Italy), **143,** *143*

Save the Brumbies Inc. (STB), 74

schooling shows, 351, 356

schoolmaster horses, 176

Schweiken (Lithuania), and Trakehner, 149

Scotland

    Clydesdale, *81,* **81–82,** 98

    Eriskay, *91,* **91–92**

    Galloway Pony, 50, 87, 110, 148

    Highland Pony, 48, *48, 103,* **103–4**

    Scottish Galloway, 94

    Shetland Pony, 84, 89, 113, 130, 131, *138,*
        138–39

SDF (synchronous diaphragmatic flutter "thumps"),
  429

seat, independent, 303

seedy toe (white line disease), 431

selenium, 239

Selle Français (France)

    about, *136,* **137**

    and Belgian Warmblood, 70

    and Danish Warmblood, 78

    and Hackney Horse, 99

semen, shipping, 386–87

Seminole tribe, 95

senior horses

    about, 38, 401, 403, 411

    and blankets, 267

    end of life decisions, 413–15

    fitness, 407–9

    investing in, 409

    nutrition and care, 404–7

    retirement process, 411–13

    signs of aging, 403–4

senses, 33, 36

septic arthritis, 396

sets back (tie-down roping), 166, 167

Sewell, Anna, 25–26

Shagya-Arabian (Austria, Hungary), *137,* **137–38**

sheath cleaning, 250

shell out the ears, 359, 367

Shetland Pony (UK, Scotland)

    about, *138,* **138–39**

    and Dartmoor Pony, 89

    and Falabella, 84

    and Miniature Horse, 113

    and Pony of the Americas, 130, 131

shipping boots, 206–7, *218*

shipping fever (respiratory disease), 215

shipping semen, 386–87

Shire (UK, England)

    about, *19, 139,* **139–40**

    and Gypsy Vanner Horse, 98

show hunter, 157

show jumping, 157–58

showmanship (western show class), 357

show premium, 355, 356

show-ring strategies, 372–73

shows

    about, 354–55, 369

    arriving at, 369

    braiding for, 366–67

    checklist, 364

    competing, 372–74

    grooming and tack for, 359–62